East Texas Mill Towns & Ghost Towns

Volume I

(Angelina, Chambers, Jefferson, Nacogdoches, Newton, Orange, Polk and Tyler Counties)

By

W. T. Block

ISBN 1-878096-31-1

Best of East Texas Publishers
515 South First • P. O. Box 1647
Lufkin, Texas 75901
(409) 634-7444

Funded by a grant from
The Pineywoods Foundation
Lufkin, Texas

Vast park-like stands of pine were commonplace throughout East Texas in the 1880s when the railroads opened up the region to sawmills and loggers.

Table of Contents

Foreword

For decades, historians in East Texas have dreamed of completing an anthology of the hundreds of sawmill towns and ghost towns that resulted from the great timber boom between the 1880s and 1920s.

Some writers have made a few attempts at collecting and recording some of the towns, but their works have been limited and, for the most part, were focused on the more famous sawmill towns like Browndell, Manning, Ratcliff and Bessmay.

In the 1990s, W.T. Block, fascinated by the rich history of the lumbering industry, began a methodical review of sawmill towns and ghost towns in every East Texas county with a timber relationship.

Learning of his work, the Pineywoods Foundation of Lufkin made a commitment in 1993 to publish Mr. Block's work, beginning with Volume I of "East Texas Sawmill Towns and Ghost Towns." Volume I contains Angelina, Chambers, Jefferson, Nacogdoches, Newton, Polk and Tyler counties. Volumes II and III, to be published by the Foundation later, will include the remainder of the Texas counties with sawmill histories.

The publishing of "East Texas Sawmill Towns and Ghost Towns" is a fitting endeavor for the Pineywoods Foundation, which traces its own genesis to the forest products industry.

The Foundation was created in the 1980s by a gift from the now-defunct Southland Paper Mills Foundation. In 1940, Southland Paper Mills, Inc. pioneered the production of the South's first pine newsprint from East Texas trees. Many of Southland's founders were, in fact, early sawmill owners in East Texas, including the Kurth, Temple, Henderson,

Thompson, and Weiner families.

Readers, especially those who have lived or worked in a sawmill community, will find Mr. Block's efforts fascinating and illuminating. His attention to detail and his ability to produce a comprehensive picture of each sawmill and its community has done much to further develop the history and heritage of the forest products industry in East Texas.

Bob Bowman
Administrator
The Pineywoods Foundation
Lufkin, Texas

MILL TOWNS AND GHOST TOWNS: EAST TEXAS BEFORE THE GREAT SAWMILLING AND LOGGING ERA

Wherever or whenever one reads of or speaks to someone who grew up in an East Texas sawmill town during the 1930's, the mood is generally one of extreme nostalgia, of a way of life that has long ago disappeared, but is not forgotten. Whether it was Bessmay, or Manning, or Diboll, or Wiergate, the memories are of schools without computers or even comfortable rest rooms, or visiting the community house just to hear a battery-powered radio, or of a date where a bowl of ice cream and a ten-cent movie were the principal ingredients. Memories of such a town almost always included acres of muddy or dusty streets, dozens of unpainted, clapboard houses, surrounded by picket fences; livestock running loose in the streets, or perhaps a barefoot urchin, munching a licorice stick and carrying a dinner pail to school or a gallon can of kerosene oil home. And always standing nearby, somewhat like a king surveying his wooded realm, was the company sawmill, where the shrill mill whistles emanated before dawn and at dusk. And during the heat of the day, there emanated also the monotonous screech of the big band saw as it chewed its way through a log. Sometimes those memories were even exclaimed aloud: "We were all poor, but nobody knew it!"

Indirectly, the writer became an occasional visitor to that scene of sawmilling - to towns such as Bessmay, Deweyville, or Honey island - around 1936, since he used to drive a "bobtail" truck there to pick up loads of lumber. Often he visited inside the sawmill and planing mills, where giant machines spit out oceans of sawdust and shavings, and where it seemed that dozens of whirling saw blades were ever ready to inflict bodily harm or even death to the careless soul. Despite the fact that one excellent volume has already chronicled mill life in about ten larger mill towns, the writer for a long time has felt a compulsion to write about the smaller mill towns omitted from that volume and thus were consigned to historical oblivion.

First, however, the writer prefers to introduce the reader to the East Texas of 1836, when no more than a half dozen crude sawmills dared to harvest only a handful of lumber. One needs only to look at page 164 of *Timber Resources of East Texas*[1] (a Kirby Lumber Company book of 1902, which described the earliest of the Kirby sawmills) to see what "old-growth" or virgin long leaf pine logs looked like. The picture is of 48-inch logs near the Woodville, Texas, log haulup, which beggared anything that today's forests have to offer.

One can readily imagine how early East Texas must have appeared to the pioneer cotton farmer, his family and his oxen, as they crossed over the Sabine River at Gaines' Ferry in Sabine County in 1837. As his oxen and mules wound their way along narrow paths, surrounded by the towering pine monarchs of the forest, the farmer thought little about the sawmilling scenario. His thoughts turned instead to the economic waste-land about him, where cotton should be sprouting, except that the huge trees were sapping the soil of its vital moisture. And if that farmer remained in East Texas, he most likely became as much a despoiler of the timberlands as did the sawmiller. Whereas the latter left only the stumps, the farmer most often girdled the giant trees and left them to dry out and rot in his cotton field.

One hindrance to sawmilling in the East Texas of 1840 was the fact that most of the machinery needed to harvest the big trees had not even been invented. Although probably 75 percent of the "old-growth" trees varied between two feet and three feet in diameter, fully one-fourth of the long leaf logs exceeded three feet in diameter, and that included the short leaf pines and hardwoods as well. There are some records of trees reaching five feet in diameter. In April, 1927, a huge long leaf log, 61-inches in diameter and sixty feet long, was cut at New Blox log camp in Jasper County, and it required three 8-wheel log wagons to transport it (see footnote for photograph in newspaper).[2] About 1900, Mexican shingle makers in Nacogdoches County cut a large short leaf pine log and counted

283 growth rings in it. The tree was already a sapling whenever the Pilgrims celebrated their first Thanksgiving.[3] About 1905, a huge short leaf pine tree on the W. R. Pickering timberlands in Shelby County was 33 feet in circumference.[4] And the photograph of a magnificent cypress tree of similar dimensions appears on page 140 of *The 35 Best Ghost Towns of East Texas.*[5]

The earliest sawmilling in East Texas was done in "whipsaw pits." In that method of cutting lumber, one man standing in the pit pushed the cross cut saw upward, while another man above the log pushed the saw downward. Another strange device used in some counties was the "peck mill." That contraption consisted of multiple, rotating axe or adze-like blades, energized by a mule or ox walking in a circle. The "peck mill" roughly squared small timbers for use in building log houses. It was said of that mill, that its staccato sounds, something like a giant woodpecker, could be heard for a mile through the forest.

The dissertation of Dr. H. P. Easton recorded that the first sawmill in East Texas appeared in 1819 in Nacogdoches, where a "Mr. Madden has engaged in building a saw and grist mill, and this is a strange activity for a Spanish town."[6] The second sawmill was a water-powered saw and grist mill, built in 1826 by Wyatt Hanks on Ayish Bayou south of San Augustine. By 1842 only the ruins of the mill dam and a few rotting timbers survived.[7] Stephen F. Austin wrote in a Spanish language letter that William Harris and "Roberto" Wilson had built a "molino de vapor" (steam sawmill) on Buffalo Bayou in 1829.[8] In 1833 T. F. McKinney and M. B. Menard built a steam sawmill on Menard Creek, a tributary of the Trinity River,[9] and in 1835 David G. Burnet and Norman Hurd built another steam mill on Buffalo Bayou.[10] By 1837, the Velasco Steam Sawmill Company was cutting cypress lumber on the Brazos River.[11] The saws in those mills were all upright sash or "muly" saws, moving in a up-down motion like the whipsaws.

The early sawmills cut primarily for local consumption since no rail system existed for transporting lumber to market. A few seaport towns - Beaumont, Sabine Pass, and Orange - were able to develop early lumber markets outside the county because they could ship by water. Nevertheless, there remained a huge demand in Texas for lumber that could only be satisfied by importing from other states. In 1840, the editor of Houston *Morning Star* noted that Houston had to import the "commonest varieties of lumber at the exorbitant price of from forty to fifty dollars a thousand."[12] A year later, the Houston *Telegraph* editor complained that:

"....With all the raw products right at hand, it seems the height of folly to look to Maine for lumber and to England and elsewhere for brick."[13] Invariably, lumber sold in Beaumont or Orange at $10 a thousand feet in 1855 commanded prices of $60 or $70 a thousand in San Antonio, Austin and Waco, because of the high cost of transportation. As late as 1850, some East Texas counties still did not have even one saw and grist mill, and in 1856, Jacob DeCordova, an early Texas publisher, observed that:[14]

> ...It must indeed seem strange...that we should be compelled to import so large an amount of lumber into Texas...Not only to Mobile are we indebted for timber and lumber, but large quantities are brought from the State of Maine. Even the ties for the Galveston, Houston, and Henderson Railroad had to be procured from the latter point...in consequence of the exorbitant rate of transportation...

All of the old-growth Atlantic Coast forests, except those in Maine, had been exhausted after about 1820, and had been replaced by small-to-medium second-growth timber. Before the Civil War, the nation depended upon yellow pine timber, grown principally between Mississippi and the Carolinas, and on northern white pine and hardwoods from the great forests of Wisconsin, Michigan, Pennsylvania, and Maine, much of which had been exhausted by 1870.

Although an infant rail system came to parts of Texas (including the Texas and New Orleans from Houston to Orange) before the Civil War, most of those lines were shoddily constructed and maintained, and they had to be rebuilt during Reconstruction days. Rail lines such as the International and Great Northern, Houston East and West Texas, Texas and New Orleans, East Texas Railroad from Beaumont to Rockland, Texas and Pacific, the Trinity and Sabine Railroad from Trinity to Colmesneil, Texas; the Gulf, Colorado and Santa Fe, and the Cottonbelt Road began to inch their way into East Texas after 1880, and suddenly, the huge East Texas timberland belt, consisting of about 20,000 square miles, was opened up for sawmillers to exploit.

At least a few sawmilling entrepreneurs came to East Texas soon after the guns of San Jacinto were silenced, with a view toward harvesting the big pines, but those who did so and remained usually settled for operating combination saw and grist mills, combined with cotton-ginning, to satisfy local demand. Those early antebellum sawmills were about equally divided between small steam mills and water-powered mills, and occasionally a mill was energized by "mule-power" or oxen. One who did not remain in East Texas at first was David R. Wingate of Hancock County, Mississippi. He visited both Jasper and Newton counties in 1844 for a month in company with Martin Ford and W. F. Sparks, but the sawmillers soon returned to Pearlington, when they realized the huge logs of East Texas had no means to reach market.[15]

Wingate also feared that his cutting equipment in Mississippi was also inadequate, one article noting that Wingate:[16]

> ...built a mill at Pearlington in which he put in a sash (upright) gang of twenty saws. The saws were driven by an independent upright engine...and had a capacity of 10,000 feet a day. As it was operated with slaves under a white foreman, a day was from daylight until dark, and (the mill) made as true, square-faced lumber as is made today with improved machinery...

Realizing that his Mississippi equipment was hopelessly inadequate for the East Texas forests, Wingate had watched closely as a friend, Dr. Montgomery, installed a 48-inch Page single circular saw at his mill on the Pearl River,[17] but the circular saw quickly choked down upon engaging logs larger than sixteen inches in diameter. The earliest Page sawmill was designed for the small second-growth timber of the Eastern Seaboard, and although the Page mill was on sale in Houston by 1848, it did not enjoy much success in the East Texas of that era.[18]

Nevertheless, Wingate, his father, and Alfred Farr soon returned to Cow Creek in Newton County with their families and 165 slaves in 1852. The former purchased 2,700 acres of the Linnville and Lewis leagues, where he raised cotton, and he and Farr installed small sawmills. In 1857, Wingate bought the bankrupt Spartan Mill Company at Sabine Pass, and after two years and much remodeling, he built it into the largest sawmill in Texas, cutting in 1859 around 30,000 feet daily and 2.5 million feet in that year. (He also raised 350 bales of cotton on his plantation at Belgrade.) By 1860, Wingate had built up an excellent coastwise lumber trade, and he also shipped cypress sugar vats and other timbers aboard his fleet of lumber schooners to the sugar-growing islands of the West Indies. In October, 1862, the Union Navy came ashore at Sabine Pass and burned D. R. Wingate and Company's sawmill, planing mill, sash and door factory, and 700,000 feet of rough lumber. After the Civil War, Wingate operated sawmills at Orange until his death in 1899.[19]

Sawmilling, a lack of railroad-building and maintenance, and commerce in general suffered miserably during the Reconstruction years, but by 1875, three factors came together to make East Texas sawmilling attractive to a few of the early Texas pioneers, as well as to many Northern entrepreneurs. By 1880, a number of railroads, the International and Great Northern, the rebuilt Texas and New Orleans, the East Texas Railroad, the Texas and Pacific, the Cottonbelt, and Trinity and Sabine, began to penetrate the East Texas pineries, where the long leaf monarchs stood. Also, bigger and better sawing equipment was being invented and manufactured, and the Allis 60-inch and 72-inch circular mills, made in Milwaukee, became the standard cutting tools all over East Texas. By 1878 the Eagle and Reliance sawmills at Beaumont were both equipped with new Allis circular mills. The old friction-feed log carriage was soon replaced by the steam "shotgun-feed" carriage; the steam log turner, log kicker, the log-stop-and-loader, the mechanical log haul-up, cut-off and drag saws for trimming log ends, and other improvements were soon added to the carriage and the log deck.

By 1885, some mills were also installing gang saws, varying in size from 21-gang to 36-inch gang saws. During the 1890's, the single-cutting band saw, rotating on eight or nine-foot "dogs," or wheels, became the standard milling tool for the bigger mills of 50,000 feet or more daily capacity. Some larger mills sported double-cutting bands and carriages. In 1860 the few East Texas steam sawmills utilized engines which developed from only 15 to 25-horsepower each. At the 1876 Centennial Exposition in Philadelphia, the might of the new Corliss steam engine was demonstrated, and by 1900 the biggest mills in East Texas contained power houses, where sometimes 350 or 400-horsepower engines were turning massive flywheels of up to 20-feet in diameter and weighing up to fifty tons each. Indeed, in twenty-five years time, the East Texas sawmilling industry had budded, blossomed, bloomed and flowered.

The third factor in East Texas sawmill history involved the northern sawmilling industry. By 1875, the great forests of Wisconsin, Michigan, and Pennsylvania were about cut out, and sawmillers in those areas began looking south to Arkansas, Louisiana, and East Texas for areas where supplies of saw logs abounded. In fact, most, but not all, of the East Texas counties had already developed sizeable sawmill industries long before the Western Louisiana parishes, because the lone north-south railroad there, the Kansas City Southern, was not completed until 1897. Midwestern retail lumber firms, such as Long-Bell, who were once dependent on northern timber supplies, soon turned southward to buy timberlands and sawmills in Texas and Louisiana. Another was William Cameron, a Scotch immigrant of Waco, who amassed one of the largest lumber fortunes in the South. Cameron began as a tie and timber contractor for the Missouri, Kansas, Texas Railroad in 1865, and by 1898, he owned about six sawmills and sixty retail yards scattered out between Waco and St. Louis, in addition to flour mills, banks, etc.[20]

There were others who began sawmilling in East Texas before the Civil War, and continued long after the war until they amassed sawmill empires as well. One was J. M. Thompson, who operated his first small sawmill in Rusk County in 1852, and by the time of his death, the Thompson-Tucker, Thompson-Ford and Thompson Brothers Lumber Company owned sawmills in Trinity, Polk, and Hardin counties.[21] James Long bought a small sawmill at Beaumont in 1860, and although Long died in 1873, his three brothers-in-law, W. A. Fletcher, John W. Keith, and Frank Carroll, formed an interlocking directorate over six large sawmills in Texas and Louisiana, to include Long Manufacturing Company (cypress shingles), Beaumont Lumber Company, and Texas Tram and Lumber Company, all of Beaumont.[22]

Others came from the northern and eastern states to buy up huge blocks of East Texas and Louisiana timber and to operate sawmills. Two of the earliest northern entrepreneurs were Henry J. Lutcher and G. Bedell Moore, who had been sawmillers at Williamsport, Pennsylvania, until their timber was cut out in 1876. In January, 1877, they made an exploratory visit to Southeast Texas, after which they bought 400 square miles of pineries in Southwest Louisiana, 60,000 acres of cypress swamps near Baton Rouge, and 100,000 acres of long leaf pine in north Newton County. Afterward, Lutcher and Moore Lumber Company cut 250,000 feet of lumber daily at their "Upper" and "Lower" sawmills on the Sabine River at Orange from 1877 until 1930.[23] Thomas L. L. Temple came from Virginia to Diboll and Pineland, Texas, via Arkansas, in 1893 and built a sawmill (later paper) conglomerate that logged timber over 250,000 acres in several counties. The biographies of men such as Cameron and Temple seldom have equals among the builders of East Texas.[24]

The writer made several trips through the most-heavily wooded East Texas counties during the summer of 1993, and one observation was apparent everywhere — that a small sawmill was about as rare a sight as a cotton field. Nevertheless, "King Cotton" dominated the East Texas economy until about 1890, and thereafter, sawmilling was the dominant industry until after World War II. In fact, the writer was the most amazed to discover that in 1859, many of those heavily-timbered counties, where it would seem that room for cotton plantations did not exist, produced the state's greatest quantities of cotton. Some counties included: Polk County-8,313 bales; Walker County-11,970 bales; Trinity County-2,945 bales; Jasper County-3,792 bales; Shelby County-3,389 bales; Newton County-2,091 bales; Upshur County-7,550 bales; Harrison County-20,006 bales; Sabine County-2,127 bales; Panola County-8,218 bales; and Rusk County-12,737 bales.[25] As early as 1843, Nacogdoches County was already shipping 12,000 bales annually, most of it wagon-freighted overland to the Red River, but that county was much larger then than it is now.

Although sawmilling eventually dominated the economy of the region, agriculture often furnished the earliest employees of the sawmills and logging contractors. Often, the census enumerations revealed that as one family member engaged in logging or sawmilling, the remaining family members tended the family farm. Certainly, a few months of logging in winter time could enable a farmer to recoup somewhat for crops lost to drought, excess rain, plant diseases, and insects. Hence, the writer suspects that the East Texas farmers' close identification with the timber industry would pass on to the sawmill community the farmer's rural cohesion, compassion, and concern for each neighbor's welfare.

In 1893, some of the largest owners of East Texas timberlands included the East Texas Land and Improvement Company (Kountze Bros. of Sabine Pass)-260,000 acres; Texas Pine Land Association-140,000 acres; Texas Tram and Lumber Company-110,000 acres; Reliance Lumber Company-90,000 acres; W. D. Herndon-80,000 acres; C.P. Huntington of Southern Pacific-80,000 acres; Beaumont Lumber Company-70,000 acres; and J. P. Irvin-60,000 acres.[26] Irvin would increase his holdings to about 140,000 acres before he eventually sold out. By 1900, Lutcher and Moore of Orange had purchased 100,000 acres of timber in north Newton County, land that Wier Long Leaf Lumber Company would eventually harvest. By 1902, most of the large blocks of land had been bought up by a number of area and out-of-state lumbermen, including Foster Lumber Company, T. L. L. Temple, Kirby Lumber Corporation, W. R. Pickering, E. B. Hayward, Thompson-Tucker Lumber Company, Frost-Trigg Associates, Texas and Louisiana Lumber Company (the "4-C" mill) at Kennard, Trinity County Lumber Company, and William Cameron.

In 1893, a Galveston editor listed 166 sawmills of East Texas, built on seven different rail lines,[27] but the writer knows from his own records that several sawmills were left off of the lists for three different railroads. In

1904, Circular No. 2090 of the Gulf, Colorado and Santa Fe Railroad listed 62 sawmills on its Beaumont to Navasota and Beaumont to Center lines.[28] In 1906, a list of sawmills on four rail lines, Santa Fe, Houston East and West Texas, Texas and New Orleans, and International and Great Northern, totalled 230 sawmills, with an potential annual capacity of 2.4 billion feet. The mills on the Cottonbelt, Texas and Pacific, and Trinity and Sabine (which in 1893 tallied up to 97 sawmills) were not included, as were those on the Orange and Northwestern, Jasper and Eastern, Nacogdoches Southeastern, and other short lines.[29] Although several "guesstimates" have been found, the writer believes that the total of uncut pine and hardwood stumpage in East Texas in 1880 probably tallied up to between sixty and seventy billion board feet.

Certainly the most unique of the East Texas mill conglomerates was the Kirby Lumber Corporation of Houston, which started out as the Texas Pine Land Association, owning 140,000 acres of timber in 1893. By 1895, it owned a small sawmill at Silsbee and about sixty miles of logging tram road, known as the Gulf, Beaumont and Kansas City Railroad. In 1901 Kirby began buying up sawmills. The maximum number of sawmills Kirby owned at any given time was about fifteen in 1905, including two at Beaumont, and one each at Orange, Silsbee, Call, Roganville, Village, Woodville, Fuqua, Bronson, Bessmay, Browndell, Kirbyville, Mobile, and Evadale. By 1905, three sawmills, two at Lillard and Ariola (Hooks' Switch), Hardin County, and another at Beaumont had already burned.[30] *Handbook of Texas* says that at the peak of operations, probably around 1910, Kirby employed 16,500 men, but that figure probably could be true only if it included everyone on all the Kirby payrolls from 1901 to 1950.

Other early Kirby mills which were dismantled included those at Aldridge and Steep Creek, San Augustine County. Other mills added at later dates included those at Voth and Honey Island. In 1907 the twelve Kirby sawmills cut 301,000,000 feet of lumber. Between 1907 and 1918, Kirby added the Steep Creek and Aldridge mills (the latter abandoned and dismantled in 1927), and in 1917 the Kirbyville mill burned down and another mill at Beaumont was dismantled. In 1918, Kirby Lumber Company cut only 200,000,000 feet of lumber (down by one-third), and they paid out in 1918 a payroll of $2,730,000 to 4,793 employees. The Browndell mill production dropped from 36 million feet in 1907 to only 8.7 million feet in 1918, and that mill burned in 1925. In 1918, all other mill production was down substantially from 1907. In 1953, the company, by then owned by the Santa Fe Railroad, consolidated all operations at a new, all-electric sawmill at Silsbee, and all operations at Honey Island, Bessmay and Call closed down by 1954.[31]

An article of 1927 noted that: "...there are about 425 sawmills in Texas, a large portion of them being located...within a radius of 150 miles of Beaumont..."[32] Another article of that year noted that 182 sawmills still operated in 19 of the long and short leaf counties (all of them south of Smith, Rusk, and Panola counties), 89 of which were so-called "big mills" because they cut over 10,000 feet daily. The following figures are for each county, followed by large mills, small mills, and then their combined daily output: Angelina Co.-ten, three, 779,000 feet; Cherokee Co.-six, eight, 169,000 feet; Hardin Co.-five, three, 274,000 feet; Houston Co.-one small, 8,000 feet; Jasper Co.-ten, nine, 712,000 feet; Jefferson Co.-two large, 105,000 feet; Liberty Co.-four, nine, 192,000 feet; Montgomery Co.-nine, nine, 507,000 feet; Nacogdoches Co.-six, thirteen, 314,000 feet; Newton Co.-five, three, 421,000 feet; Orange Co.-three, one, 308,000 feet; Polk Co.-seven, six, 363,000 feet; San Jacinto Co.-two large, 86,000 feet; Sabine Co.-four, five, 295,000; San Augustine Co.-five, seven, 187,000 feet; Shelby Co.-two, one, 248,000 feet; Trinity Co.-three, two, 276,000 feet; Tyler Co.-four, three, 265,000 feet; and Walker Co.- six, four, 161,000 feet.[33]

The concept of occupational safety was unheard of in the early sawmills of East Texas. Much like the early Texas oil fields, where men often suffocated from poisonous gases spewing from the earth, or the early meat-packing industry, the loss of fingers and arms to accidents or even the death of a sawmill employee was considered one of the hazards of engaging in that trade. One might say - if one wished to avoid an industrial accident, one studied for the ministry. Write-ups about accidents and mill deaths were so commonplace in the newspapers that the reader's eye barely blinked. For instance, after only the fourth month of mill operation at Nacogdoches in March, 1905, the new E. B. Hayward sawmill had already chalked up the following unenviable record:[34]

> ...Tom Weeks, while feeding a planer at the Hayward mill...Saturday accidently had his hand cut completely off. This makes the twelfth man hurt by accident at the mill {in only four months), including one that fell from a scaffold and was killed outright...
>
> ...From one cause or another, the band saw of the big Hayward mill was demolished last Thursday (March, 1905), and the edger ("bull edger" or gang saw) also was

> put out of commission for several days. When a big gang saw breaks, it is a dangerous customer and generally succeeds in cutting half a dozen men in two before it is stopped...

The greatest fear of a band sawyer or gang sawyer came when the teeth of the saw blade struck a big nail, railroad spike, barbed wire, or other metal imbedded in a log. After 1895, members of the International Workers of the World (the "Wobblies") attempted to organize and called occasional sawmill strikes at Orange and elsewhere. There were many such attempts in East Texas after 1900, and the most frequent complaints were about industrial safety, low pay or long waits for pay, and 11 or 12-hour work days. Sometimes long strikes brought on industrial sabotage, and the most common expressions of such worker unrest were the nailing of rail spikes into company-owned logs or timber and plant arson. In August, 1911, there were thirty East Texas sawmills closed down by labor troubles, and the burning of the Aldridge, Texas sawmill on August 25, 1911, was known to have been a case of intentional arson although no one was ever prosecuted. In 1914, there was a second attempt there to burn the sawmill. The fire was put out before it did much damage, but an empty kerosene barrel, the contents of which had been poured out among the stacks of dried lumber, was discovered.[35]

The writer hopes to assume an impartial stance while discussing the early mill town social conditions. Just as sawmills were free to hire and fire employees, the latter were free to change jobs at will. However, as Dr. H. P. Easton pointed out in his dissertation, "The History of The Texas Lumbering Industry," about all a logger or mill worker could accomplish by leaving one job for another was perhaps to rid himself of a particularly obnoxious boss, because wages, work hours, and work conditions were about the same everywhere. An examination of the wages and work hours at thirteen Beaumont and Orange sawmills in 1880 revealed that $1.50 daily for unskilled labor and $3.00 daily for skilled labor was the industry standard. An eleven-hour work day was also the standard for the thirteen mills, although two sawmills at Orange, Lutcher and Moore and A. K. Norris, scheduled an eleven and one-half hour work day for their employees in both winter and summer. Although a sawmill work day tended to drop to ten hours after 1905, the wages and working conditions remained about the same for several decades.[36]

In 1915, George Creel published a diatribe of indictments in *Harper's Weekly*, entitled "The Feudal Towns of Texas," against Kirby Lumber

Corporation and its sawmill towns in Jasper County. In particular, Creel condemned the "company towns," such as Kirbyville, Browndel and Bessmay, where the Kirby company paid off in "mill checks," good for exchange without discount only at the company store, and the Kirby paymaster, who came around at very irregular intervals of about two or three months. Since he compared East Texas sawmilling only to the company-owned coal and ore-mining towns elsewhere, Creel rebuked particularly a lifestyle whereby an infant was patted on the buttocks at birth by a company doctor, lived in a company house, was fed by the company store, barbered by the company barber, educated and churched in the company-owned church and school house, and at the time of death, was buried in a company coffin in a company-owned cemetery. In addition, Creel objected to the collection of both a hospital fee and a workman's compensation fee from sawmill employees, which Creel claimed was illegal, as well as the company "black list."[37]

The writer readily recognizes the inhumanity of the early days of sawmilling, the occupational hazards, lack of environmental concerns, long hours, low pay, etc., but those factors were common to many manufacturing industries during that age of unbridled capitalism, which recognized no government right to control it. One need only to read Upton Sinclair's scathing attack on the meat-packing industry (*The Jungle*) and the other "Muckraker" writings to know that American industry of that age was less than perfect. A few years ago, the writer researched and wrote a book about the inhumane working conditions and hazards to life in the Sour Lake boom town of 1903-1904, a period when for every barrel of oil recovered and sold, another barrel was wasted, washed into the mud, or burned as waste oil. Also, the writer recalls around 1925 when every inch of space between Lumberton and Woodville or Jasper was a cutover wasteland of stumps before reforestration came into vogue. The writer's parents also told him of a day before World War I when a man labored 84 hours a week (seven 12-hour days) to earn a living in Jefferson County's oil refineries. Hence, if some one chooses to become "judge and jury" of the age of East Texas sawmilling, he should do so with the eyesight of 1890, and not with the hindsight of 1990.

Just as every one with a pocket full of dollar bills rushed to Sour Lake or Spindletop, Texas, to drill an oil well in 1902, some one else ran into the East Texas woods to build a sawmill. And they did it because everyone else was doing it, and in spite of the fact that the market might already be over-saturated with crude oil and lumber. Between 1893 and 1897, the

East Texas sawmiller had to operate with the barest margin of profit, and when the market for lumber dropped even lower or ceased to exist, he operated his mill three days a week, or shut down entirely, or went bankrupt. Many elderly people recall when those conditions emerged again during the 1930's. The one person that the writer cannot speak so fondly of was the absentee or non-resident sawmiller, who was not present to witness or share his employees' work layoffs, or other sorrows and joys.

In writing these chapters, the writer's most heart-warming experience was to read the stories of the 1930's by the former residents of Call, Manning, Deweyville, and Honey Island — of how they helped cushion the sorrows of and cared for one another during the Great Depression when the sawmill was shut down. As an example, John C. Gee wrote of those days in Call, Texas, with the following comments:[38]

> ... Our ease was interrupted when the Great Depression that hit the United States in 1929 filtered on down...Call was able to survive those terrible times, I believe, better than any other place in the world. The sawmill shut down, of course, and all working for wages came to a halt. But the great sense of love that always held the community together naturally led to a great sharing of the little that each person had. When a person killed a deer..., it was shared with the neighbors. To top that off, Doctor Beavers dispensed rice from his silos in China, Texas, to every family each week...The final blow came in 1953 when the sawmill was closed for good. Life, as we had known it, ...had come to an end...

When John Gee wrote those words, he spoke for every former school kid who grew up in a sawmill town during the 1930's. With that statement, the writer feels he has prepared the reader for the pages ahead and is inclined to end this introductory chapter. Since beginning these sawmill chapters in 1991, he has already been addressed by many persons, seeking information about grandparents or other relatives who long ago lived in various mill towns and worked in various sawmills. While the reader will most likely not find anything about a particular relative, perhaps the reader may learn something about the mill towns and sawmills where the relatives lived and worked.

— W. T. Block
Nederland, Texas

Angelina County

High-wheeled logging carts, pulled by mule teams, were used to log early timber stands in Angelina County.

ANGELINA COUNTY

THE GEOGRAPHIC AND TOPOGRAPHIC DESCRIPTION

Beautiful Angelina County is one of Southeast Texas' heavily-timbered counties, with its county seat located at Lufkin. Roughly fifty by twenty-five miles in size, it contains 940 square miles, and is also unique in that about seventy percent of its boundaries consists of two rivers, the Neches and Angelina. About forty miles of its borders to the north and southeast are shared in common with Jasper and Cherokee counties, being the only part of its confines not determined by water courses. Generally, the county's neighbors consist of Cherokee and Nacogdoches counties on the north, Tyler and Polk counties on the south, Trinity and Houston counties on the west, and Jasper and San Augustine counties on the east.

Other than the Neches and Angelina rivers, Angelina County also contains many other water courses in the form of creeks, as follows: Shawnee, Popher, Stanley, Odell, Durasno, Jack, Cedar, Biloxi, Buck, and several others. Like the rivers, the creek beds help to determine the undulating soil surfaces, which generally are near the 250-foot level. And perhaps Angelina County enjoys a greater amount of level or near-level meadow and timber surfaces than some of its neighbors, particularly Jasper, Newton, and San Augustine counties, where the surface undulations are particularly marked and irregular. Soils vary somewhat, but are generally brown and gray sandy loams, intermixed with red and black clays.[1]

The timber areas of Angelina County can be identified best as a long leaf pine belt in the southern region and a short leaf and loblolly pine sector in the northern region. It is not known what the amount of uncut pine stumpage (log measure) may have totaled in 1870 or earlier, but it must have exceeded three billion feet. One inaccurate source estimated the amount in 1900 as one billion feet.[2] but a more accurate estimate in 1904 placed the long leaf stumpage at 1,340,800,000 board feet and the short leaf and loblolly stumpage at 1,190,400,000.[3] Interspersed among the pines, particularly in the river and creek bottoms, are a great variety of hardwoods, much of which were not marketable timber. One source described Angelina County's saleable hardwoods as follows:[4]

> ... white oak...red oak...sweet gum, black gum, beach, birch, locust, ash, magnolia, cypress, black walnut... cottonwood, sycamore, elm, cedar, hackberry, basswood (linden), china, tupelo gum, dogwood, persimmon, cherry, and hickory...

A 1937 estimate of farm and meadow lands amounted to fifty square miles, but the writer assumes that those areas too were forest lands before they were cleared for plantations in antebellum days. The virgin forests of pine lands usually harvested from 8,000 to 10,000 board feet per acre, but some areas were known to harvest as much as 22,000 feet an acre. Angelina short leaf pine was generally larger in diameter, up to five feet on occasion, but the marketable trunk of short leaf was much shorter than long leaf. Angelina long leaf pine was smaller in diameter, usually between two and three feet, but the saleable trunk length often reached as much as ninety feet long.[5] A long leaf pine tree cut in Jasper County in 1926 was 61 inches in diameter by 120 feet long.

ANGELINA COUNTY'S EARLY TRAMS AND RAILROADS

The first of the railroads to reach Angelina County was the Houston East and West Texas, a narrow gauge line, chartered in 1875, which ultimately reached Logansport, Louisiana (before reaching Shreveport under a different name and charter). It was converted to standard gauge in 1894. Most of the line was built to Angelina County by 1882, and it eventually reached Nacogdoches by 1884. Not only did the road lay out the county seat town, but it also named it after a probable railroad inspector, Abraham P. Lufkin. Burke, another Angelina County town, was named for another railroad official. Ultimately, the railroad passed through Burke, Emporia, Diboll, Lufkin, and Michelli (also known as

Angelina), all of them in Angelina County.[6]

Founded in 1890, the Angelina and Neches River Railroad, was a 30-mile long logging road and common carrier chartered in August, 1900, by the Angelina County Lumber Company of Keltys. The line extended from Keltys to its terminus at Chireno, Nacogdoches County. Two of the Keltys sawmill's logging "fronts" were at Acol and at Camp Nancy near Zavalla, and often the mill had to pay "trackage rights" in order for its log trains to pass over the Texas and New Orleans tracks or other rail lines before reaching the Angelina and Neches River rails.[7] Keltys and Ewing were two of the Angelina County towns which lay on that route.

The Kansas and Gulf Short Line Railroad, also known as the Cottonbelt or St. Louis Southwestern Railroad, was chartered in February, 1880, to build from Tyler, Texas, to the Louisiana line. That railroad was absorbed or changed hands several times. About 45 miles of narrow gauge trackage were built from Tyler to Rusk by 1882, later being extended to Lufkin and eastward to the San Augustine County line. Several Angelina County sawmill towns — Pollok, Clawson, Keltys, Lufkin — were located on the Cottonbelt Railroad.[8]

The Texas and New Orleans Railroad was originally chartered as the East Texas Railroad in April, 1880, and by 1882, had completed 103 miles of trackage from Sabine Pass to Rockland on the Neches River. After the railroad's sale to C. P. Huntington of Southern Pacific, the Texas and New Orleans line was completed through Angelina County to Nacogdoches, and eventually to Dallas. The town of Huntington was named for the new owner, and the sawmill town of Manning was connected by a private tram road to both the Texas and New Orleans and the Cottonbelt rails.[9] The towns of Zavalla, Huntington, Baber and Retrieve were scattered along the Texas and New Orleans route.

The Houston, Shreveport and Gulf was a short line railroad built in 1906 to connect the sawmill town of Manning with Huntington. The Carter-Kelley Lumber Company sawmill at Manning had no other means to transport their finished lumber, and the short line or tram railroad afforded that mill with excellent outlets to both the Cottonbelt and the Texas and New Orleans rails.[10]

Another standard gauge line was the tram road of the Tyler Car and Lumber Company (later William Cameron) sawmill at Michelli. Since that mill logged in the Redland vicinity, north of Lufkin, it was a matter of convenience to extend the tram road to Clawson to connect with the Cottonbelt and thus shorten their products' journey to market. The ten mile Michelli tram was built about 1893, and the rails were eventually abandoned and removed about 1903. For years, the citizens and mills of Nacogdoches had hoped that the Cottonbelt would extend service to their town via the Michelli tram, but fate proved otherwise. When Tyler Car and Lumber Company owned Michelli, all their lumber was shipped to Tyler, via Clawson and the Cottonbelt route, for disposition, and the Cameron Company also shipped over that route to Waco and points north on the St. Louis Southwestern line. Although the Houston East and West Texas railroad passed through Michelli, the sawmill shipped none of its products over that road.[11]

Begun in 1894, the Texas South-Eastern Railroad was chartered in 1900 by the owners of the Southern Pine Lumber Company as a logging and common carrier railroad to connect the sawmill at Diboll with the logging camps at Blix and Vair. A part of the trackage was abandoned in 1908, but a 21-mile track was added to connect Diboll with Vair and Lufkin with Blix, thus connecting with both the Cottonbelt and the Texas and New Orleans Railroads.[12]

There were many more logging trams in Angelina County, often ranging from two to twenty-five miles in length, and some of them will be discussed along with their respective sawmill owners. And believe it not, some of them used wooden rails, over which tram cars were drawn with mules. This occurred usually between 1885 and 1895, and was common all over East Texas.

THE EARLIEST STEAM SAWMILLS

Timberwise, the East Texas of 1840 was much like a person floating in the sea without a single drop of water to drink. Despite three billion feet of virgin timber standing within its boundaries, Angelina County had neither the steam equipment needed to cut it nor the railroads needed to move lumber to market. How extremely sad, for in 1841, a Houston editor complained that "with all the raw products right at hand, it seems the height of folly to look to Maine for lumber..."[13] As late as 1856, a Texas realtor observed that:[14]

> ...It must indeed seem strange...that we should be compelled to import so large an amount of lumber into Texas... Not only to Mobile are we indebted for timber and lumber, but large quantities are brought from the State of Maine. Even the ties for the Galveston, Houston, and Henderson Railroad had to be

> procured from the latter point...in consequence of the exorbitant rate of transportation...

According to her 1937 M. A. thesis, Effie Maddox Boon noted that Angelina County lacked the spring-fed creeks, so common to Jasper and Newton counties, which abetted the construction of water mills. Without spring-fed creeks, water mills needed a mill pond and mill dam for impounding the winter rains. Nevertheless, Angelina County still had some early water mills, most notably those of White's mill near Lufkin and Slater's mill near Zavalla.[15] Actually, the oldest sawmill in Texas was built not far from the county's boundaries. In 1826, Wyatt Hanks built Texas' first water-powered combination gin, grist mill and sawmill on Ayish Bayou, a few miles south of San Augustine. By 1842, only the ruins of the mill dam were still visible.[16]

Without a doubt, the most common early method for sawing lumber in East Texas was the whipsaw pit. Such a pit enabled one man to stand beneath the log to push a crosscut saw upward, moving it in an upright or vertical position. Another much less common method was the "peck mill," which was in use in nearby Jasper County by 1843. A "peck mill" consisted of a multiple swinging axe or adze-like contraption, powered by mules or oxen, which roughly squared a small timber for the building of log houses. It was often said that the sound of a "peck mill," much like a big woodpecker, could be heard for a mile through the forest.[17]

Perhaps lumber from Angelina County's first whipsaw pits and water mills was used to build the first keelboats that floated cotton down the Angelina River to Sabine Pass. Thomas F. McKinney floated the first keelboat loaded with cotton from Pattonia on the Angelina River in 1830.[18] Such boats were often broken up and sold as lumber upon reaching their destination. Captains Robert S. and Moses L. Patton built a number of such keelboats at Pattonia, beginning with the *Thomas J. Rusk* in 1844.[19] In 1846, they built the first Neches River steamboat there, the *Angelina*, which Captain Moses L. Patton sailed to the coast on several cotton-hauling voyages each year. In February, 1850, the cotton steamer *Angelina* struck a snag and sank near Evadale, Jasper County, while returning to Pattonia with a load of lumber and freight.[20]

In her master's thesis, Effie Boon also recorded that "steam sawmills were not introduced (into Angelina County) until after the Civil War," which is incorrect.[21] There were two steam sawmills there that were recorded in the county's 1860 Schedule V, Products of Industry, census. Their exact locations, however, are unknown to the author, and the names of their owners are also omitted from Angelina County's most notable history, *Land of The Little Angel.*

During the census year ending July 1, 1860, James M. Anderson's sawmill (it was not a combination gin and grist mill) was valued at $3,000 and used during the census year 1.5 million feet of log stumpage and 500 cords of wood fuel, worth $2,100. His steam engine was rated at twelve horsepower. His mill employed six hands, paid a total of $200 wages each month. In that census, no cutting equipment was recorded, but the writer assumes that Anderson used one or two upright sash saws. Anderson produced 1.3 million feet of rough lumber, valued at $12,000, all of which was sold to satisfy local demand. Elsewhere in the county, Sam Cole's sawmill was worth $5,000, and it also used 1.5 million feet of pine stumpage, worth $1,800, during the census year. He utilized an 18-horsepower steam engine to power his sash saw(s). Cole also employed eight men to whom he paid a total of $300 monthly. He produced during the year 1.2 million feet of lumber, worth $10,000.[22] Most likely, there were many other early Angelina County sawmills recorded in the manufacturers/products of industry censuses, but the author has not checked them all out. The 1850 Schedule V, Products of Industry, for Angelina County is also on microfilm at the Lamar University Library. Two Schedules V, Products of Industry, censuses for Angelina County are on microfilm at the Texas State Archives. Angelina County's sawmillers of 1870 are recorded on Microfilm Reel No. 46 there, and those for 1880 are recorded on Microfilm Reel No. 48.

Two other early steam sawmills were reported as operating in Angelina County in 1867, according to Boon's thesis and *Land of The Little Angel.* One was operated by Dr. W. W. Manning and the other by Charles L. Kelty; and both were in the general vicinity of Homer. Kelty's mill was the first sawmill in the county equipped with a planing mill. About 1870, Kelty moved his sawmill to the northwest side of Lufkin, a place which still bears his name, where he sold out in 1887 to the Angelina County Lumber Company. No statistics survive for either mill unless they were recorded in the censuses mentioned above.[23]

Homer V. McMickle, a Confederate veteran of General R. E. Lee's army, built a small sawmill during the middle 1880's at McMickle, a Houston East and West Texas rail stop on the north side of Burke. Since McMickle gave up his appointment as postmaster at Burke in 1885, that may have been the year he began sawmilling. The only known reference to that mill appeared in that railroad's list of sawmills, published in 1893, and it noted that the H. V. McMickle Lumber Company cut 12,000 feet of lumber daily.[24] No other statistics were furnished. McMickle was no longer listed on the railroad's list of rail stops of 1898, nor on another list of the Houston East and West Texas sawmills published in 1900.[25]

According to one record, a man named Richard Blair owned a sawmill at Pollok during the 1880's.[26] By 1893, Richard Blair's sawmill at Lufkin was cutting 10,000 feet daily, and in 1900, there was still a Richard Blair and A. J. Caruthers sawmill somewhere near Lufkin. The 1900 Angelina County census, Precinct 1, identifies the following families at just three residences, as follows: at residence 477 was Richard Blair, lumber manufacturer; at res. 478 was Dr. John Reagan, mill physician; boarder at res. 478, Fred Evans, mill bookkeeper; boarder at res. 478, Ashley Stroud, mill shipping clerk; and res. 479, A. J. Caruthers, lumber manufacturer. While it is generally not prudent to make historical suppositions, there was also a Blair Lumber Company sawmill in 1900, located at Blair, Texas (a. k. a. Gallagher), between Garrison and Timpson in Shelby County. At the same time there was also a Bobo Lumber Company located at Bobo (a. k. a. Meldrum), Texas, between Timpson and Teneha in Shelby County. According to one source, a railroad porter used to chant out the upcoming rail stops of "Tenaha, Timpson, Bobo and Blair" as the passenger car moved toward the southwest. Could Richard Blair have unwittingly lent his name to the World War I Doughboy crapshooters' chant of "Tenaha, Timpson, Bobo, and Blair?" While unproven, it seems highly likely and plausible to this writer.[27]

THE WANDERING BOYNTON AND JEANS BROTHERS

After large, well-financed sawmill interests, usually from out-of-state, moved into East Texas during the 1880's and 1890's to buy up huge tracts of railroad and privately-owned forest land, it became increasing harder for the smaller sawmill companies to purchase more than a 5,000-acre tract for a timber reserve. Such a company was Boynton Brothers Lumber Company sawmill, located at Boynton, ten miles southeast of Huntington on the Cottonbelt Railroad. Albert and Walter Boynton erected tbeir sawmill and planer there in 1904 to harvest their small tract of about 4,500 acres of timber near the Angelina River. Two sources noted that the Boynton brothers milled good quality lumber at the rate of 25,000 feet daily and worked 75 men. They also owned about five miles of steel rails, one locomotive, several tram cars, ox and mule teams, but they used no steam skidders or loaders in their forest operations before their timber was cut out in 1908.[28] In January, 1904, one news account rated Boynton Brothers as one of Angelina County's seven largest sawmills, cutting 50,000 feet daily. A year later, another newspaper article observed that they cut only 40,000 feet of pine lumber daily.[29]

In 1908, Walter and Albert Boynton dismantled and reassembled their sawmill at Logtown, Newton County, where they owned a 5,000-acre tract of timber. The Boynton brothers carried no fancy mill titles, the census of 1910 identifying Albert only as "owner and planer foreman" and Walter as "owner and sawmill foreman." During their four-year stay, Logtown grew to about 700 persons, with houses, commissary, church, and schools, and one hundred men working in the mills and forests.[30]

In 1912, Boynton Brothers again dismantled their mill and moved to Sturgis in San Augustine County. In 1918, the Lufkin Foundry and Machine Company designed some equipment for their Sturgis mill. About 1920, Boynton Brothers Lumber Company moved back to Angelina County, where in 1927 their hardwood mill was one of five mills still operating in that county.[31]

About 1898, Will F. and Tom J. Jeans erected their sawmill in Angelina County. The 1900 Angelina County census enumerated them as "sawmill lumbermen" at Justice Precinct No. 2, residence 218, which apparently was the town of Ora since Tom J. Jeans was Ora's postmaster in 1901. No information is available about the size of the Ora sawmill, but it probably didn't exceed 20,000 feet daily. Since the Jeans sawmill was east of Huntington, its tram probably intersected the Cottonbelt Railroad.[32] By 1902 the Jeans brothers had moved their mill to Clevenger in Nacogdoches County, and by 1905, the Jeans sawmill was cutting timber east of Rush in Sabine County.[33] Apparently, Will and Tom Jeans prospered modestly and did not suffer from the frequent moves and the lack of a large timber reserve. By 1906 the Jeans Brothers' milling interests were divided into three sawmilling companies, as follows: T.W. Jeans and Son, a 40,000-foot pine sawmill at Bronson, Sabine County; E. F. Jeans and Co., a 20,000-foot hardwood mill, also at Bronson; and T. W. Jeans and Co., a 20,000-foot pine mill at San Augustine.[34]

THE SAWMILL AT MICHELLI

Almost immediately, the writer will have to deal with the correct spelling of Michelli, since he has added an extra 'l' to the spelling which appears in Effie Boon's M. A. thesis, as well as in *Land of the Little Angel*. Boon's only source of information about Michelli was a letter from Anderson Jasper Peavy of Shreveport in July, 1934, which of itself would seem sufficient. Yet two of A. J. Peavy's obituaries used the "double-l" spelling for Michelli, one from Lufkin reading as follows: "...Mr. Peavy began his career (ca. 1886) working for the W. H. Bonner General Store, Lufkin; and several years prior to 1893, logged for the Michelli Mill, near the Angelina River. Soon thereafter, he made a contract to log the Angelina County Lumber Co.'s mill at Keltys..." A Shreveport obituary of Peavy added that: "...He (Peavy) later engaged in the logging business at Michelli, Texas, where he worked for the Tyler Car and Lumber Company..."[35] Certainly, A. J. Peavy did not log at Michelli after 1893, since each obituary described his activities for each year thereafter. Regarding the spelling of Michelli, a list of the Houston East and West Texas Railroad sawmills for April, 1893, spelled the name "Michelli," as did a Beaumont *Journal* history of the abandoned town in November, 1904. And many records in the Angelina County archives use the double-L spelling. Hence, the writer believes that Effie Boon adopted the spelling of the original Spanish land grant holder, Vicente Micheli, as she found it in the Texas Land Office archives.

There are other discrepancies in Boon's article, notably her statements that; "...This mill was put down about 1900 by the Tyler Car and Lumber Company..." and "...The chief contractor for logging was A. J. Peavy, who began the logging operations in 1893 and continued until the mill cut out in 1906..." Actually Peavy's logging at Michelli ended in 1893 rather than in 1906. Boon added that: "...In 1906, the mill went into the hands of a receiver and ... the members of the logging company rented the mill and cut the logs it had in the river..." Again, that statement is incorrect, for Michelli and its mill had ceased to exist as early as 1903, probably by 1902, and by 1904, with all its mill machinery, tram tracks, and most of the housing removed, Michelli was already a ghost town.[36] Actually, A. Jasper Peavy had been connected with so many different sawmills and logging operations in Angelina County between 1888 and 1906, the year that he moved to Louisiana, that the writer fears he had become confused regarding his activities of 45 years earlier.

Nevertheless, the writer must accept some of Effie Boon's statements as being factually correct, namely, that A. L. Clark was manager of the mill; that three-fourths of Michelli's 300 woods and mill hands were Negroes; that the sawmill had a single circular saw of 90,000 feet daily capacity; and that Michelli was probably the only Angelina County sawmill to use the Angelina River as a mill pond. And certainly, Michelli had a population of about 1,000 people in 1900, along with a commissary, church, and schools.[37]

The writer believes that the Tyler Car and Lumber Company sawmill was built at Michelli no later than 1890, and most likely one or two years earlier. In April, 1893, it was one of the two largest sawmills in Angelina County, cutting 50,000 feet daily (as did Keltys). The same company had a 60,000 sawmill and its headquarters at Tyler, Texas.[38] In 1900, when Angelina County Lumber Company at Keltys was cutting 75,000 feet daily, Michelli (by then named Angelina) was cutting 85,000 feet daily.[39]

Jonathan Gerland, an archivist at Texas Forestry Museum during the summer of 1993, made the following comments about Michelli, taken from the Angelina County archives, in his letter to the writer, dated July 1, 1993, namely:[40]

> ... Michelli: Deed Records, (Vol.) Z: (pp.) 104-127, 130-131. It appears that Tyler Car and Lumber Company properties went into receivership CA 1897, because of a law suit filled by Paramore Investment Company of St. Louis about 1897. From that, William Cameron (and Co., Waco) emerged with the Michelli operation on March 14, 1898. Among properties conveyed: 13 miles of RR tracks (from Durst Station on the Cottonbelt to Michelli on the HE&WTRR) and the sawmill plant at Michelli, including land and timber. The sawmill engine was a Rankin and Futch Corliss (22"x42"), powered by a battery of 52-inch x 16-foot boilers. The saw equipment consisted of one carriage (3-block Allis), one W. F. Hill steam log turner, and the largest saw blades listed were 54-inch diameter. The planer mill consisted of a Land and Bodley (16"x24") engine, two 50-inch by 16-foot Huntsville boilers, a 126-foot line shaft, double-surfacer, three matchers, picket header, inside and outside moulders, Hoyt resaw, table edgers, and a swinging cut-off saw. Dry kiln measured 34x 94 feet, and the sawmill building proper measured 56x150 feet. From later entries, I found that Cameron began selling plats of land almost immediately after purchasing property. From maps and shipping guides at SFAS University, Nacogdoches, Michelli mail was sent to

> the town of Angelina. Angelina town became Kurth Ca. 1925...

As of 1900, William Cameron and Company of Waco owned four other sawmills in extreme Southeast Texas, namely, Saron, Trinity Co.; Carmona, Polk Co.; Rockland, Tyler Co.; and Nona, Hardin County, and most likely it did not need an additional sawmill. While there always remains the possibility that Michelli's timber reserve may have been nearly exhausted, the writer believes that the Cameron Company bought the Michelli mill expressly to increase its own timber reserve and to dismantle the mill.

Nevertheless, there is much evidence that William Cameron operated the Michelli sawmill for at least four years before abandoning it. The railroad bridge and the Michelli sawmill were located at the north end of Michelli Island, a mile-long island in the Angelina River, which enabled the mill to "boom up" (a river fence built of logs chained together) the south side of the river for a log pond. Jonathan Gerland of the Texas Forestry Museum located a 1900 history of the Michelli (by then known as Angelina) mill in the *American Lumberman*, a part of which follows verbatim:[41]

> ... At Angelina, Texas, 125 miles north of Houston...is one of the big yellow pine mills purchased by Mr. Cameron and still owned by the company. It is managed by S. Wilkins...This plant was purchased in March, 1898...Mr. Cameron...refitted it, overhauled it and made it a modern mill, and now it always has plenty of stock on hand to draw from...The equipment is an Allis circular with a capacity of 18,000,000 feet a year and the timber that is owned in connection with the plant is sufficient to supply the plant for ten years. The mill is rather scattered as to location. It covers considerable ground from the railroad down to the mill, which is situated on the river bank, where the river can be used as a log pond by booming it in.
>
> ...To avoid any possibility of loss of timber by flood, the company has cut a canal across a bend and left the loop which is used for a mill pond as slack water by turning the channel of the Angelina river down through the canal across the neck of the bend.
>
> ...The plant consists of a sawmill, a planing mill of good

> capacity, and a good dry kiln and plenty of sheds. There is always about 5,000,000 feet of lumber carried here in stock...There is a railroad roundhouse here which is owned by the company, where the locomotives are taken care of...To handle this end of the business properly, the company has put in eighteen miles of broad (standard) gauge track and uses four locomotives and thirty cars. The logging equipment is modern and complete...In connection with this, there is also a repair shop, where repairs are made to locomotives in case of breakage while in use.
>
> ...The commissary is well-equipped and carries a stock to fill the needs of all of the people employed...There are quite a number of men quartered in good substantial buildings, though not fancy...The houses for the men are built in comfortable style, airy and roomy, and the location being a good, healthy place, there is consequently very little sickness at this mill...

The last swan song about the Michelli sawmill was published in November, 1904, and the writer will include its almost verbatim, as follows:[42]

> ...The Passing of Michelli—The fact that a big sawmill happens to locate in a certain section of the country is no reason why that particular section of the country is going to become permanent...Such for instance was Michelli, on the Angelina River, ten miles north of Lufkin, on the Houston East and West Texas Railroad. Ten years ago it was the seat of one of William Cameron and Co.'s largest mills in Texas. It was beautifully situated on the banks of the Angelina River, and the river was used for booming the logs.
>
> ...The tram road ran into the vast piney woods for ten miles, and when all the timber on a certain line of tram was consumed, the tram rails were picked up and laid in another direction. The traveler or hunter in the woods bordering the Angelina River will come on these old abandoned tram roads occasionally, as they thread their way deviously through the forest.
>
> ...Michelli had every promise of becoming a town of some magnitude, and the railroad company, believing that per-

> haps it would, established a depot, ticket and telegraph office there. A standard gauge railroad was built by the company westward ten miles to Clawson on the Cottonbelt Railroad, and the company shipped all their lumber over that route.
>
> ...The Nacogdoches people believed for a time that the Cottonbelt Railroad would use the Cameron road from Clawson to Michelli, and thence come to their town over the Houston East and West Texas Railroad from Michelli. It was only ten miles to Michelli...but such steps were not taken. And with the passing of Michelli, the road from Michelli to Clawson also passed, and there is hardly a trace left in the country through which it ran.
>
> ...With hardly any warning, the mill was torn down and moved away, the lumber was all shipped out, and the shacks sold to people in the surrounding country. A number of shacks are still standing, also the old depot telegraph office, and these are inhabited by bats or other vermin of the forest.
>
> ...The only person now living at Michelli is the bridge keeper of the railroad. The town had a large hotel, several stores, a splendid church and school, and a number of other business enterprises. They have all gone, and...even the name of the place has been changed to Angelina. Since Nacogdoches County went "dry" last year (1903),...a saloon has been put up on the Angelina side of the river in the suburbs of what was once Michelli.(1904)

And thus ended the story of the old Michelli sawmill in Angelina County.

THE EARLY SAWMILL AT KELTYS

Certainly one of Angelina County's greatest assets was Joseph H. Kurth, Sr. and his partners, who organized the Angelina County Lumber Company in 1887. Kurth was born in Bonn, Germany, in 1857 and emigrated to Galveston, Texas, in 1878. For three years, he labored at a variety of odd jobs at Seguin and Willis and spent a few months working in and studying the lumber business at Beaumont. About 1882, he became associated with Sam Allen in the operation of Allen's mills at Corrigan and Asia on the Trinity and Sabine Railroad. In 1884, Allen helped finance Kurth's first small sawmill at Kurth Station, east of Corrigan, and along the way Kurth met and became the partner of Simon W. Henderson of Corrigan, with whom he would remain associated in business for the remainder of his life.

By 1886 Kurth and Henderson could see that timber along the Trinity and Sabine Railroad was already being rapidly depleted by a dozen sawmills. In 1887 they turned to Angelina County, where they learned that Charles L. Kelty of Keltys had operated a small sawmill and the first planing mill in Angelina County since 1870. After some negotiation, they bought out Kelty's business in 1887, at which time the Angelina County Lumber Company was born, with its third and newest partner being a Jewish merchant of Corrigan, Sam Wiener. From its beginning, the new Keltys company chose to develop a long-term plan of forest preservation on its lands in north Angelina and Nacogdoches counties and avoid the disastrous "cut and run" techniques usually practiced by most East Texas sawmillers.[43]

In 1890 the first history of the new Kurth sawmill was published, and the writer has chosen to reprint most of it verbatim because of the story's wealth of information, as follows:[44]

> ...Lumber Business at Keltys, April 6—Prominent among the overshadowing industries of East Texas, the lumber business will properly take first rank...To illustrate the volume of business being done at Keltys, a little mill burg of some 300 souls, on the Kansas and Gulf Short Line, one and one-half miles north of Lufkin, I will furnish you with a few dots concerning the fine new sawmill located there and owned by Messrs. Henderson and Kurth, young men of rare push and vim, and recognized leaders in their business. ...Their plant is one of the finest and best-equipped in this section, and the facilities for handling lumber are splendid. The mill has a cutting capacity of 50,000 feet a day; the Prescott steam carriage feed is used. J. A. Ewing, a man well up in his profession, is mill foreman; J. A. Snodgrass...is sawyer; while Mr. James Smith is the clever fireman. The output is strictly first-class, yellow pine long leaf lumber. The mill and dry sheds, which are in tiptop order, are protected by corrogated roofing. R. W. Scurlock is the yard foreman. The planer is furnished with surfacing, flooring, and picket machines, resaws, etc. A large No. 2 sizer is now

being put in, and other improvements are in contemplation, all the outgrowth of a steadily increasing trade. B. F. Fite is the foreman of this department.

...The present dry kilns will dry 10,000 feet per day, but much larger ones are now in process of erection and will soon be in operation. A fine system of waterworks is being added that will render the plant and its valued surroundings comparatively safe from the dreaded fire fiend.

...The logs are brought to the mill on a tram road owned by the firm, and laid mostly with steel rails, extending to a distance of five and one-quarter miles in the magnificent pineries. The gauge is the same as that of the Kansas and Gulf Short Line. The Edna Ewing is the name of the locomotive that pulls the cars over this road, and Mr. J. T. King handles the throttle in an efficient manner. Mr. King has five years experience in this exciting calling, being three years at the Morris mill (Bowers, Polk County) on the Gulf (Trinity) and Sabine road and two at this point. We are indebted to him for a safe and pleasant ride over this road, that gave us a fine opportunity to note the large operations in the woods, presided over by Mr. T. Y. Faulkner as (woods) foreman.

...While a large and well-selected stock of lumber is carried, still more orders are on file than can be filled. A large amount is furnished to the railroad here, and heavy shipments are made to the Texas trade, as well as to Kansas, the Indian Territory, Missouri, Illinois, and Nebraska. The mill gives employment to fully one hundred men, most of whom live here with their families in homes furnished by the company.

... Messrs. Henderson and Kurth have a contract to supply the Kansas and Gulf Short Line with 60,000 oak ties, and for the purpose of handling this material in an expeditious manner, they are now constructing a transfer switch from the main line to the log road over which the railroad's cars can be run direct to the yard.

...The mill commissary, which does a business of about $120 a day, is in charge of Mr. Melvin Glenn, a young gentleman of refined manners and pleasing dress, while Mr. J. E. West...has charge of the books. Mr. West is a

> native of Harris County and is very popular here and at Lufkin, where he is widely known.
> ...Mr. S. W. Henderson, one of the leading members of the firm, is also connected with the mercantile firm of J. L. Henderson and Co. of Corrigan. Mr. J. H. Kurth, the active and efficient managing partner of the firm, resides here with his charming and affable wife in a neat and pleasant cottage where the wayfaring man always meets with a cordial reception and elegant entertainment.
> Mr. Kurth is a noble example of what can be accomplished by push, energy and sound judgment. Leaving his faraway home in Germany when a mere youth, he first touched American soil twelve years ago, a stranger to our language and customs, and commenced the battle of life in which he has...worked himself up until he now stands in the front ranks of our millmen.

THE EARLIEST YEARS OF THE KELTYS SAWMILL

In October, 1890, Angelina County Lumber Company was incorporated at Homer, Texas, with a capital stock of $100,000, with J. H. Kurth, Sam Wiener, and Simon Henderson as the principal officers.[45] The decade of the 1890's was especially turbulent, both financially and weather-wise, for the East Texas lumber industry for a variety of reasons - a national depression in 1893-1894, and in 1896-1897, by lumber overproduction, depressed markets, high interest rates, flooded timberlands, or prolonged droughts. Many mills went bankrupt; others operated on half-time or three-quarter time, or even closed down for long periods of time, as the Keltys mill did in 1896. In May, 1896, a Keltys mill spokesman reported that: "...a few orders are coming in...but they are so few and far between, they are beginning to look to a mill man's eyes like gold dollars would to a pauper's (eyes)..." In November, 1896, a Galveston editor observed that "...the Angelina County Lumber Company is making preparations to start up its mill, which has been closed down for several months..."[46] Lack of lumber demand was the obvious reason; it appeared that prosperity did not return to Keltys until about 1898.

The Keltys census of 1900 revealed who many of the key plant personnel at Angelina County Lumber Company, were as follows: (pct. 1, res. 292) Ben McWhorter, bookkeeper; James Butts, yard foreman;

Thomas J. Bonner, log contractor; J. H. Kurth, lumber manufacturer; John Prosser, filer; Geo. W. Morris, scaler; James Chambers, sawyer; Edgar Jones, locomotive engineer; Martin Goodman, locomotive engineer; Alex Hollinger, locomotive engineer; Curtis Grimes, blacksmith; M. Peavy, hotel keeper; Joseph A. Henderson, shipping clerk; Dr. Eddie Clark, mill physician; Eli Wiener, lumber manufacturer; Wesley West, planer foreman; and (res. 407) Simon Henderson, lumberman manufacturer.[47]

By 1904, the Keltys sawmill had been enlarged and modernized sufficiently to enable it to cut 125,000 feet daily.[48] In 1906, while J. H. Kurth was on a visit to his native Germany, the Keltys mill fell victim to the dreaded fire demon and it burned to the ground, a total loss. The mill, however, was quickly rebuilt, with a combination of circular and band saws which returned it to its earlier capacity of 125,000 feet daily or 3,000,000 feet monthly. About the same time, capital stock was increased to $225,000.[49] Modernization at Keltys, though, was always a continuing process. In 1908, the owners added a new pair of twin steam engines, size sixteen-inch by 22-inch stroke, to the mill equipment, and in 1909, they upgraded the planing mill by replacing old equipment with new matchers, moulders, sizers, dressers, picket machine, resaw and edger, as well as a fan and blowpipe system designed to clear all shavings immediately.[50]

A letter from Jonathan Gerland at the Texas Forestry Museum reported the following equipment changes over a long period of years from 1906 until 1955. It appears verbatim, as follows:[50a]

> ...After the 1906 fire, the (Keltys) mill was built as a single band and 'Dixie circular.' A 1928 appraisal listed one 18"x24" Houston, Stanwood and Gamble twin horizontal side valve steam engine (1919 model), and three 72'x18' Casey and Hedges boilers. Sawmill proper building measured 50'x186' and included an 8 foot band mill and 60-inch circular saw. Also included were one 66" eight-saw edger, a 20-foot six-saw overhead slab slasher with 40-inch saws, and a 27-inch undercut trimmer with twelve 22-inch saws. After the 1939 fire, the plant was built as a double band saw rig (rated at 20,000 feet per hour powered by 4 boilers producing 1,050 horsepower. The main sawmill engine was a 28"x48' stroke Filer and Stowell. Dry kilns totaled six in number, with nine dry sheds, and seven planing machines. By 1948 (before the electric mill), a

> small hardwood mill was added to cut ties for the A&NR railroad. It was rated at 15,000 feet and was diesel-powered. The main mill went electric in 1955, and was rated at 30 million board feet a year...

By 1900 Eli Wiener, a brother of Sam, was living at Keltys and taking a very active role in mill management. It seemed likely too that Sam Wiener, the secretary-treasurer, may have become less active since in 1895 he moved to Monroe, Louisiana to enter business there. By 1934 Eli Wiener had advanced to company president.[51]

By 1910, many people occupying key positions at the mill had changed hands since sawmill labor always experienced a high turnover rate. Personnel included W. A. Smith, checker; Black Powell, assistant planer mill foreman; Frank Powell, planer mill foreman; Lee Whitman, bookkeeper; Martin Goodman, locomotive engineer (one of the few employees left over from 1900); M. A. Moses, planing mill foreman; Abb Jones, blacksmith; Charlie Fuller, blacksmith; Charles Beckford, sawyer; Russell Beckford, filer; George H. Beckford, sawyer; John Struber, filer; Dock H. Nelson, engineer of steam loader; Joe Beard, mill pond foreman; Robert Lowry, dry kiln foreman; William H. Jellison, shipping clerk; Hugh Cavin, stenographer; R. J. Smith, sawyer; Eli Wiener, lumber manufacturer, Mary Hornsby, boarding house owner; Simon W. Henderson, sawmill manager; Jack Henderson, commissary clerk; Percy Douglas, commissary manager; Dr. Eddie Clark, mill physician; Ernest L. Kurth, bookkeeper; and Leonard Cotton, locomotive engineer.[52]

The Keltys log tram from about 1890 until 1900 was a short tram road, consisting in part of four by four wooden rails over which a few log cars were pulled by a small Shay or "dinky" engine or by ox and mule teams. Oxen and mule teams were also the standard fare for pulling logs on eight-wheel wagons and loading the tram cars. Within a few years, mules and oxen had to be left in corrals in the forest, where railroad spurs crowfooted out in all directions in order to reach the uncut stands of timber. Such locations were called "fronts," and in time the Keltys logging camps of Acol and Camp Nancy needed housing and even a church and school for the loggers in the same manner that the sawmill town did.

The Keltys tram road was incorporated separately as the Angelina and Neches River Railroad, a common carrier, in August, 1900, although the tram operated much earlier. The tram eventually ran for over thirty miles to Chireno in Nacogdoches County. By 1923, Keltys was logging at a new front, called Camp Nancy, near Zavalla in south Angelina County, and by

1933, when that camp was cut out, it moved farther south into Tyler County. Hence, it became necessary for Angelina County Lumber Company to pay "trackage rights" for its log trains to pass over the Texas and New Orleans rails to the intersection of the Angelina and Neches River Railroad. The last action of the log train crews was to unload each train load of logs into the Keltys mill pond. At all sawmill ponds, some logs were referred to as "sinkers" because they were too heavy to float, and a special dredging machine operation had to be utilized to raise them to the surface to be milled.[53]

THE KELTYS SAWMILL EQUIPMENT

Mill operations at Keltys were divided generally into four parts, as follows: the sawmill; the planing mill, which included flooring and picket machines; the dry kilns, where green lumber was steam-dried in about 36 hours of time; and the yard, dry sheds, and loading docks, where box cars for shipment were loaded. There were also a dozen auxiliary units which will be discussed later. In 1934, the planing mill engine was a Corliss, which weighted 69,000 pounds. By 1948, there were two sawmills at Keltys—a hardwood mill, which cut 15,000 feet daily; and a pine mill. Each of the two sawing units in the pine mill consisted of endless band saws, fourteen inches wide and fifty feet long, as well as the 55-foot, shotgun-feed log carriages. Each saw was held in place on two eight-foot circular "dogs," and each saw was removed about every three hours for sharpening. Other mill cutting equipment included a drag saw to cut logs to length, an endless chain "haul-up" for bringing logs from the pond to the carriage, steam kickers and steam log turners, edger saws, trimmers, cut-off saws, shashers, resaw, perhaps a Wickes gang saw, the long line shaft with its power take-offs, and line rollers and conveyers for moving slabs and boards. The cut of the band saws was 10,000 feet an hour.

In 1948, the power house at Keltys consisted of three horizontal boilers each of 150 horsepower, and one vertical boiler of 600 horsepower. The boilers in turn powered a single 800 horsepower Filer-Stowell engine with a twenty-eight by sixty-sixty inch stroke. The 50-ton flywheel was twenty feet in diameter, with a 40-inch face. The belting from it to the long line shaft was made of 12-ply rubber, 41 inches wide by 127 feet long; the belt cost $1,800 and lasted ten years.

Green lumber was loaded onto iron dollies, which were pushed directly into one of the plant's six dry kilns, each a brick building sixty-feet by 120-

feet in size. Some lumber was allowed to air-dry for between four and six months. After passing through one of seven planing machines, finished lumber was then stored in one of the nine large Keltys dry sheds or moved straight to the loading dock for shipment. There were many more auxiliary units at the Keltys sawmill which will not be discussed in detail, namely: the office, boarding house, post office, filing room, blacksmith shop, carpenter shop; round house and machine shop, where the locomotives, tram cars, loaders and skidders, and all steam equipment were repaired; two mill ponds, two elevated water tanks, a water hose and fire plug protection system; and a power house where a 100-kilowatt dynamo supplied lighting to the mill and mill town. Many large mills also operated an ice plant, but since there was no mention of one at Keltys, ice was probably hauled from Lufkin. There was also a second fire at Keltys about 1948, not nearly as serious as the one of 1906, but the electrical plant building was one of those that was badly damaged.[54]

THE SOCIAL LIFE IN THE MILL TOWN OF KELTYS

Lita Mayberry's M. A. thesis about Keltys zeroes in on the social life of the sawmill town during the year 1948. She divided the town into four units, that is, the business and public houses centering around the mill office, commissary, and community hall; and three or four styles and divisions of housing, one area for supervisory personnel, two other areas of cottages for white workers, and a section of housing for Negro employees across the Cottonbelt tracks from the mill. Lita Mayberry further subdivided the cottages into Class A, B, C, and D housing. The Class A homes were larger houses for higher-salaried personnel, were painted white, fenced, were spacious with eight to ten rooms, all conveniences, garage and garden. Type B houses were less spacious, four to six room painted cottages with electricity, water, gas and sewer connections. Type C housing had four or five rooms, porches and fencing, electricity and water, were painted, but no garages, gas or sewer connections. Type D housing were two or three rooms, were unpainted with electricity and water, but no garages, sewer or gas. All houses were screened. Of course, when Effie Boon and Lita Mayberry wrote about Keltys, strict social segregation was still a way of life in East Texas.

In 1948, twelve hundred people lived in Keltys, and 380 men worked for the sawmill company, 330 of whom were white and fifty were Negroes. Another 35 families lived in the company logging camps. Those houses not equipped with gas cooked and heated their homes with free fuel from the company slab or scrap wood pile, or else with kerosene stoves. Mayberry wrote that Keltys' Negro employees owned reasonably good clothes and furniture, and that in 1948 some of them owned cars and electric appliances.

Where once the Keltys commissary had been a cracker barrels, coffins and yard goods affair, it had by 1948 developed into a modern general store, some 45 feet by 100 feet in size, which employed 12 people and displayed all packaged and canned goods, staple and fancy groceries, with a general wearing apparel department, hardware, a drug department with cosmetics, notions, and patent medicines, and a feed store attached. Next door was the mill physician's office. Employees paid $1.25 monthly as a hospital fee for a family of two.

Before 1900, Keltys had only a "community church," but by 1934 it had separate Methodist and Baptist churches for whites. The community hall was used for a variety of meetings, parties, dances, lodge meetings, games, skating, movies and other amusements. In 1934 Keltys had one grammar school for white children. By 1948, however, a new Mabel Malear Elementary School had been built at Keltys, whereas secondary students were bused to Lufkin, since Keltys was in the Lufkin School District. Negro children attended segregated schools in Lufkin or at Brandon, an elementary school. In 1948, white churchgoers still attended the Methodist and Baptist churches, and Negro churchgoers attended a Baptist church of their own, organized in 1892, which was in a new building free of debt.[55]

As the old company management had mostly passed on by 1930, new officers succeeded them, and by 1934, Eli Wiener was president of the company; Ernest L. Kurth was vice president and general manager; and D. W. Thompson had become secretary and treasurer. During the depression and World War II years, E. L. Kurth was the company's driving force, who brought startling advances and even a paper mill to the area. Kurth had a serious bout with illness, however, losing his vocal chords during the late 1940's, and following his death in 1960, the Angelina County Lumber Company and its huge acreage of timberlands were soon sold to the Owens-Illinois Company, who a few years later closed down the mill permanently.[56]

SAWMILLING AT BABER, ZAVALLA, RETRIEVE, HUNTINGTON

There were some sawmills in south and eastern Angelina County that Effie Boon's thesis classified as "small sawmills," meaning those cutting 25,000 feet daily or less, and she included the towns of Baber and Boynton as fitting that category. Apparently, neither town met her criteria exactly, since in 1904 the Boynton Brothers sawmill cut 50,000 feet daily and in 1909 the Baber sawmill cut 60,000 feet daily.[57]

When Samuel Fain Carter, Sr. of the old Emporia Lumber Company retired from active milling in 1906, the mill superintendent and secretary-treasurer of the old Emporia company was J. Press Carter, his younger brother. Press Carter promptly sold his large block of Emporia stock in 1906 in order to buy large stands of timber in Angelina, Nacogdoches and San Augustine counties. He then organized and built the Attoyac River Lumber Company, an 80,000-foot mill, at Mayo, Nacogdoches County and the J. P. Carter Lumber Company, a 60,000-foot mill, at Baber, Texas, located four miles southeast of Huntington on the Texas and New Orleans Railroad. Press Carter furnished the following information about his Baber sawmill when he was interviewed by *Southern Industrial and Lumber Review* in Houston in 1909, as follows:[58]

> ...The Carter Lumber Company also owns a circular mill of 18,000,000 feet annual capacity at Baber, on the Texas and New Orleans Railroad. This mill is equipped with a planer, steam dry kilns, etc. The general offices of the Carter Lumber Company are at Houston...through which the output of both the above mills is marketed...

Some of the key plant personnel in the 1910 census of the Baber sawmill included Fred Selman, saw filer; Edgar T. Hickman, mill superintendent and postmaster; Carl Hickman, bookkeeper; Dr. Byron L. Jordan, mill physician; William A. Lindsey, shipping clerk; Bony Harville, sawyer; Thomas Maxey, planer mill foreman; William H. Wisenbaum, yard foreman; Seth Wisenbaum, checker; and Shelby Price, dry kiln foreman.[59]

Usually, a 60,000-foot sawmill required at least one hundred mill hands and loggers to operate in a normal fashion. About 1910, Baber probably reached its maximum population of about 400 persons, complete with commissary, a church and a school, before the timber was exhausted in 1913, and the mill was dismantled and the site abandoned.[60]

Huntington, Texas. is not mentioned either in Effie Boon's "History of

Angelina County," or in *Land of The Little Angel* as being a significant lumbering community, and apparently no large lumber firm ever located there, even though it was the junction of the Cottonbelt and Texas and New Orleans railroads. Nevertheless, there were a series of small sawmills there after 1900, and there were some crosstie and piling industries there as well. It could have been that most of the timber surrounding the town had already been bought up in large tracts by the large lumber companies, but whatever the reason, it certainly appeared that Huntington developed an agricultural and cotton gin economy much earlier than any other Angelina County town did.[61]

Named for C. P. Huntington, the Southern Pacific rail magnate, Huntington was first surveyed in June, 1900.[62] Joseph H. Burnaman built the first sawmill there about 1901, but lacking access to an adequate timber reserve, his mill was already cut out by 1904. One news account of 1904 observed that: "...The mill of Joe Burnaman at Huntington has closed down permanently on account of being out of reach of timber..."[62]

Burnaman's mill was small, probably 25,000 feet daily. Nevertheless, the loss of its payroll was sorely felt in Huntington, as another article revealed:[63]

> ...The mill of J. H. Burnaman at Huntington is not running at present. It has no timber in reach and will be moved to Bronson in San Augustine County (sic: Bronson is in Sabine County), sometime this year...The removal of this mill from Huntington will cause the town to lose several hundred dollars' worth of trade every month. Huntington is surrounded by good agricultural, fruit, and truck-farming land. But in the immediate vicinity of the town, the people have not as yet turned their attention in that direction. The main thing that has sustained Huntington as a trading point has been the tie and piling and lumber business combined, and now since the mill...is going to move, and the tie and piling business is dead, Huntington finds itself in the dullest period of its existence...

Despite the seeming inaccessibility of marketable timber near Huntington, three other sawmills were located there by September, 1906. In 1906, the Nunnally-McConnico sawmill cut 25,000 feet of pine daily and gave employment to about 35 or 40 men. Their logging camp was located at a front named McConnico, south of Huntington. Also in 1906, Charles H. Angell's sawmill cut 25,000 feet a day at Huntington. Since Angell was postmaster at Krub for five months in 1905, perhaps Krub may have been the name of a village which grew up around his sawmill. Sebron J. Cowart's small mill was credited to Huntington in 1906, but it may actually have been located at Ivy (a. k. a. Odell), a nearby village. S. J. Cowart had served as Ivy's postmaster in 1898, and in 1906, Cowart's mill cut only 3,000 feet of lumber daily. The writer believes that the life of these three sawmills, and Huntington's history as a lumber center, ended about 1910.[64]

Hardly anything is known by the author about Retrieve, Texas, beyond what is available in either Boon's thesis or *Land of The Little Angel*. Located six miles southeast of Huntington on the Texas and New Orleans Railroad, it was known to have had a 15,000-foot sawmill and to have employed about thirty mill hands, but apparently even the name of the lumber company is lost. All of the writer's railroad lists of sawmills are dated between 1893 and 1906, and none of them list an Angelina County rail stop named Retrieve. The mill probably was founded and cut out between 1907 and 1912.[65]

The best report on Zavalla in the early years appeared in a lumber journal in 1904, as follows:[66]

> ...Zavalla is twelve miles south of Huntington on the same road (T&NO) and is kept up almost entirely by the local trade from the various lumber, stave and piling camps in the vicinity. The Frost Lumber Company of Lufkin has two spurs that penetrate the heavy piney to within a few miles of Zavalla. A large stave mill is in operation on the Neches River, south of Zavalla, and has local headquarters in town. The finest pinery now to be found in Angelina County extends from Zavalla eastwardly to the Angelina and Neches Rivers. This timber belongs partly to the Kirby (Lumber) Company and partly to the (Wm.) Cameron Company...The timber interests in the vicinity of Zavalla will keep the town up for five years, after which the people will either have to turn their attention to fruit or truck farming or move...

Undoubtedly, one or more small sawmills operated at Zavalla around 1900, but if so, the name(s) do not appear on any list of Texas and New Orleans Railroad's list of sawmills up until 1906. As of that year, no nearby timber was available to small sawmillers anyway. The Kirby Company loggers cut timber in the extreme southeast Angelina forests for their mill at Woodville, Tyler County. William Cameron and Company logged its Rockland mill from that region too, and so did Hal Aldridge for his Aldridge sawmill in Jasper County.

EARLY SAWMILLING AT LUFKIN

As was previously stated, the first sawmill of record at Lufkin was that of Richard Blair about 1890. It was a small mill of 10,000 feet daily capacity and was soon dismantled and moved away.[67] Among surviving papers of the Angelina County Lumber Company are the impressive letterheads of the Lufkin Manufacturing Company, "Manufacturers Of Yellow Pine Lumber," at Lufkin, Texas. The letters carry dates only of July and November, 1896, and list the company officers as being: Judge E. J. Mantooth, president; J. T. Maroney, vice president; J. V. Burch, secretary; W. H. Bonner, treasurer; and W. H. Barron, general manager. No other mention of Lufkin Manufacturing Company has ever been located by the author, and he concludes that it must have been a short lived enterprise which perhaps went bankrupt.[68]

Lufkin acquired its first large sawmill as a result of the organization of the Lufkin Land and Lumber Company by Enoch W. Frost, Edwin Ambrose Frost, George Alfred Kelley, and Thomas L. L. Temple in May, 1899, with a capitalization of $70,000. A letterhead from Lufkin Land and Lumber Company, dated December 1, 1899, confirmed the purchase of materials from Angelina County Lumber Company for use in building the new sawmill, and a list of the Houston East and West Texas Railroad sawmills, dated January 1, 1900, bore the notation that: "...A modern mill is also being erected at Lufkin and will have a daily capacity of about 75,000 feet..."[69]

E. W. Frost and his son, Edwin A. Frost, were timber barons who owned a sawmill conglomerate in Arkansas and Louisiana with headquarters at Texarkana and Shreveport. A native of Canada, G. A. Kelley worked in sawmills from age fifteen and came to Arkansas in 1897 as a representative of E. P. Allis and Company (now Allis-Chalmers), a sawmill manufacturer of Milwaukee. Kelley was a first-rate sawmill builder and engineer, who held patents on both the "Kelley steam log kicker" and the "Kelley steam log turner."[70]

By 1904, the Lufkin Land sawmill was cutting 125,000 feet daily and was located on the east side of town. Although in Lufkin's corporate limits, the village that sprang up around the sawmill became known as

"Lufkin Land." The town's 1900 census revealed who held many of that company's key positions, as follows: George S. Prestridge, mill foreman; S. W. Prestridge, log contractor; Clarence Brown, sawyer; William F. Green, engineer; L. Hobbs, planer foreman; Absalom Smith, dry kiln foreman; Jas. D. Gillis, yard foreman; G. A. Kelley, owner and superintendent; Dr. Thomas W. Large, mill physician; Silas Gardner, checking clerk; Joseph P. Dunn, shipping clerk; Edwin A. Frost, owner-manufacturer; Abby Thompson, machinist; and M. Christopher, scaler.[71]

In 1905, the following history of the sawmill appeared in a Beaumont newspaper, as follows:[72]

> ... A PAYING PROPOSITION —A Description of The Plant Of The Lufkin Land and Lumber Company-Lufkin, Texas, April 5—About one-half mile from Lufkin is the plant of the Lufkin Land and Lumber Company. This company was organized in May, 1899. A complete plant was built and equipped. Since its incipiency, it has been a paying proposition, largely due to the fact that it has been well-officered and well-managed. Manager G. A. Kelley has everything well in hand...One of the features of this fine plant is its water protection. Not only has it an elevated tank, but also pumps and a connection with the city water works, which gives ample pressure and an unlimited quantity of water in case of fire. It also has several big ponds, one of which is being dredged, and there are two dredge barges for that purpose. This will insure them an abundance of water. The loading docks will accommodate 20 cars. The lumber sheds are very large and well-constructed.
>
> ...Roominess is a feature of the plant. The yards are well-arranged and easy of access. The class of labor is above average. Officers: E. W. Frost, president; G. A. Kelley, vice-president; and E. A. Frost, secretary-treasurer.
>
> ...Operating department: G. A. Kelley, manager; J. P. Dunn, shipping clerk and yard foreman; C. G. Gribble, bookkeeper; W. G. Banks, stenographer; Calvin Cochran; sawmill foreman; F. O. Hollenback, planer mill foreman; D. S. Prestridge, J. W. Prestridge, J. W. Buchanan, woods superintendents; O. O. Clark, sawyer; W. L. Fisher, filer; M. Rosenbaum, Allen Ferr, checkers; D. C. Payton, commissary manager; Walter Fountain, chief locomotive engineer; Joe Giles, Lewis Rigsby,

> Henry Walker, Claude McCrowden, locomotive engineers; M. S. Wood, master mechanic; M. M. Cowsert, sawmill engineer; S. V. Fuller, assistant engineer; W. L. Bishop, filer; George Babers, A. A. Spears, Dave Smith, planer engineers.
> ...General Information: Sawmill capacity, daily-200,000 feet; stumpage-10 years; feet in yard-2,000,000; output in 1904—49,000,000 feet; saws-two single-cutting bands; sawmill engine-350 H. P.; boilers-four 60-inch by 18 feet; kilns-six rooms, standard, brick; water protection tank-55 feet high, 15,000 gallons, pump, and city water.
> ...Planer Mill: Engine-18x22 inch stroke, 75 H. P.; boilers-two 60-inch by 16 feet; 1 matcher, 5 Hall and Brown machines, 1 moulder; one resaw; one edger; capacity-200,000 feet daily; output in 1904—50,000,000 feet.
> ...Electric Appliances: Atlas automatic engine, 150 H. P.; sheds—two rough, two dressed; commissary stock—$47,000; men employed—200; average wages per month-$12,000; artificial ponds—one large, two small; locomotives-seven; tram cars—ninety; employee houses-120.

Early in its existence, Lufkin Land and Lumber Company acquired 53,000 acres of timberland, mostly in eastern Angelina and San Augustine counties, and the company quickly chartered and built the Texas and Louisiana Railroad, which as a tram road, extended eastward from Lufkin (crossing the Texas and New Orleans rails at White City) to Broaddus in San Augustine County. Much of the tram trackage was later absorbed by the Cottonbelt Railroad. As the steel rails moved eastward, Lufkin Land used a steam loader to place logs onto tram cars and later added a steam skidder to its logging equipment. By 1904, the mill was tramming 28 miles to the Attoyac River and eventually 35 miles to the vicinity of Broaddus in San Augustine County, while some of its tram spurs reached the vicinity of Zavalla.[73]

A newspaper editor reported in 1904 that the Lufkin Land sawmill accepted the long log haul from San Augustine County stoically for its several daily log trains, as witness the following quotes:[74]

> . . . The Lufkin Land and Lumber Company says they can tram their logs 50 miles, and it will be a whole lot cheaper to do so than to move the mill to the timber in San Augustine County...The...company will thus be tramming their logs a distance of thirty-five miles, twenty-three of which are over the

Cottonbelt Railroad...

Like many East Texas sawmills, Lufkin Land had problems keeping its mill ponds in tiptop shape, and one editor observed that:[75]

> ...The big mill at Lufkin is building a dredge boat to be used in raising sunken logs in their pond at the mill. Besides raising ("sinker") logs, the boat will also be used in dredging the pond and keeping it clean so the water supply will not be interfered with. The pond covers an area of about five acres and is ten feet deep in some places. The mill has been closed down for a few weeks to make some necessary repairs; it will start up again soon...

The citizens of Lufkin were startled in June, 1905, to learn that the huge Long-Bell Lumber Company of Kansas City, Missouri, had purchased their local sawmill. Long-Bell was principally a large retail concern, with about forty lumber yards scattered out over Kansas and Missouri, but the company owned vast timberlands in western Louisiana and two sawmills at Lake Charles and DeRidder. One news account recorded that:[76]

> ...The main feature of the past ten days...was the purchase by the Long-Bell Lumber Company of Kansas City, Missouri, of the big sawmill of the Lufkin Land and Lumber Company, together with 53,000 acres of fine timberland, ... that about 22,000 acres are what is known as cutover land ... As an industrial proposition it has been the biggest thing that ever happened in Angelina County, and has done more to building up Lufkin as a real live city, as well as extending the Cottonbelt Railroad eastward...
>
> ...The Long-Bell people are making great improvements to their equipment at the mill, and also in the matter of rolling stock which is to be used in handling logs over the Lufkin-Broaddus extension...The Long-Bell people by the terms of the agreement came into a vast quantity of timber in San Augustine County...

Once again, the census of 1910 revealed significant changes among key personnel at the Long-Bell sawmill, as follows: Belle Gates, boarding house owner; Walter Sims, planer foreman; Luther McClerkin, shipping clerk; Elmore Sparks, Simon McAdams, steam engineers; William Threet, assistant shipping clerk; Eugene Toy, commissary manager; R. W. Swank, sawmill foreman; Louis Ripley, Charles Cameron, locomotive engineers; W. R. Davis, yard foreman; Oscar Eddington, Bennie Lindsey, Jim Yancey, graders; E. Baxter, machinist; and W. B. Stewart, bookkeeper.[77]

As Long-Bell Lumber Company, the big Lufkin sawmill enjoyed a long 25 year span of prosperity until an exhausted timber reserve, coupled with the Great Depression and a plummeted lumber market, forced the company to shut down the mill permanently. Long-Bell operated at various time a number of logging camps — Broaddus, White City, Hamburg, Lakeview, Donovan, and Banister — mostly in San Augustine County, and the loggers' housing in all but the latter location consisted of two small rooms bolted onto a rail car. In 1923, as Long-Bell experienced a shortage of virgin forests and was recutting its second growth timber, the Lufkin plant erected a hardwood mill in connection with the pine mill.

The Long-Bell mill was also one of only about three Angelina County sawmills that operated a night shift. The surrounding mill community, known as Lufkin Land, developed its own facilities — a company-owned commissary, employee housing, drug store, community hall, and grammar school. And when the sawmill suddenly closed in 1930, it left a great void in the city as 250 men were unemployed. And in the financial mood of the year 1930 in Angelina County, there was almost no hope of finding other employment.[78]

Another Lufkin mill that has been totally ignored in Angelina County's historical writings was the Martin Wagon Company's hardwood sawmill. The wagon factory made a variety of wooden log carts and trailors, but its specialty was an oversized, eight-wheeled log wagon, built to carry the largest logs of East Texas, and shipped also to all points in the timbered southern and northwestern states. Founded early in 1910 by D. W. Martin, the wagon company's long-time president was Bennie Lee Zeagler, and apparently at its beginning, the Martin Wagon Company sawmill was privately-owned by Wilbur Brainard, an arrangement that soon ended.[79]

In the 1910 census for Lufkin, the following Martin Wagon Company's key sawmill personnel were as follows: (beginning from Pct. 1, res. 10) Bennie Lee Zeagler, manager, wagon factory; Harvey Burnett, sawmill filer; Clarence Burnett, assistant filer; William Couffer, William Thatcher, machinists; Reese Vinson, locomotive engineer; Marvin Lewis, checker; John H. Cochran, sawmill foreman; Joe P. Dunn, yard foreman; Joe L. Lapoint, sawyer; Lee Mantooth, machinist; Ira Holbrook, sawyer; Josh Cochran, manager, planing mill; Sebron J. Cowart, manager, dry kilns; Ben Edwards, stenographer; Lyman Bettisworth, lumber office manager; Ralph Muckleroy, bookkeeper; and (res. 137) Wilbur Brainard, owns mill.[80]

Certainly the Martin Wagon Company survived as a major Lufkin manufacturer and sawmiller for at least twenty years. The wagon plant was absorbed into the Lufkin Foundry and Machine Company in 1939, but the writer knows nothing about the hardwood mill after 1927. The mill probably didn't survive long during the Great Depression. A 1927 article, entitled "Lufkin Thrives With Pine Mills," revealed that:[81]

> ...Lufkin and lumber are unconsciously thought of as being as closely related as ham and eggs...Two of the largest pine mills in the south are located there, the Long-Bell Lumber Company and the Angelina County Lumber Company...There are two large hardwood mills there, one operated by Boynton (Bros.) Lumber Company and the other by the Martin Wagon Company...
>
> ...Not quite so large (as Lufkin Foundry), but no less interesting is the factory of the Martin Wagon Company, where the famous eight-wheel, heavy duty (log) wagons are made. These too are shipped to all parts of the United States and to a number of foreign countries...

Another sawmill generally credited to Lufkin was the Jack Creek Lumber Company. A letterhead in the Angelina County Lumber Company papers, dated October 12, 1909, indicated Lufkin was the plant's headquarters and that E. T. Clark was president and J. H. Kurth, Jr., was secretary-treasurer. *Land of The Little Angel*, page 171, located the "small mill on Jack Creek, four miles southwest" of Clawson, and the letterhead indicated "mills located on Eastern Texas Railroad," perhaps as much as eight miles west of Lufkin on the rail line to Kennard. Since the Jack Creek mill was a Kurth enterprise, its daily capacity was probably forty or fifty thousand feet daily.

With the depletion of the forests around 1930, the value of the sawmilling industry to Lufkin's economy declined rapidly as factories manufacturing trailors, paper, and oil machinery came to replace the old-time sawmills as an employment resource for the county's citizens.

OTHER SAWMILLING AND LUMBERING CENTERS

Clark's Ferry, also known as Miami or Neches City, was an early Angelina County river crossing south of Diboll. It was noted in *Land of The Little Angel* as being the "present terminus of the H. E. and W. T. (Houston East and West Texas) Railway" in the year 1882. The *Southern Industrial and Lumber Review* observed in 1908 that a company was

"...engaged in building and installing a small sawmill plant at Clark's Ferry on the line of the Houston East and West Texas Railway on the Neches River. The mill is of circular equipment..." This was probably one of the many crosstie or stave-making enterprises in Angelina County.[82]

Nancy or Camp Nancy, south of Zavalla, was for many years a permanent logging site of the Angelina County Lumber Company after 1923, located on the Texas and New Orleans Railroad. A commissary and 200 houses for the loggers was maintained there, and one source added that a small sawmill was moved there from San Augustine County. The loggers were moved about 1930 to a new logging front in Tyler County named Acol.[83]

Ora, located east of Huntington, was the location in 1900 of the Jeans Brothers Lumber Company. Will F. and Tom J. Jeans were listed in the 1900 census as "sawmill lumbermen," and in 1901 Tom Jeans was postmaster at Ora. In 1902 the Jeans Brothers moved their sawmill to Clevenger in Nacogdoches County, but other small sawmillers replaced them at Ora at intervals.[84]

Manton, the site of the 12,500-acre Angelina Orchard Company, was the location of at least one small sawmill, located near Huntington. A news editor of 1905 observed that: "...The Manton Orchard Company in Angelina County is building their sawmill during the present year. The mill will be used for cutting small timber into crate and box material..." The Manton sawmill lasted only about five years.[85]

Durant, located between Clawson and Pollok on the St. Louis Southwestern Railroad, was a sawmill town from about 1907 until 1915, and was named for the first sawmiller there, the Durant Lumber Company. By 1909, the Odum and Chambers Lumber Company operated a 25,000-foot mill there and employed forty men, even though *Handbook of Texas* credited Durant with only five residents in 1910. The Durant 1910 census listed some of the mill employees as including: Oscar Weaver, sawmill engineer; Dr. Claude Alexander, sawmill physician; Louie Odum, sawmill owner and postmaster; James T. Chambers, part owner and sawmill foreman; Joe E. Odum, part owner and planer mill foreman; and James B. Abercrombie, sawmill engineer.[86]

Baker was a small sawmill town, also on the St. Louis Southwestern Railroad, near Pollok. Baker was not listed in an 1893 list of Cottonbelt sawmills. However, a 1900 list reported a 20,000-foot mill there, but did not give the owner's name. Since the reactivated post office there lasted only from 1894 until 1901, those were probably the dates that the sawmill was also operated. And since a small sawmill owner was usually the town's postmaster as well, the mill probably belonged either to Martin W. Townsend or John S. Matthews.[87]

The early sawmill town of Aldridge, although perhaps only one or two miles south of the Angelina County line, was situated wholly in Jasper County. A check by the writer of six or seven different maps indicated that, EAST of Highway 69, although the Angelina County line runs roughly parallel to and only one or two miles removed from the Neches River, no point on that line touches the river. Although Boykin Springs, a public campground (14 miles southeast of Zavalla on Highway 63) is probably on or nearly on the county line, the old Aldridge ghost town, with its massive and vine-covered, concrete boiler and engine rooms, lies about two or more miles south of Boykin Springs on the banks of the river. *Handbook of Texas* also places Aldridge in Jasper County.

Krub, in Angelina County, was apparently the location of Charles H. Angell's 25,000-foot sawmill in 1906. Although that mill was credited to Huntington in a railroad list of sawmills, Angell was also the postmaster of Krub for five months during 1905 (in fact its only postmaster before it was discontinued).[88]

Donovan was the logging camp of Long-Bell Lumber Company prior to 1915. Afterward, the company's logging fronts moved into San Augustine County, until all of their virgin timber there was exhaused. In 1926 the logging front was moved back to Donovan so the Long-Bell loggers could cut the second growth timber on about 22,000 acres of cutover land owned in that region. Donovan was abandoned in 1930.[89]

Hoshall, according to *35 Best Ghost Towns*, was located between Lufkin and Burke. It was the site of a commissary, a boarding house, as well as a 20,000-foot sawmill owned by Luke Wright. Named for W. E. Hoshall, the sawmill existed for only a few years.[90]

Davisville, on the Houston East and West Texas Railroad, was the location of the Wright-Jones Lumber Company sawmill for a few years between 1908 and about 1915. The town had a commissary and a population of 150 men in 1910. The sawmill was small, about a 20,000-foot cut daily. A plat of part of the Michelli Grant, drawn in October, 1909, showed Davisville plainly located about midway between Lufkin and the Angelina River. The plat also showed the location of the Davisville Lumber Company, which the writer presumes to have been the site of the Wright-Jones sawmill.[91]

Burke, on the Houston East and West Texas Railroad 8 miles south of

Lufkin, was one of Angelina County's earliest sawmill towns, although the industry there did not last long. *Handbook of Texas* noted that "in 1885 the village had three sawmills, three cotton gins, a church, school, and a population of five," which seems a mighty few people for so much industry. In 1888, R. W. Haltom noted that Burke had "three stores, a general merchandise store, a drug store, and a sawmill," along with several houses and fifty scholars in school. A railroad list of sawmills in 1893 observed that the Burke Lumber Company operated a 20,000 foot mill there, but a similar list of 1900 did not reveal any sawmilling or lumber production at Burke. Jonathan Gerland of the Texas Forestry Museum could ascertain from deed records only that lumber production there after 1900 centered around T. J. Treadwell, W. F. Eden, and E. G. Prudhomme. *Land of The Little Angel* reveals that T. J Treadwell was an early Burke merchant, but makes no mention of the others.[92]

Caruthers was started around 1900 as the logging camp for the Clawson Lumber Company. It was named for the mill's general manager, A. J. "Jack" Caruthers, who was also a native of McClennan County and an official of the Waco Lumber Company, that owned a group of Central Texas retail lumber yards. Logging at Caruthers ended about 1908. Members of the Peavy family operated a small sawmill there about 1900.[93]

Mott, was an early settlement on the Angelina River around 1870. Named for J. H. Mott, who owned a cotton gin and grist mill, Mott also had a small sawmill during the 1880's. The businesses there sounded very much like those often surrounding a water mill. The town declined rapidly after 1890.[94]

Platt was an early logging camp in north Angelina County that was generally mentioned in connection with Old Platt and North Platt, also logging camps. There was a store, a one-room school, and a post office there between 1902 and 1913. All logging operations there had ended by 1949.[95]

Alecdo was also a logging camp on the Texas and South-Eastern Railroad for the Southern Pine Lumber Company of Diboll. At one time, the camp stood on the bank of the Neches River, but later it was moved farther inland. Logging ended there about 1920.[96]

Collwood was a small settlement of World War II days. J. V. Lankford and Son was the operator of a small sawmill there, which he later sold to Luke Wright. Sawmilling ended there about 1946.[97] There were other logging and small sawmilling sites in Angelina County, which were of minor importance. One such place was Durst's Mill, on the Cottonbelt near Clawson, which was the western terminus of the 13-mile Michelli tram road. At one time around 1890, it may have been the site of a small sawmill operated by the Durst Lumber Company.[98]

SAWMILLING AT CLAWSON AND POLLOK

Clawson, northwest of Lufkin, had a checkered lumbering history over about thirty years of time. The first sawmiller there about 1887, according to R. W. Haltom, was T. R. Bonner and Company, which in 1888 was cutting 30,000 feet daily and employed about one hundred mill hands. The town took its name from T. W. Clawson, the mill manager. A deed record of 1890 conveyed 145 acres of land to T. R. Bonner and Company.[99]

As a result of research in the Angelina County Deed Records, Jonathan Gerland, archivist of the Texas Forestry Museum, furnished the writer with the following information, by letter dated July 1, 1993:[100]

> ... Clawson: (This mill changed hands under the names of T. R. Bonner, S. C. Olive and A. J. Caruthers {the latter two as Clawson Lumber Company}, and Foster Lumber Company...Deed (Vol.) S, (p.) 384 conveyed 145 acres, 6 miles northwest of Lufkin to T. R. Bonner and Co. on Oct. 23, 1890. Deed of Trust (Vol.) C, (pp.) 95-99; Deed Records: (Vol.) S, (pp.) 384, 426-427, 428-431; Vol. 11, (pp.) 35-40. (This is the same tract later conveyed with the mill property of the Clawson Lumber Co. Apparently, Bonner moved and rebuilt a larger, 75,000-foot mill here.) A fire destroyed the sawmill and dry kilns on Aug. 23, 1891. A. L. Clark of the Tyler Car and Lumber Co. became trustee and sold the burned-out mill to A. J. Caruthers, S. C. Olive, J. W. Mann, and J. K. Rose, all of McClennan County on Dec. 30, 1891. By Jan. 1903, Clawson appeared on Foster Lumber Co.'s price lists (as 65,000 feet, "actual, not imaginary"). Not until April 30, 1903, was the mill sold to Foster. My interpretation of Bonner moving is open to debate, but in any case, the mill of Clawson Lumber Company was not the same as the Bonner mill mentioned by Haltom—see letterheads, also note that A. J. and R. A. Caruthers are described in 1902 as "agents" of Foster Lumber Co. The Caruthers brothers

were also involved with a certain Pine Island Lumber Company of Angelina County.

In 1893, a list of sawmills on the Cottonbelt Railroad, dated April, 1893, observed that the Clawson Lumber Company was cutting 40,000 feet daily, and another list of 1900 revealed that Clawson was cutting 50,000 feet daily.[101] Caruthers, Olive, Mann, and Rose, all of Waco, also owned the Waco Lumber Company, which owned a series of retail lumber firms in Central Texas. Gerland has also furnished to the writer five letterheads of the Clawson Lumber Company, two dated in 1897 and three in 1899. Each letterhead bore notations of daily mill capacities as follows: Clawson sawmill, 75,000 feet; planing mill, 75,000 feet; dry kiln, 35,000 feet. One letter from Jack Caruthers to Simon Henderson in Keltys read as follows: "...Simon, I would like for you to come up tonight. We are going to cut a water melon..."[102]

The 1900 Angelina County census, precinct 1, residences 477 to 535, listed several persons who held key positions at the Clawson Lumber Company, as follows: (res. 535) Robert A. Caruthers, lumber manufacturer; — Richards, planing mill foreman; Dora Gibson, boarding house owner; John Sargent, grader; Nathan Evans, sawmill foreman; Theo Methveys, filer; (res. 479) A. J. Caruthers, lumber manufacturer; Ashley Stroud, shipping clerk; Fred Evans, bookkeeper; Dr. John Reagan, mill physician; and (res. 477) Richard Blair, lumber manufacturer.[103]

There can be no doubt that the above enumeration was that of Clawson because Robert A. Caruthers was the Clawson postmaster in 1900. The writer is wholly unable to explain one circumstance, however. Richard Blair, Jack Caruthers, and Robert Caruthers, all listed as "lumber manufacturers," lived side by side among the mill's other key personnel, but neither Robert Caruthers' nor Robert Blair's name appeared on the letterheads of Clawson Lumber Company in 1899. The writer has only one guess or premise for that fact. Since Gerland noted that the Caruthers brothers were also linked to a Pine Island Lumber Company somewhere, could the latter's sawmill have been located nearby, and R. A. Caruthers and Robert Blair were actually linked to it, but residing in Clawson? The answer most likely will never be known.

In the same year (1900) and precinct, at residences 689 and 792, Caton B. Powelsy was listed as a "lumber manufacturer; also Sam Wilkings appeared in the census as a "sawmill manager," but the writer can not identify the mill or location. Also at precinct 3, residence 220, the names of "James Bonner, superintendent of lumber mill;" and "Charles L.

Bonner, assistant superintendent of lumber mill," appear in the same census, and again the sites are unidentifiable. Hence, the writer wonders if T. R. Bonner and Company, or some of his relatives, were still sawing timber as late as 1900?

The editors of *Land of The Little Angel* published the following quote from the *Lufkin Tribune* of 1902, as follows:[104]

> ...Clawson is...a live sawmill village. The place has a large store, church and school house. The big sawmill, known as Pine Island, is under the direction of A. J. Caruthers and is running to its fullest capacity...

Exactly what happened to Clawson Lumber Company in 1902 is unclear to the author. It probably amounted to a bankruptcy and public sale of assets, which resulted in the Foster Lumber Company owning the big Clawson mill. Jonathan Gerland of Lufkin determined in Angelina County deed records that the Caruthers brothers were "agents" of Foster Lumber Company in 1902, even though the sale to Foster was not dated until April 20, 1903. Foster owned many of its nine Southeast Texas sawmills through its subsidiary, the Trinity River Lumber Company, and Foster's price list of 1903 revealed that its Clawson sawmill was cutting 65,000 feet daily.[105]

There was another sawmill at Clawson, contemporary with the Caruthers and Foster mill. Henderson Land and Lumber Company operated a sawmill at Clawson from about 1902 until 1906, of which Anderson Jasper Peavy was the plant superintendent. Jasper Peavy left the Clawson sawmill in 1906 when he left Lufkin to superintend the Frost-Johnson sawmill at Mansfield, Louisiana. Nothing else is known about the Henderson Land and Lumber sawmill.[106]

About 1905, two of A. J. Caruthers' sons operated a small mill at Clawson before selling out to Ashley Stroud and J. M. Burnett in 1907. In 1900 Stroud had been Jack Caruthers' shipping clerk, and Stroud and Burnett sold out to the Bivins and Blythe Lumber Company in 1909. When the latter mill cut out its timber in 1911, the sawmilling era at Clawson effectively ended.[107]

The Bivins and Blythe Lumber Company at Clawson cut about 25,000 feet of lumber daily and employed 34 mill hands. As of 1910, the following names are a few of the key mill employees as taken from the census enumeration, as follows: Keyser Blythe, planing mill foreman; Dr. William W. Dunn, mill physician; Joseph H. Edwards, sawyer; Edward W. Blythe, part owner and postmaster; Frank H. Bivins, part owner and sawmill foreman; Lawson Bargainer and James A. Denton, locomotive engineers.[108]

The writer has been left so confused about the sawmilling epoch at Pollok that he is almost tempted to omit that story, except that that action would accomplish nothing. There are three secondary accounts about Pollok, and *The Handbook of Texas* noted that:[109]

> ...Pollok...was established in the early 1880's by the Harris-Lipsitz Lumber Company...The station on the Kansas and Gulf Short Line...was also called Bodan for the Bodan Lumber Company...Population was 150 in 1910...

Effie Boon wrote in her "History of Angelina County, Texas" that Pollok was established:[110]

> ...By the Harris-Lipsitz Lumber Company during the early nineties...The mill had a daily capacity of around 90,000 feet and employed about 250 men (error). This number of laborers usually gave a mill town a population of eight hundred to one thousand...The sawmill at Pollok operated for about fifteen years...

In *Land of The Little Angel*, the editors told a quite different story which will require a lengthy quote, as follows:[111]

> ...The town's inception goes back to the 1880's when Richard Blair built a small sawmill...Later the J. A. Young Lumber Company built a larger mill at Pollok, and in 1899, the firm sold its holdings to the Bodan Lumber Company for $25,000...The Bodan Company...increased the output of lumber to about 30,000 feet a day. Pollok became a flourishing sawmill town, consisting of many tenant houses, a post office, a commissary store...After the mill was destroyed by fire, however, the town declined.
> ...The stock holders then dissolved the corporation and sold the (burned) mill to Louis Lipsitz and A. Harris, two of the principal stockholders in the Bodan company. The two then built a small mill in order to cut the timber still standing on the land...

Immediately, it becomes evident that the account as told in *Land of The Little Angel* corresponds very closely with that of the available primary sources, and that in fact, the Harris-Lipsitz Lumber Company was the last sawmiller at Pollok, not the first. Bowman noted in *35 Best Ghost Towns* that the Pollok sawmill was dismantled in 1922 and moved to Forrest. Along with the reference to Forrest, the Pollok sawmill's ties to Cherokee County and the Chronister Lumber Company at Forrest will quickly become evident.[112] And instead of fifteen years as one writer noted, the sawmilling age at Pollok lasted for about thirty years.

The first primary source about Pollok was R. W. Haltom's remarks of 1888 which acknowledged the existence of two stores, "doing a general mercantile business;" a post office, and a sawmill, "the latter employing a number of hands and having a fine trade."[113] Everything points to the fact that the first mill indeed belonged to Richard Blair. By 1893, a list of the Houston East and West Texas sawmills recorded that Richard Blair's 10,000-foot mill had been moved to Lufkin, but a list of the Cottonbelt's sawmills on the same page indicated that there was no sawmill at all at Pollok in 1893. The name of the owner and size of the mill was omitted from the 1900 Cottonbelt list of mills, the only comment being - "and a small mill at Pollok..."[114]

In January, 1897, another name arose among the sawmillers there. Galveston *Daily News* recorded in January, 1897, that: "...The mill at Pollok, owned by H. Y. and C. W. Fisher, which is now in a trustee's hands, has resumed operations..."[115] From the old papers of the Angelina County Lumber Company, the writer recently obtained a copy of a letterhead from Jonathan Gerland of Lufkin, which reads: "Office of Bodan Lumber Company, Pollok, Texas, November 22, 1899..." Except for the notation, "Telephone Office: Wells, Texas," there was unfortunately no other inscriptions, names of company officers, mill statistics, or anything else. Even the signature of the letter has been clipped off by the photocopying machine.

A History of Cherokee County acknowledged that "Dr. C. W. Fisher was an Alto citizen...(who) in addition to his dental practice, his sawmill and farming interests...also had extensive sawmill interests in Pollok..."[116]

Some of the mystery about the Pollok sawmill is explained by Dorothy Young's article in *The History of Angelina County, 1840-1991* and in a letter received from Jonathan Gerland of the Texas Forestry Museum in August, 1993. As was witnessed by the Galveston *News* article above, the mill that Dr. Fisher built at Pollok in 1894 was in financial distress by 1896, and Dr. Fisher was forced out (but not very far). After the Young Lumber Company assumed charge of the Pollok mill, Fisher's successors were his first cousin, John A. Young and the owner's wife, Mary A. Fisher, probably the result of a sheriff's sale. M. A. Fisher may have transferred her share to her mother-in-law, Elizabeth Fisher, but not because of her death (she was still alive in 1934). However, Young became sole owner of the Fisher mill on July 6, 1897.

As a result of the company charter of Bodan Lumber Company on file in Austin, J. A. Young transferred ownership for $25,000 on February 10, 1899, to a trio of new partners, which included himself, J. J. Carter of Pollok, and J. Lipsitz of Tyler, Smith County. Although Lipsitz was not one of the original organizers of Chronister Lumber Company of Forrest, Cherokee County, that company came under his control in 1901 when J. Lipsitz became its president, and he served until his death in 1910; his son, Louis Lipsitz, succeeded him. A. Harris of New York was L. Lipsitz' brother-in-law; Harris also owned the A. Harris Lumber Company of Dallas County and A. Harris and Company of Smith County, and Dr. and Mrs. Fisher sold some of their remaining timber lands in Angelina County to Harris' companies.

Thus, except in name, Bodan Lumber Company became a subsidiary of Chronister Lumber Company, Louis Lipsitz becoming president of both Chronister and Bodan and S. W. Littlejohn, who was mill superintendent of both Chronister mills at Forrest and Wildhurst, became secretary-treasurer of Bodan. Dorothy Young in her article observed that the Pollok mill burned both in 1915 and 1917, but was rebuilt each time. In June and July, 1917, the Bodan Lumber agreed to a dissolution and sale to Harris and Lipsitz, but the transaction appeared to be only a paper shuffle since Harris and Lipsitz already owned it. The sawmill epoch in Pollok ended in 1923 when the Lipsitz mill was dismantled and moved to Chronister's mill in Forrest, Texas.[117]

An 1895 map of the Cottonbelt Route of the St. Louis Southwestern Railway (from Kurth Records at S. F. Austin State University, Box 1, Vol. 5) listed only two rail stops in Angelina County—Lufkin and Bodan. Neither Clawson nor Keltys were listed.

The key employees of the Pollok sawmill, as taken from the 1910 census, were as follows: Jesse G. Driver, blacksmith; John W. Gilchrist, sawmill superintendent; George Abernathy, shipping clerk; Charles Boles, sawmill foreman; James R. Cain, sawmill engineer; R. Robertson, Earl Robertson, graders; Samuel H. Prosser, filer; Lee Robertson, assistant planing mill foreman; Rutledge Weems, locomotive engineer; Henry J. D. Willge, yard foreman; Robert L. George, bookkeeper; Elijah Pruitt, planer mill foreman; and William S. Watson, postmaster of Pollok. The Harris-Lipsitz mill employed 78 men and cut about 40,000 feet daily. The sawmilling epoch at Pollok ended after the mill was dismantled in 1922.[118]

SAWMILLING AT EMPORIA AND EWING

Emporia and the Emporia Lumber Company sawmill were located one mile south of Diboll and existed from about 1893 until 1906. The company began as Jones and Carter in 1892 when Samuel Fain Carter and M. T. Jones, both of Houston, began their sawmilling partnership together. M. T. Jones was a well-known and wealthy retail lumber dealer in Houston, but he also owned two large sawmills at Orange, the D. R. Wingate Lumber Company and the Orange Lumber Company (he was also the uncle of Jesse Jones, who founded the *Houston Chronicle*).

S. F. Carter was born at Huntsville, Alabama, in 1857, but he and his younger brother and partner, J. Press Carter, were raised at Sherman, Texas in Civil War days. S. F. Carter settled at Beaumont in 1882, and within a short time owned an interest in the Village Mills Company, a subsidiary of the Texas Tram and Lumber Company of Beaumont. By 1888 Carter was secretary and business manager of Texas Tram, as well as superintendent of its Village sawmill. In 1892, Sam Carter sold his stock in Texas Tram Company and moved to Houston. And Press Carter went to work for his brother as commissary manager at Village in 1888, but by 1890, he was bookkeeper and shipping clerk for M. T. Jones' Wingate sawmill at Orange. [119]

Jones and S. F. Carter purchased 5,755 acres of prime pine timberland from W. H. Bonner of Angelina County in November, 1892. Although Emporia was not listed as a rail stop on the Houston East and West Texas Railroad in April, 1893, the writer believes that the mill began sawing timber soon thereafter. The town was awarded a post office in July, 1893, and when the Emporia Lumber Company was chartered on November 1, 1894, Jones and Carter deed 9,186 acres of timber to the new corporation. The first postmaster at Emporia was Joseph P. Carter in 1894. In 1900 Joseph P. Carter was also recorded as mill superintendent at Emporia. Other key mill employees included: F. Hollenbeck, planer mill foreman; Gus Clemons, steel gang foreman; S. Perkins, hotel keeper; and Arthur Asplet, shipping clerk. About 125 employees worked for Emporia sawmill in the mills and woods, and one page of the 1900 Precinct 2 census was titled "Emporia Log Camp."[120]

The Emporia Lumber Company sawmill burned down in July, 1897, and two Galveston news accounts reveal that S.F. Carter almost did not rebuild. They are as follows:[121]

> ...S. F. Carter of Emporia was here (Galveston) on Tuesday. Mr. Carter has not yet decided when he will commence work on his mill to replace the one recently burned down there, as he finds he can buy lumber just as he needs it, cheaper than he can make it...
>
> ...Mr. S. F. Carter of the Emporia Lumber Company informed The News reporter this week that arrangements were being perfected for the rebuilding of this mill at Emporia within a short time...

Effie Boon recorded in her thesis that Emporia mill began in 1893 with a daily cutting capacity of only 40,000 feet and employed fifty men. The railroad's list of its sawmills reported in January, 1900, that the Emporia mill was cutting 75,000 feet daily. Also a photocopy of an Emporia Lumber Company (original is in the Angelina County Lumber Company archives) letterhead, furnished to this writer by Jonathan Gerland and dated August 28, 1897, revealed that the sawmill's specialty was "bridge timber and crossties, car decking, car roofing, and car siding," and the owners' "paid up capital" was "$100,000." Elsewhere the letterhead revealed that the daily cutting capacity of the Emporia sawmill was 75,000 feet and the planing mill was 100,000 feet. In 1897, the company officers were S. F. Carter, president and general manager; M. T. Jones, vice president; and J. Press Carter, secretary.[122]

Press Carter revealed in his biography that when he arrived at Emporia in 1893, he served at first as bookkeeper and commissary manager, but eventually he worked his way up to mill superintendent. About 1902, M. T. Jones withdrew from Emporia Lumber Company and sold his interests to his partners, but S. F. Carter retained the majority ownership. However, J. Press Carter and E. L. Crooke bought a considerable interest in Emporia Lumber Company at that time. The new company officers became S. F. Carter, president; J. Press Carter, vice president; and E. L. Crooke, secretary. By 1905, all three men were living in Houston. About 1900, Emporia Lumber Company built a new 75,000 foot mill and acquired a sizeable block of timber at Doucette, Tyler County. At first the new mill carried the Emporia name as well, but it was eventually chartered as the Sunset Lumber Company.[123]

No record of Emporia's cutting equipment has ever been located by the writer, but it was probably equipped with an Allis circular mill of about

60-inch diameter. All that is known about the Emporia tram road was that two locomotive engineers were employed; hence, the tram probably had about six to ten miles of steel track, with probably one locomotive and one Shay engine on the spurs, and probably twenty tram cars in use for logs. Boon's thesis observed that, although a steam loader was eventually added, most of the forest labor was done with horses, mules and oxen, and two-wheeled carts. Car houses, bolted onto rail flat cars, were used on the spurs to house the loggers because of the ease of movement.

Apparently, Boon's interview in 1934 was with S. F. Carter, Jr., because his father died in Houston in 1928. By most standards, Emporia, because of its relatively short life as a mill town, lacked most of the conveniences of the older and larger mill towns. Employee housing, except for a few cottages, were of unpainted, clapboard construction with no inside conveniences. There was a commissary, a community church and hall, and a school, and the town's population probably reached about 500 people.[124] A few things are certain about the mill's demise. The company sold its Doucette mill and timberlands at a sizeable profit in 1906 to Thompson-Tucker Lumber Company. With apparently nothing left in Angelina County except 10,000 acres of cutover lands, Emporia Lumber Company dismantled it Emporia sawmill in 1906, abandoned the site, and the owners divided up their profits. S. F. Carter retired from active lumber pursuits, organized Houston's Lumbermen's National Bank in 1907, and served as its president and board chairman until he died in 1928. In 1906, Press Carter organized the J. P. Carter Lumber Company with his profits and built sawmills at Baber, Angelina County, and in Nacogdoches County. And the ghost town of Emporia returned to the forests that had spawned its birth.

Certainly a news article of November, 1904, did not reveal that the Emporia mill had exhausted its timber or was planning to close, having read as follows:[125]

> ...Water is reported very scarce with the Emporia Lumber people at the mill on the H. E. and W. T. Railroad in Angelina County. In order to run on full time, water has been hauled in tanks from the Neches River, a mile distant...

Ewing, Texas was founded by the Bohlssen Manufacturing Company as a hardwood sawmill in 1920. The founder was Henry G. Bohlssen, but the principal investors were members of the Henderson and Kurth families of Keltys, who had bought a sawmill in Arkansas for Bohlssen to use. The town was named for another of the investors, James E. Ewing. Neither Boon's thesis nor *Land of The Little Angel* are explicit about what the founding Bohlssen's full name was, noting only that John Bohlssen was the first postmaster and that the mill owner named Bohlssen was killed in a sawmill accident about 1922. In 1910 Henry G. Bohlssen owned a hardwood sawmill in Liberty County on the Gulf, Colorado and Santa Fe Railroad, near Fuqua and Milvid. In 1915 the Henry G. Bohlssen Hardwood Company was operating a 20,000 foot mill in Jasper, and in 1919, Bohlssen was sawmilling in New Caney, Montgomery County, when Simon Henderson corresponded with him about moving to Ewing. After Bohlssen's death, the Lufkin Foundry and Machine Company drew two technical drawings, Nos. 1990.015.0031 and 0032, for the H. G. Bohlssen Manufacturing Company in 1924, but soon afterward the firm's name was changed to Angelina County Hardwood Company.[126]

Ewing was located on the Keltys tram road, known as the Angelina and Neches River Railroad, near the hardwood bottomlands of the Angelina River. After Bohlssen's death, the mill was renamed the Angelina Hardwood Lumber Company after its sale to the Keltys associates. Boon rarely furnished any mechanical details or social history of any sawmill, noting only that that the Ewing hardwood plant cut about 60,000 feet daily in 1934 and employed about 300 men. *Land of The Little Angel* noted also that Ewing reached its zenith about 1929, the year that the veneer plant was installed there, and the town contained about 1,000 people, along with commissary, stores, churches, and schools. Although Angelina County had had three hardwood sawmills in 1927, Boynton Brothers, Martin Wagon Company, and Ewing, the latter hardwood mill was the only one that survived; and it too closed down when the timber was exhausted about 1938. The veneer plant cut material for crates, boxes, etc.[127]

Boon seemed more intrigued with the types of trees a sawmill cut than with any economic, mechanical, or social details of a mill town. Her thesis noted that Ewing cut 26 kinds of hardwood, with 55% of its cut being sweet gum and most of the remainder red and white oak. Most hardwoods were air-dried. Some varieties of gum trees were used for furniture, veneer, cheap lumber, car decking, crates and boxes. Tupelo gum made very expensive lumber for a variety of uses. White oak was used for wagon stock, flooring, tool handles, farm implements, decking and bridge timbers. Some cypress was cut for well-curbing and boat timbers. Ash, magnolia, elm, and hickory were in great demand for furniture and barrel staves, especially in England and Europe.[128]

The Ewing sawmill did not maintain a logging front, but carried its woods crews back and forth daily on the work train. The company used a steam loader in the forest, as well as a locomotive, tram road, and tram cars, but most of its conventional logging equipment was done with mule teams, oxen, trucks, and two-wheeled carts. Boon also observed that Ewing had a "distinctive industrial atmosphere" to it, so much so that one might forget it was a town as well.[129]

THE ORGANIZERS OF CARTER-KELLEY LUMBER COMPANY

In 1906, a very unusual circumstance developed which would result in the building of the sawmill town of Manning. During the 1920's, when the population of Lufkin was 5,000, the population of Manning reached 1,500, but it declined rapidly thereafter. Its origin stemmed from the fact that about 1898, W. T. Carter and Brother, a sawmill concern of Camden, Polk County, purchased several alternate sections of prime long leaf timber (school lands) in south Angelina County at $5.00 an acre in order to increase his timber reserve. The other circumstance occurred in June, 1905, when the sale of the Lufkin Land and Lumber Company to Long-Bell Lumber Company of Kansas City, Missouri, left George A. Kelley, one of the Lufkin Land founders, with a surplus of $200,000 from the sale that he needed to reinvest. Being a master machinist and probably the most knowledgeable man of sawmill machinery ever to reach East Texas, Kelley's partnership with W. T. Carter of Camden was fortunate for both, not to mention Angelina County's economy, and perhaps inevitable.[130]

Two Master of Arts theses and a book have made the histories of the Keltys and Diboll mill communities easy to write. Some school essays about the history of Manning, written as senior class English assignments, were collected by R. L. Poland in 1985, and with the memoirs of some oldtimers added to them, they became the backbone of a 29-page published pamphlet, entitled *Were You At Manning?* And if each East Texas mill town's teachers had likewise assigned such essays and preserved them, the writing of East Texas sawmill history would not have become such an unsung, forgotten, and difficult project to write.

The original stockholders of the Carter-Kelley Lumber Company included G. A. Kelley, W. T. Carter, Sr., Jack Thomas, Judge E. J. Mantooth, Tom Collins, C. G. Gribble, L. C. Carter, Earnest Carter, and G. C. Garrison. Nevertheless, W. T. Carter and Kelley owned the lion's share, amounting to about 85 percent. W. T. Carter, Sr. was born in Smith County in 1856, but spent much of his youth in Angelina County, and later in Trinity County, where he learned the sawmilling business under his father's tutelage. In 1882 he bought huge blocks of timberlands in western Polk County and eastern Tyler County, later adding to his timber reserve with additional tracts in southern Angelina County. In addition, he owned the sawmills at Bowers and Barnum, on the Trinity and Sabine Railroad in Polk County, and when those two mills either burned or cut out prior to 1897, Carter built his big mill at Camden, which would became the show place of W. T. Carter and Brother Lumber Company for the next seventy years. W. T. Carter died at Houston in 1921.[131]

George A. Kelley has been discussed some previously. Born in Canada in 1861, he became a master machinist, who acquired two sawmill patents (the Kelley steam log kicker and log turner) while employed by E. P. Allis and Co., a sawmill manufacturer of Milwaukee. In 1897, he became a traveling representative of the company in Arkansas and Texas, when he joined with E. W. and E. A. Frost to found the Lufkin Land and Lumber Company in 1899. He had just sold out in June, 1905, when he joined W. T. Carter, Sr. in founding the Carter-Kelley Lumber Company of Manning. Kelley sold out his mill interests in 1926 and after he retired, he dropped dead on a Houston street in 1931.[132]

Jack Thomas, Tom Collins, C. G. Gribble, and G. C. Garrison had been Carter or Carter-Kelley mill managers who acquired blocks of stock, but later sold them back to the company. Judge E. J. Mantooth was a Lufkin attorney and business man, who still owned company stock at the time the mill burned. L. C. and Earnest Carter were W. T. Carter, Sr.'s brothers, who by 1930 were either dead or incapacitated. G. A. Kelley sold out his interests to Aubrey Carter about 1926. By that year, nearly all of the Carter-Kelley Lumber Company stock had passed to Aubrey Carter (W. T. Carter, Sr.'s only son), W. T. Carter, Jr. (Carter, Sr.'s nephew and adopted son), and Aubrey Carter's four sisters.[133]

MANNING, TEXAS: THE EARLY DAYS

According to one writer, the first sawmill at Manning in 1906 was a small circular rig, installed expressly to cut lumber for the company-owned buildings and housing and to cut crossties and timbers for the tram, the "dollyways" (elevated walkways connecting sawmill, kilns, etc.), and the company-owned Houston Shreveport and Gulf Railroad to Huntington. One account noted that the first employees resided in tents; then in the "shotgun" housing, built on one side of town for the Negro families, but used by whites at first until cottages for white families were completed. When the railroad to Huntington was completed, the first load of sawmill and power plant machinery arrived to be bolted down to the new concrete foundations.[134]

The first permanent sawmill at Manning was also small, was personally installed by Kelley, and was probably an Allis single circular with about a sixty-inch saw and a 45,000 foot daily capacity. Effie Boon noted in her thesis that "...The (Manning) mill was never operated at night...," but on the contrary, that first mill was designed precisely to run at night, as follows:[135]

> ...The sawmill of the Carter-Kelley Lumber Company at Manning, Texas has been operating for some time on the night and day shift. Mr. W. T. Carter...states that the Carter-Kelley mill was made to run nights...The mill was made exactly one-half the capacity of the planer, kilns, and the shipping department, and was built purposely for double-shift operation. The night and day capacity of the mill is 90,000 feet...[136]

Boon also reported that three-fourths of the Manning employees were white and one-fourth were Negroes.[137] In the early days (1910), there were also fourteen immigrant Italians and 21 immigrant Mexicans in the woods crew. Early race incidents may have provoked the low percentage of black employees, as one record of 1908 revealed:[138]

> ...Cases of trouble between the sawmill companies and the farmers owning land close to the mills...have been the cause of a great deal of trouble during the past year...In most cases the trouble seems to be purely malicious on the part of the farmers. They sell the timber on their lands to the sawmill companies, and when these concerns send their crews to cut the timber, the farmers refuse to allow the Negro workers of the company to venture on their land to cut the timber.
>
> ...During the past month, there have been reports of trouble of this nature from the mill of the Carter-Kelley company at Manning...It was found necessary to get officers to protect the colored laborers while they were doing their work...

One historian of Manning wrote that Tom Collins was the second plant superintendent at Manning and that Charles G. Gribble was the fourth. If that were indeed the case, then the Manning sawmill had a different superintendent every year because Gribble was the plant manager in 1910 when Manning sawmill was only four years old. Gribble, who had been Kelley's bookkeeper at Lufkin Land, followed his old boss to Manning in 1906, where he became the town's first postmster. The following are some of the Manning key personnel as were recorded in the 1910 census: John M. Smith, Dock Malone, Edward Hicks, Walter Parker, Samuel Morris, machinists; Clyde Woodard, William H. Jones, P. H. Scroggins, W. V. Fountain, Willis E. Furlow, locomotive engineers; William M. Gibbs, woods foreman; Dave Medlock, John R. Medlock, Lord Rayburn, millwrights; Dickson Jones, Lonnie Johnson, sawyers; Isaac Tisdale, sawmill foreman; Cleve Bateman, dry kiln foreman; Henry Comb, yard foreman; E.L. Culpepper, sawmill engineer; Clem Wright, shipping clerk; C. G. Gribble, superintendent; and Dr. Jesse W. Hawkins, mill physician.[139]

THE MANNING SAWMILL AND EQUIPMENT

The sawmill town of Manning was to witness several company disasters over the years. The Carter-Kelley planing mill burned on the night of December 19, 1911, and the remainder of the plant was saved by cutting loose the wooden, elevated dolly ways. Manning endured three windstorms (perhaps cyclones or hurricanes) during the early years. One blew the roof off the north side of the sawmill, and another storm blew off the roof on the mill's south side as well as the commissary roof.

At noon on March 25, 1916, the Manning sawmill and kilns burned down, but the planing mill was spared, again by cutting loose the dolly ways. After the fire, the entire sawmill was rebuilt and modernized, a project requiring three months to complete. In 1932, the commisary building burned, a conflagration during which "one could read a newspaper in almost any part of Manning." The mill store fire was considered as having been arson. The final fire which destroyed the entire sawmill occured in 1936.[140]

The rebuilt sawmill of 1916 was a double band rig with a planing mill and dry kilns equally as modern. The main mill engine was a Corliss Valve steam engine, turning a 28-foot flywheel with a 40-inch face and a long belt which powered the line shaft. Boiler fuel was sawdust and shavings, but an 84-inch "Harvey Hogg" ground up some of the slab waste into chips for the burners.

The Carter-Kelley company operated one train - a passenger car and loaded or empty box cars - one round trip daily over its ten-mile Houston, Shreveport and Gulf Railroad to Huntington. A work train carried the loggers (called "flatheads") to the forest and returned them daily. Manning had one Shay engine for use on the tram spurs and five locomotives, each of the latter having "cabbage head" smoke stacks to restrain the cinders. Manning also had a locomotive round house and a well-equipped machine shop. As early as 1910, the sawmill employed five machinists and three millwrights to keep the mill machinery, locomotives, power house, skidder, loader and all steam equipment in tiptop shape.[141]

The owners built two mill ponds at Manning, known respectively as the "upper pond," which was a reserve reservoir and public swimming hole containing 35 acres, and the "lower pond." The latter, containing 25 acres, was adjacent to the sawmill and all log trains unloaded there before the logs reached the "haul-up" going into the mill.[142] A map of Manning indicated that a dam had been thrown up across Brushy Creek in order to create the lower pond. Nearby was a large, elevated water tank, and at all vulnerable points of the mill property, hose carts, water hydrants, buckets, and barrels of water were deposited to combat fires.

Boon wrote in her thesis that the Manning woods crew did not use a steam skidder, citing her interview with W. M. Gibbs, the mill superintendent. However, R. L. Poland wrote that at one time the woods crews used both steam skidders and log loaders, but because the cables and logs pulled by the skidder tore up young trees so badly, use of the skidder was abandoned. Poland compared the log loader operator's handling of the controls and levers to "Van Cliburn (the Moscow festival winner) with the keys of the piano." The two key positions among the loggers were the sawyers, two of whom had to drop each tree with a crosscut saw, and the "mule skinner," who pulled the logs to the rail siding with his mule team. R. L. Poland considered Charlie King to be "the best mule skinner at Manning ever to hitch up a team of mules." Loggers used two types of carts at Manning, one being the "piling cart," with its two high wheels, to which one end of the log was attached. The other was the large Martin eight-wheel, heavy duty cart, used for the larger logs.[143]

Although the fire of 1936 was the compelling reason for the demise of Manning and its sawmill, the town was already on the decline at any time after 1929, the year the Great Depression began. And lumber demand and production dropped acutely for many years thereafter. Although the Manning sawmill had a daily capacity of over 100,000 feet and an annual capacity exceeding 30,000,000 feet, the mill generally cut only about 50,000 feet daily. Lumber demand fell rapidly after 1930, and the awful spector of a mill shutdown was always present. As Angelina County's virgin forests were mostly cut out by 1932, the Manning mill was still cutting some timber in the Neches River bottoms, but it began tramming most of its logs from Camp Ruby, the logging front of W. T. Carter and Brother's mill at Camden, Polk County. While the company-owned railroad ran north to Huntington, it also ran south to the Camden mill (where it connected with the Moscow and Camden short line), and thence fourteen miles farther south to Camp Ruby. And the logs from that front were second-growth, lower quality stuff, with a log rarely exceeding twelve inches in diameter.[144]

MANNING—ITS BUSINESS DISTRICT AND MILL TOWN

Daily life in Manning was probably no better, no worse than any other East Texas mill town. The Negro quarter, with its own church, school, hotel, barber shop, etc., was totally segregated to conform to the laws and customs of Texas in that age. Streets consisted only of dirt, with dust everywhere in dry weather, and houses were all unpainted except for the mill superintendent's home and a few business houses. Water came from the mill pond, which was purified by a condensing plant before it was pumped into the water tower. Mill wages were paid in brass "mill checks," which served as legal tender at all the local businesses.

Only a few mill supervisors' homes had running water in early days. Most houses had only a hydrant in the front yard. The company's electricity was direct current, generated by two dynamoes. Most employee houses did not have electricity before 1930 and used only kerosene lamps for lighting. Eventually, about half the company houses were wired for electricity. Both cooking and heating were done with wood stoves, using wood fuel supplied by the company for a dollar a wagon load.

There were also no refrigerators in Manning before 1930, and everyone bought ice from the company ice house in either 25 or 50-pound blocks. Ice for the ice house arrived in 250-pound blocks and was unloaded from a refrigerator boxcar about twice weekly. The Carter-Kelley company had built another commissary building to replace the one that burned in 1932. It had several departments; one was the post office and mill check office. Others were the grocery and dry goods departments, as well as the meat market. Manning operated its own slaughter house, where animals were killed about twice weekly. The mill even owned some cattle, but purchased some animals from nearby farmers. Other departments were the ice house and the drug store, which had its own soda fountain.

The office headquarters was the prettiest building in town, being painted white. In 1913, a movie house, known as the Lyric Theatre, was built. It was equipped with only a silent machine, which was turned with a hand crank, and the movie house was never converted to 'vitaphone' for sound pictures. There was also a two-story Masonic Hall, one floor of which was for public amusements and entertainments.

Manning had one community church building, which was used alternately by the Methodists and Baptists. There were also Missionary Baptist and Pentecostal congregations. There was also a church in the Negro quarter. Manning's schools consisted of a grammar and high school for whites as well as a Negro school. A Mr. Day was superintendent until 1930 when M. M. Flournoy arrived. The schools were open a full nine-months term for about 200 students, taught by seven teachers.

There were only a few farmers at Manning. Some of them raised watermelons, potatoes, corn, and other garden produce, but about seventy-five percent of them were cotton farmers. About fifteen men and women were employed in the commissary, drug store, postal and mill check office, barber and pressing shops, ice house and slaughter house, and elsewhere in the business district.

There were also a variety of diversionary activities. The Manning Country Club was a nice, clean place for adults and teenagers to entertain and enjoy themselves. There were a Masonic Lodge and other fraternal organizations, as well as church and singing groups. Two mill physicians, Drs. Denman and Clements, practiced in Manning. Sometimes a "medicine show" would come to town. W. M. Gibbs was the last mill superintendent at Manning, and when he died, he directed that his home and 1,200 acres of land be sold to M. M. Flournoy, who decided to abandon his school and become a farmer when the sawmill burned for the last time.

Most of the people moved away after the sawmill burned in 1936, but Manning did not become a ghost town in the usual sense of the word because Flournoy and several other farmers stayed on. And the lovely Gibbs home (the mill superintendent's two-story home) became the Flournoy home. Everything that was movable was quickly torn down and moved away. Even the scap iron left over from the fire was sent to

Beaumont, where it was shipped to Japan to be made into munitions for World War II. Nevertheless, Manning was still a ghost town to most people, for no longer could one see the clouds of steam rising from the boiler house, or witness the activity everywhere in the dirt streets, or hear the screech of the big band saw as it ripped its way through a log, or smell the rosiny freshness of the sawdust and green lumber as it crawled out of the machines. And without those items, any mill town had already been reduced to a ghost town.

One other small sawmill at Manning in 1915 was the Turner-Cummings Hardware Company. The mill, which was probably connected with the C. R. Cummings Export Company in Houston, owned a single ciruclar mill which cut only 8,000 feet daily of white oak, red oak, gum, walnut, and hickory, but owned no planing mill. The mill cut wagon and other dimension timbers mostly for export.[145]

THE EARLY DAYS OF DIBOLL, TEXAS

The amazing fact about Diboll, Texas, is that it has survived until the present day as a sawmill town. It was founded about twelve miles south of Lufkin in 1894, a few months after Thomas Louis Latane Temple organized the Southern Pine Lumber Company at Texarkana in 1893. Diboll lies on the route of the old Houston East and West Texas Railroad, now the Southern Pacific. In conjunction with mill operations, a company-owned and operated tram road, the Texas South-Eastern Railroad was chartered in October, 1900, to connect Diboll with its early log camp at Blix, thence to Lufkin and the St. Louis Southwestern Railroad.[146]

Temple, when age 34, arrived in Angelina County in 1893, where he bought a 7,000-acre tract of long leaf pine from the Copes heirs, one of whom was J. C. Diboll, who gave his name to the mill town.[147] He was born in Essex County, Virginia, on March 18, 1859, and his brother and partner was born there in 1862. The brothers were orphaned at an early age, and at seventeen, Thomas Temple moved to Texarkana, where he was associated at first with varied mill interests. In 1889, he organized the first Southern Pine Lumber Company at Kingsland, Arkansas, which soon went bankrupt, and one of Temple's first chores after opening the Diboll mill was to pay off the liabilities left over from that business failure. William Temple followed his older brother to Arkansas in 1880, but he soon settled at Fulton, Arkansas, where he became a very successful merchant.[148]

Apparently at all times, T. L. L. Temple retained the lion's share of Southern Pine Lumber Company stock, and during the earliest years, the only company officers listed on a January, 1896, company letterhead were T. L. L.Temple, president; and William Temple, vice president and treasurer.[149] A photocopy of another letterhead in the writer's possession added the names of L. D. Gilbert, secretary and general manager; H. A. Carpenter, treasurer; and J. E. Hintz, general sales agent. The same letter listed mills at Diboll and Pineland; a capital stock of $1,250,000; and an annual mill capacity of 150,000,000 feet of lumber.[150] By 1923, another letterhead revealed that Arthur Temple had replaced Hintz as sales manager, and annual mill capacity had increased to 200,000,000 feet.[151]

According to one source, the Diboll mill started up in November, 1894, and Will Carlisle was the first plant manager. Temple's next timber purchases included 12,000 acres in Angelina County from M. T. Jones; 13,000 acres in San Augustine County; and timber purchases would continue in Trinity, Houston, Cherokee and Anderson counties until the company's timberlands totaled 200,000 acres. In 1899, Tom Temple acquired a one-third partnership when he, E. A. Frost, and G. A. Kelley organized the Lufkin Land and Lumber Company.[152]

Construction of the first small mill began at Diboll in March, 1894, and it had a 50,000 feet daily potential, the mill probably being an Allis single circular. The first planing mill built soon afterward contained five Hall and Brown Machinery Company matchers and moulders. Apparently, additional machinery was quickly installed, because when the Houston East and West Texas Railroad published its list of sawmills in January, 1900, the Diboll cutting capacity had increased to 75,000 board feet.[153]

The 1900 Diboll census revealed who occupied the key positions at the sawmill, as follows: Robert Green, pond foreman; W. P. Rutland, Robert P. Rutland, commissary clerks; William Ashford, planer mill foreman; Watson Walker, general manager; Dr. Lee Matthews, mill physician; C. Bates, mill engineer; Frank Dunlap, shipping clerk; Robert J. Beam, Jack Pipkin, Tom Sikes, locomotive engineers; C. H. Daugherty, yard foreman; S. Treadwell, commissary manager; D. McCall, telegraph operator; and C. G. Brooke, bookkeeper. A. J. Crager and — Arrington were merchants, and Clemon Appleyard, Sarah Treadwell, Eva McCall, and Pearl Brooke were school teachers.[154]

On March 1, 1903, the Southern Pine company began erecting a second sawmill there, and by June 12th, the new mill had already begun sawing timber. Equipment was installed in a new mill building two and one-half stories high and sixty feet by 208 feet long. A description of the new equipment follows:[155]

> ...The band mill on the short side has a carriage 20 feet in length and on the long side 32 feet in length. The long length carriage is run by a 12-inch "shotgun" (steam) feed...A 500 horsepower Filer and Stowell 24x40-inch (stroke) Corliss engine runs the sawmill and lath mill. Four boilers are employed to generate the necessary power. Besides the lumber output—250,000 feet in 24 hours — there is produced at this mill 60,000 lath daily.
> ...The dry kilns are located 200 feet from the sawmill...They consist of six rooms, 20x120 feet, and built entirely of brick...These kilns were erected by National Dry Kiln Company of Indiana. The kilns will hold 300,000 feet of lumber when filled. It may be said they will turn out 100,000 feet daily...
> ...The planing mill...is 80x180 feet in area and the machinery (Hall and Brown) consists of one outside moulder, one inside moulder, one double sizer, one 18-inch machine (matcher), three 15-inch machines, and six 9-inch machines, two resaws and two edgers. Power to run the planing mill is generated in three boilers, 60-inch by 16 feet, and is transmitted to the machinery by a Giddings and Lewis (steam) engine. The blower system consists of one double 70-inch and one double 50-inch blower and necessary piping...The planing mill at Diboll has a daily capacity of 250,000 feet ...

Diboll had two mill ponds, a main water reserve pond and another where logs were skidded and floated before being carried up the "haul-up" to the carriage. The elevated tank stood 90-feet high and held 33,000 gallons. Water protection consisted a Morse pump with 8-inch suction, along with necessary piping, hose carts, and hydrants located at all vital points. The sawmill also turned a large dynamo which energized 18 arc lights and 200 incandescant bulbs. Diboll also had a telephone circuit which connected the office to the log camp, the railway office, the city of Lufkin, and all connections of the Southwestern Telephone Company.

Many personnel changes had occurred during the four years after 1900. Watson Walker was still the plant's general manager, and others included: Charles Fredericks, sawmill foreman; William Ashford, planing mill foreman; Sam Lingard, shipping clerk, John A. Massingill, woods foreman; Emmett Massingill, log scaler; R. B Tucker, yard foreman; and W. J. Raef, general manager of the log tram, the Texas South-Eastern Railroad.[156]

With Diboll's population and mill capacity ever increasing, the number of people holding key positions in the company also grew. Among those enumerated during Diboll's 1910 census included: Watson Walker, mill manager; William Effinger, O. H. Welse, bookkeepers; Harry Temple, shipping clerk; Robert Effinger, John W. Vaughan, checkers; Ed Jennings, Henry T. Mooney, William E. Baucom, locomotive engineers; Louis G. Hardin, Jesse B. Fuller, millwrights; Dr. David A. Mann, mill physician; W. H. Agee, mill druggist, Marion H. Rogers, sawmill foreman; Harry E. McCloy, Ben Tucker, Moses Pruitt, yard foremen; E.A. Blount, hotel keeper; Mary Ballinger, boarding house owner; David Adams, sawmill engineer; Mathew Clemons, lath mill foreman; Monroe Smith, scaler; Willie Thomas, Bascom Weeks, Warren Spivey, Robert B. Strait, Claude Hodges, graders; Moses W. Green, Fred H. Chambers, blacksmiths; A. Ballinger, John Balzer, Harrison Cook, saw filers; Wilfred Eubank, telegraph operator; James Prewitt, steel gang foreman; William Edwards, carpenter foreman; and R. L. Weeks, planing mill foreman.[157]

Another Diboll article of 1912 revealed that, in addition to mills No. 1 and No. 2, a third sawmill for hardwood had just been built at Diboll, a mill which styled itself under its news photograph as "the largest Texas sawmill plant." A description of the Diboll mills of 1912 follows:[158]

> ...Diboll...mill is located on the Houston East and West Texas Railroad, and its cutting equipment is three large and modern sawmills, standing side by side on one great mill pond. Viewing (the photograph) from left to right will be found mills Nos. 1, 2, and 3. No. 1 is the original Diboll mill. It is equipped with two single-cutting bands and a heavy gang (saw, all of which had been added after 1900), and is the largest of the three mills. It is used for cutting pine only...
>
> ...Mill No. 2 is equipped with a single-cutting band (saw) mill of heavy design. It was built in 1906 (error-1903) as a hardwood mill, but was later converted into a yellow pine mill for which it is splendidly equipped. Mill No. 3 was completed about two months ago (April, 1912), and has been in operation only about that length of time. It is a hardwood mill strictly...It is equipped to cut logs into boards, and to further convert them

> into various dimensions for the making of furniture, wagon material, hardwood specialties, etc., and the equipment permits the saving of a great deal of lumber that in the ordinary hardwood mill goes to the slab pit to be burned. It is one of the best hardwood mills on earth, with a capacity of 40,000 or 50,000 feet (daily)...
>
> . . . The combined cutting capacity of these mills is greater than that of any other milling institution in the state of Texas. The rest of the equipment in the Diboll plant is in excellent keeping with the large milling capacity. The planing mill is one of the largest in the south; the dry kilns are of huge capacity; the sheds are of tremendous capacity; and the logging and shipping departments are equipped for all necessities. The product of this plant is sold through the general sales office of the company in Texarkana.....

On April 8, 1915, the new hardwood mill just mentioned went up in smoke, a total loss, as the following news account reveals:[159]

> ...The big hardwood mill of the Southern Pine Lumber Company of Diboll, Texas, was destroyed by fire on the night of April 8th. The combined efforts of the mill crews saved the remainder of the plant. The hardwood plant stands side by side with the two yellow pine sawmills of the same concern, and the entire plant was threatened for some time...
>
> ...The hardwood mill was built about two (actually, 3) years ago...and was claimed to be one of the most modern sawmills in existence — certainly the best-equipped in Texas. It was equipped with a single-cutting band (saw) mill and with a variety of special machinery for working hardwood lumber. The fire loss is estimated at $90,000, fairly well-covered by insurance. The mill will unquestionably be rebuilt, as the Southern Pine Lumber Company...has developed their hardwood end of the business to a very considerable degree...

LOGGING THE DIBOLL SAWMILL

Logging began at Diboll in 1894 under contract to William M. Atwood, who built the first narrow-gauge tram road into the nearby pinery. In 1896, the contractor was James J. Bonner, but in 1897, the sawmill bought the tram road, and converted its rails to standard gauge. In 1900 the tram road was incorporated as the Texas South-Eastern Railroad, and W. J. Raef was appointed general manager. Late in 1903, the company bought a McGiffert log loader for use in the woods and immediately branded its use by the woods crew a huge success.

By 1904, the log tram ran from Diboll Junction to connect with the Cottonbelt line at Lufkin. The distance from Ashford to the log camp was 21 miles, giving the tram a total length of 31 miles. In 1908 the original trackage was abandoned, because the timber was cut out, so the company built new trackage to the log camp at Vair.[160]

After Diboll loggers "cut out" in Angelina County, the tram road was moved to Trinity County until the loggers cut out there in 1906. Then they moved to Houston County until the reserve timber there was exhausted in 1912. Logging was then transferred to the 13,000-acre tract in San Augustine County until the timber there was cut out in 1921. Afterward the loggers moved into Cherokee and Anderson counties, where logging continued until about 1940.[161]

From the log camp at Vair, the Texas South-Eastern tram road moved up the east bank of the Neches River for fifty miles until it reached the site of Fastrill, a semi-permanent logging camp southwest of Rusk in Cherokee County. Along the way, the woods crews left a string of abandoned logging locations between Diboll and Fastrill, namely: Rutland, Blix, Gilbert, Vair, Rayville, Heff, Pisgah, Walkerton, Anderson Crossing, Kenleyville, Bluff City, Dredville and Lehane, before reaching Fastrill.[162]

One history of Cherokee County recorded what community life was like at one of Diboll's main logging camps around 1932, as follows:[163]

> ...Since 1922 the Southern Pine Lumber Company has maintained a camp at Fastrill, 12 miles southwest of Rusk, which serves as a base for its extensive logging operations in Cherokee County. Today Fastrill has a population of more than six hundred, two churches, and a four-teacher school. The company furnishes the land, teams, and tools for its men to farm during off-hours. and a community canning plant to aid the conservation of their produce. (This was in the heart of the Great Depression.)
>
> ...When business was at its peak, the Fastrill monthly payroll was $30,000; the annual output fifty million feet of lumber. Fifteen hundred trees have been cut in one day. According to Superintendent F. Goetzman...officials estimate that, operating at the present rate, it will take eight or

> ten years (that is, until about 1942) to exhaust the Southern Pine holdings...

Although Southern Pine loggers profited enormously in labor costs from use of the McGiffert log loader for loading the tram cars, the woods crews had mixed success in the use of rehaul and horse skidders. The latter were used during World War I when labor was extremely scarce. Effie Boon observed in her thesis that the skidders were very destructive of seedling and young pine timber during the 1920's at a time when Southern Pine Lumber Company was becoming the most cautious about protecting its second growth forests of timberland. In 1934, Boon noted that 200 men were working at the logging camp, where there were 75 cottages, a boarding house, a school house, all screened with electric lights and water, but lacking sewers. In that year, although Southern Pine still owned a ten year supply of old-growth timber, the company also owned 200,000 acres of cutover lands, on which a young stand of timber would have to become the mainstay of Diboll operations for decades to come.[164]

In 1927, assistant mill manager P. A. Strauss reported on the reforestration program then in progress on the company's cutover lands, adding that:[165]

> ...We are taking care of the young trees on all our holdings. We give them a better opportunity of growing by keeping the undergrowth cut. It was discovered by accident that when we put cattle to graze on the cutover lands...that this improved the growth to a great extent...
>
> ...A feature of the Southern's mill holdings is the cattle ranch, situated fourteen miles from Diboll, and upon which there are grazed 2,000 head of beeves. A substantial investment was made in the best Brahma stock, and this blood is being bred into the company's ordinary range cattle, improving their size and quality to a great extent...

As of 1927, Diboll's two mills consisted of a double band saw in one mill unit and a double-band and gang saw in the other unit. The saws and planers cut 200,000 feet of pine and hardwood daily, and they gave employment to six hundred men. Diboll had a 1927 population of 1,500 people, and it had been an incorporated city since 1902.[166]

THE DIBOLL SAWMILL OF 1940

When Hamilton P. Easton studied two of the largest sawmills left in East Texas (Diboll and Wier Long Leaf sawmill at Wiergate, Newton County) in 1940, he was impressed with some of the differences in the ways the two mills functioned. He noticed, for instance, that a majority of logs arriving at Diboll on tram cars were unloaded by an overhead derrick rather than logs being skidded directly into the mill pond. Some logs on the old type of tram cars were still skidded directly into the mill pond if they were known to be "floaters." At that time, a Diboll guide informed Easton that "modern timber doesn't float as well" as the logs of earlier days. Another cause was brought about by the change in tram cars. Before 1930, all logs arriving at the mill pond were held in place by chains. When the log train arrived at the pond's edge, chains were loosened, and the logs ran down a skidway into the pond. Invariably, a few pine logs were so heavy that they sank (and were known as "sinkers"), and they required a dredging barge to raise them to the surface. All hardwood logs were "sinkers," and they had to be unloaded by the derrick.

The change from chain cars to "pole cars" required that all logs be lifted by the derrick and deposited onto an incline upon which they rolled down to the endless chain "haul-up," which carried them up to the drag saw and carriage. On the "pole cars," there were eight or ten-foot poles or saplings "projecting vertically" along the bed of the flat car to hold the logs in place, and if the vertical poles were suddenly removed, logs scattered in such a quick descent that it spelled death and injury to the unloading crews. (Wiergate logs were all pine and went straight to the mill pond. Wiergate used no "pole cars.")

Logs were debarked before being band-sawed on two surfaces to a predetermined width. Normally, an endless band saw varied from 35 to 55 feet in length, about fourteen-inches wide, and were held taut on two rotating "dogs" eight-feet in diameter. However, one band saw at Diboll may have been much larger. Easton wrote that it "goes through the roof and through the floor in one giant, continuous oval." If one-inch boards were desired, the log remnant might then pass through the 32-gang saw, which could cut thirty or more boards at one time. Again Easton wrote that four small logs could pass through Diboll's gang saw at one time, whereas Wiergate had no gang saw. If four-inch or larger timbers (including railroad stringers and crossties) were being cut, they would be sliced by the band saw as the log rode on the carriage. After that point, boards passed through the edgers, trimmers, and cut-off saws, resulting in many slab pieces that might or might not be large enough for further processing. If slabs were too small for "re-manufacture," they were chipped by the "Harvey fuel hog" to be used as fuel for the boiler house. If slabs were large enough for "re-manufacturing," they might pass on to the lath mill, box factory, handle factory or elsewhere.[167]

Effie Boon wrote that the Diboll installation differed from the Angelina County Lumber Company sawmill at Keltys only in that it had a box factory instead of a picket machine.[168] H. B. Ellison wrote in his 1947 history of Temple Lumber Company at Pineland that that plant's lath mill produced "18,000 squares per day, used principally in making handles for brooms and mops. Most of the squares turned out by the Pineland lath mill are shipped to Diboll, where they are converted into the finished product at the rate of about 1,000,000 per month."[169]

Diboll's most recent history book, *The Cornbread Whistle*, made ample note of Tom Temple's aversion to waste wood of a size that he thought could be remanufactured at a profit. Sometimes Mr. Temple walked along the dollyways which connected the planing mill to other points, and if he found "even a short piece of lumber" that had fallen off a truck, he would carry the stick or sticks back to the office and place them on the shipping clerk's desk.[170] Hence, both the Diboll plant and the Temple Lumber Company at Pineland separated and re-manufactured waste wood in a manner almost unknown to the other East Texas sawmills.

Dr. Easton noted as well that green lumber removed from the sawmill was then graded and stacked on "small-wheeled platforms called trucks," which were then moved over tracks into the kilns, where the lumber was steam-dried for about forty or fifty hours (or more). Again, this process differed from the Wiergate mill, which used mules to pull the dollies to the kilns instead of trucks. After a second grading process, lumber at the Diboll plant was passed through the moulders, matchers, etc., in the planing mill in order to give it a smooth finish. From that point, it passed either to the dry shed or to the loading dock for shipment in box cars. Lumber fresh from the planing mill might included "S-4-S" (right-angle surface on four sides), "tongue and groove" like flooring or centermatch, "lipped," such as shiplap or various types of siding, moulding, or other.[171]

Effie Boon observed too that the Diboll mills cut everything from the smallest moulding to railroad stringers and walking beams, size either 8-inch by 16-inch by 28-feet long or (for beams) 14-inch by 26-inch by 26-feet long. Obviously, such large timbers required overhead chain hoists to handle them. In the hardwood mill, 65 percent of the products were made from red gum or black gum. The remainder was mostly made from oak or a great variety of other hardwoods found throughout East Texas. And Southern Pine woods crews logged all along the Neches River banks and bottomlands, where cypresses, oaks, and other nut-bearing hardwoods thrived. Pine lumber could be either kiln-dried or air-dried, but to season most (but not all) hardwoods properly, they needed to be air-dried for many, many months.[172]

At the end of his article on Diboll, Dr. Easton left the following description of one segment of the milling industry there that seemed to intrigue him the most, as follows:[173]

> ... In connection with the grading, a large mill such as the one at Diboll makes use of short lengths and otherwise off-standard lumber in its "re-manufacturing plant" and for end-matched (oak) flooring. Some of the lumber goes to the kilns to be dried for handles, which are made at the Diboll handle factory. The Diboll mill organization is especially large because of the inclusion there of box, crate, and handle manufacturing divisions. The re-manufacturing plant takes care of the waste lumber which is too good to be thrown away. This is utilized in the form of "shorts," chair stock, and other pieces. Another plant works over the "crooks' and broken pieces. The shorts are made into toys, lawn furniture, and so on, and are derived from broken pieces, bark ends, and the like...

DIBOLL—FROM VILLAGE TO MILL CITY

It is certain that the first chronicles of the city of Diboll never envisioned a sawmill city of any length of permanency. Such just did not exist in the old East Texas pattern of sawmilling, and by 1906, the neighboring town of Emporia had already "cut out and run." Michelli in north Angelina County had also "cut out and run," as had also Barnum, Bowers, Valda and Wakefield in neighboring Polk County.

The earliest writings described Diboll as a small mill town that even the trains rarely stopped at. It was customary to board and exit the train at Emporia, a mile to the south, in the earliest days. One lady's earliest recollection of Diboll was of "a few scattered shacks, about six houses on th east side of the S. P. (originally H. E. and W. T.) Railroad tracks, no schools and no churches.[174]

The earliest mill town housing was likely to be a box-like, "shotgun" structure of two or three rooms, built of upright clapboards, the cracks of which were covered with one by four-inch battens. Robert L. Weeks recalled that: "...We lived in a company house and the rent was low then. The first house I lived in was a little box house, and I think we paid something like three or four dollars a month." Mrs. J. W.Taylor recalled that her family had helped built the first Baptist church building in Diboll in 1897, and the first school was built there about the same year.[175]

Somewhat like the houses described at Keltys, the housing provided for Diboll's key personnel and mill supervisors were considerably more spacious, were screened and inside a fenced yard. By 1904, an American Lumberman reporter observed that Diboll: "...has about 1,000 inhabitants who live by the saw and about the saw...It is safe to estimate that Diboll will be a sawmill town distinctively and particularly for twenty-five years to come..." The same 1904 article revealed that at least some houses had running water and were wired for electricity.[176]

The editors of *Land of The Little Angel* noted that by 1908 houses which contained both wallpaper and electricity were for rent from $6 to $12 monthly. The company charged nothing additional for electricity, outhouse cleaning or garbage collections, but electricity was usually turned off at 10:00 P. M. each night.[177] One oldtimer wrote of organizing a baseball team and playing Lufkin in 1897. When Tom Temple learned of the rowdiness and drunkeness that developed at the game, he closed down that activity and had a farmer plow up and plant their ball park.

Like most any mill town, social life in early Diboll centered very much around the mill commissary, managed by W. P. Rutland. There the latest bear-hunting stories, gossip, and mill town news were bantered around among the shoppers, visitors, and whittlers, who might purchase only a dime's worth of snuff, candy, or tobacco, or shop for the latest in men's and ladies' apparel. In the drug department could be found all the latest patent medicines and elixirs. And nearby was the ice house, where chunks weighing either 25 or 50 pounds could be bought for home use in summertime.[178]

In 1927 assistant mill manager P. A. Strauss was proud of Diboll's many achievements, observing that Diboll:[179]

> ...the cozy little town, boasting one of the largest mills in the pine territory, sits back confortably in the knowledge that it has a long life. It is not believed that Diboll will meet the fate of those other sawmill towns which, when the pine was cut, closed their doors altogether...
>
> ...Diboll today is distinctly on the East Texas map...Great shade trees guard the main street and the company office

> buildings and a group of mercantile houses at the center of the picture...Diboll has a population of 1,500 ...

The onslaught of the Great Depression created havoc for Southern Pine Lumber Company at a time when lumber demand and prices plummeted, and many nearby sawmills were shutting down. One bank wanted to plunge the company into insolvency. On one occasion, a Diboll man named John Oliver "loaned the (Southern Pine) company $40,000 or $50,000," and Tom Temple had to sell 80,974 acres of cutover land to the National Forest Service at the ridiculously low price of $2.75 an acre in order to help keep his company afloat. Reduced work schedules and a short work week were quickly implemented; some personnel were cut; mills and housing were left without repairs; and other drastic cost-reducing measures were inaugurated in order to avoid bankruptcy.[180]

In 1934, Effie Boon left the following evaluation of Diboll, the mill, and the town's racial composition, as follows:[181]

> ...It (the mill) employs six hundred men. Forty-nine percent of those men are native-born white men, of which about twenty-five percent were born in Angelina County, and the greater part of the remaining twenty-four percent in East Texas. Forty-nine percent of the laborers are Negroes, of whom probably twenty-five percent were born in Arkansas, Mississippi, or Louisiana, and the balance were reared at the sawmills in East Texas. About two percent are Mexicans...
>
> ...Diboll...has a population of 1,500 inhabitants. The employees live in houses that are owned by the company. The houses are well-screened and kept in good repair and are furnished with lights and water. The town maintains three churches and has the second-best school system in the county. The white scholastic population is 450 and the Negro (population) 150. The Diboll schools have first class affiliation...

The depression lasted until 1940, and was followed by another "phenomenal onslaught," World War II, which also created significant changes in Diboll and the lumber industry. In 1941 lumber demand and prices skyrocketed overnight, while hundreds of military bases were being built, and soon afterward a severe labor shortage developed in the East Texas sawmill towns as people moved to better paying defense jobs in Lufkin, Houston, and Beaumont, and young men were called to military service.

Suddenly, Southern Pine Lumber Company decided to divest itself of its employee housing, and all who so desired were allowed to buy the homes they lived in. Although a rent-free health clinic was sponsored by the sawmill, the former mill physician entered private practice in Diboll, and he charged for each visit since he was no longer paid by the sawmill.[182] Diboll has undergone a thousand other changes since 1940, much too numerous to relate in this sawmill history, and the reader will find these well-documented either in *The Cornbread Whistle* or in *Land of The Little Angel.*

A SUMMARY OF ANGELINA COUNTY SAWMILLING

Sawmilling was the principal industrial asset of East Texas for six decades, from 1880 until 1940; and without any doubt, during that time frame, Angelina County gave up to the axe and the circular saw about two and one-half billion board feet of its sylvan monarchs. (Recall that in 1904 the county had 2.53 billion feet of stumpage still standing.)

In 1927, as the East Texas timber era was almost drawing to a close, there were still 182 sawmills left in twenty East Texas counties (there had been about 500 at one time). Of that number (182), Angelina County still had ten large sawmills, cutting 755,000 feet daily, and three small mills, cutting 24,000 feet daily.[183] One can readily recognize that of the first number, Keltys, Long-Bell, Martin Wagon, Boynton Hardwood, Ewing, Manning, and Diboll were among the ten largest mills, and Long-Bell and Diboll owned more than one mill each. How many sawmills cut timber in Angelina County in the course of a century is absolute guess-work, but the figure must have equalled or exceeded one hundred.

Other than the sawmills, the wealth of Angelina County's banks, the Lufkin Foundry, the numerous mercantile and supply houses also fed off the forests and milling interests, which still applies to some degree even in 1993. The dollar value of Angelina County's timber assets were calculated during all the decennial censuses, but the exact figures are not available to the writer at present. Along the way, perhaps a dozen Angelina County mill towns — Michelli, Boynton, Ewing, Retrieve, Manning, Clawson, Pollok, Baber — all blossomed into respectable mill towns, and each then withered and died as the timber was cut away. The unsightly aftermath of the sawmilling era was the countless millions of acres of cutover, often burned over, stump lands that scarred the East Texas timberlands.

Nevertheless, sawmilling did provide a livelihood for probably 4,000 or 5,000 families (if not more) in Angelina County during the heyday of lumbering. Although work hours were long and arduous, wages were low, and daily living knew no frills or ornamentation, there were some oldtimers who swore that those were the better years—when life was simpler, folks were friendlier, and criminality was almost unknown. And one or two of those oldtimers out around the ghost towns of Manning or Michelli still swear that, provided one's nostalgia is geared to fever-pitch and his ear is cupped well to windward, one might still discern the faint murmurs of yesteryear's sobs and laughter or catch the distant echoes of the big band saws' screech as each day's last arc of sunlight disappears in the west.

Chambers County

The use of crosscut saws and axes were the principal means to fell trees in early Chambers County.

CHAMBERS COUNTY

The writer is grateful to Kevin Ladd of Wallisville for information generously provided.

THE EARLY AND SMALLER SAWMILLS

Chambers County, in extreme Southeast Texas, is one of only three or four East Texas counties bordering on the Gulf of Mexico and still containing some sawmill history. Lumbering in that county was only a major industry between 1893 and 1915, although some small sawmills existed prior to those dates and others continued to operate in that county until after World War II. Most of the mills were dependent upon logs floated down the Trinity River.

Chambers County, like neighboring Jefferson County, is located in a geological zone known as the Beaumont Clay; it is also a part of the Fayette Prairie, a 12,000 square mile stretch of almost treeless coastal prairie. Hence, logging could hardly be said to have been a major industry there at any given time, despite the fact that perhaps 300 men worked in sawmills there after 1900. Also like Jefferson County, Chambers County is sparsely covered with forests except along the Trinity flood plain, mostly on the east bank, where mixed pine and hardwood forests extend eastward in places to a depth of about ten miles. Before 1900 some sawmilling also took place on the west bank in the vicinity of Cove and Old River. In 1900 the county was credited with a one-quarter billion feet of merchantable pine timber and a large quantity of marketable hardwoods as well as cypress. Hence, most of the county's sawmills depended on logs floated down from upstream on the Trinity River.

One source noted that the "trees in this section are the same as those along the Atlantic and Gulf Coast Plains. These trees thrive in an abundant rainfall. East of the Trinity River, the forests are of long leaf, while west of the Trinity River, the timber is mainly post oak, black jack, and live oak."[1]

The first sawmill along the Trinity River was probably on Menard Creek, north of Liberty, rather than in Chambers County, although its exact location is no longer known. The sawmill, founded in 1833, three years prior to the Battle of San Jacinto, was owned by M. B. Menard, a founder of Galveston, and Thomas F. McKinney, who was Galveston's first merchant and financier of the Texas Revolution. The steam mill was of the upright sash, or muly, variety, capable of cutting about 3,000 feet daily, and some of its saws were used to cut boards and timbers for McKinney's steamboats being built at Galveston's Lufkin Shipyard. In February, 1842, the boiler of Menard and McKinney's sawmill exploded, killing two slaves, but it was soon repaired and sawing again. The Menard mill cut timber for fifteen years, being dismantled about 1848.[2]

The second sawmill on the Trinity River and probably the first in Chambers County was that of Robert Wiseman. A deed conveyance of November, 1846, from Wiseman to Thomas McDonald, noted that Wiseman owned a 400-acre tract on Old River, near Cove, "known as the mill tract, where a steam sawmill was heretofore erected."[3]

The next sawmill known to have been erected in Chambers County was built by Clarke and Kilgore about 1850. One source observed that:[4]

> ...About six miles up the river from the mouth, (there) is a large sawmill, owned my Messrs. Clarke and Kilgore. It affords a considerable profit to schooners and other small craft...as they can get there and return with loads of lumber for Galveston. The logs, chiefly cypress, are rafted from above, down the river at a trifling expense, and when sawed into lumber, sells for $20 per M. (thousand feet) at the mill....

Two years later, the same Galveston editor reported that:

> ...The steam sawmill of Messrs. Clarke and Kilgore, about six miles above the mouth of the Trinity, blew up the other day. Fortunately...no lives were lost and very little damage was done. The accident will soon be repaired, and the mill put in operation again...[5]

Kilgore's mill lasted at least until 1860 when the Chambers County Schedule V census, Products of Industry, was made. In 1855, the sawmill was advertised for sale, but it still remained unsold several months later as one early news account revealed:

> ...The subscriber (Kilgore) offers for sale the Union Steam Sawmill...This mill is in good repair and will be sold very low on immediate application to the subscriber on the premises...[6]

The Chambers County census of 1860, Products of Industry Schedule, listed two sawmills as being active in Chambers County. One was that of R. O. W. McManus, located about "five miles upriver" and probably in the vicinity of Lake Charlotte. The sawmill represented a "capital investment of $3,000" and cut around 50,000 feet of lumber annually. According to Miriam Partlow, McManus resided at Moss Bluff, but he also had large land grants in the Lake Charlotte area.[7] However, *The Handbook of Texas* noted that Moss Bluff had "grist and sawmills" as early as 1840.[8]

One source revealed that a sawmill owned by Charles H. Moore of Galveston and located on the east side of the Trinity River appeared on George W. Grover's map of Chambers County in 1885. Moore, for forty years the largest retail lumber dealer on Galveston Island, also owned sawmills in Jefferson and Orange County.[9]

An early deed record revealed on March 1, 1883, that H. A. Wallis leased five acres of land "on the waters of Miller's Lake Pass" to A. B. Chamberlain and Company "for shingle and sawmill uses and purposes." The lease included the right for Chamberlain to build a "log boom way along the entire shore of said lake pass."[10] Dr. A. B. Chamberlain was a well-known physician of Galveston, who during the 1870's was U. S. Immigration and Public Health officer at Sabine Pass and in 1906 was grand master of the Masonic Lodge-Texas District. Another deed record observed that on December 10, 1883, Chamberlain and Co. sold the steam sawmill on Miller's Lake Pass to Leroy Jones and W. J. Swilley.[11] Evidently the transaction with Jones and Swilley was not completed, for on January 7, 1884, Chamberlain sold for $4,000 "the saw and shingle mill...together with the engine and boilers and machinery therein," to Charles H. Moore of Galveston and William Markham of Orange, Orange County.[12] No other information is available about the Chamberlain and Company mill, later the C. H. Moore sawmill, which was located about one and a half miles east of Wallisville.

Another sawmill on the west side of the Trinity River in 1885 was that of John R. Wooten, the county clerk of Chambers County. As of January,

1884, the steam sawmill known as the "New Hope Sawmill" belonged to the firm of Brooks and Perkins, and it was located "on Dorr's Island, and one and a half miles above the town of Wallisville." On February 14, 1884, Perkins, Augusta Lemuel, and H. F. Levy of Harris County sold the "New Hope Sawmill" to Wooten for $1,600. Equipment included one 52-inch circular saw, probably capable of cutting 10,000 feet or more daily.[13] Soon afterward, Wooten sold the mill to William Gallon and C. E. Parker. A subsequent partnership, comprised of Parker and Henry S. Mackin, operated the Dorr's Island sawmill under the firm name of the "Wallisville Mills." It is unknown how long the mill remained in operation or its subsequent fate.[14]

In December, 1884, a Galveston editor noted that the two sawmills were very important to the future of Wallisville, adding that:[15]

> ...Two sawmills are going up in Chambers County, near Wallisville, and a correspondent says the town is destined to become a great lumber market...The mills will get their log supply from the Trinity River, which affords plenty of cypress and a 'right smart' of pine timber of the loblolly variety...

It is unknown whether the 'Wallisville Mills" emerged into the Wallisville Lumber Company or not.[16] The Wallisville Lumber Company sawmill was owned by W. E. Stephens, and in March, 1897, a local editor called it the city's "chief manufacturing enterprise."[17]

In 1897, the Liberty *Vindicator* editor paid a compliment to that company, noting that:[18]

> ...The Wallisville Lumber Manufacturing Company has put in a new edger and cut-off saw and is making all the lumber it can make, running ten hours a day. During the past week, ships have been carrying lumber to Old River and Double Bayou, and farmers in localities not reached by water transportation have been hauling lumber from the mill in their wagons...

In December, 1896, one editor observed that: "...The Wallisville Lumber Company is putting in a substantial boom at this place to stop their logs, which they raft down the Trinity..."[19]

On June 14 and 15, 1899, two advertisements of the Wallisville Lumber Company, each different from the other, appeared in the Wallisville *Age*. As a dealer in "dressed and undressed" lumber, the ads suggested that the mill had dry kilns and planer in addition to circular saws, but the author otherwise has located no reord of the mill's cutting and planing capacity. The ads added that the sawmill cut ash and cypress lumber in addition to pine lumber.[20]

One other record about the Wallisville Lumber Company has been observed by the author, although no record of the mill's demise has ever been located and it certainly sawed timber for some years after 1900. A local editor of 1897 revealed that:[21]

> ...The Wallisville Lumber Company has made more clear money this year than any other lumber company in the state on the amount of capital invested...
> ... Situated on the bank of the Trinity River, with cheap transportation to Galveston and in easy reach of the bountiful forests of pine, cypress, ash, and walnut timber, the Wallisville Lumber Company (W. E. Stephens' mill) can deliver at Galveston at the same price and can make a half-dollar a thousand (feet) on it...

Another sawmill was removed from Turtle Bayou to Wallisville at the same time that the Wallisville Lumber Company sawmill was in operation. Turtle Bayou, where James Taylor White settled in Spanish Texas about 1818, is perhaps the longest bayou in Chambers County, as it meanders its way through the pine and hardwood forests east of Wallisville. It appears that three men who were operating sawmills settled at Wallisville. In April, 1896, J. H. Higginbotham closed down his Turtle Bayou sawmill and relocated it. One source noted that "Major J. H. Higginbotham has torn down his sawmill at Turtle Bayou and will remove it either to Wallisville or San Jacinto, where he will put up a shingle mill."[22]

Another source recorded that William A. Beckwith was operating a sawmill on Turtle Bayou in 1896.[23] The editor of Liberty *Vindicator* carried the following quote during that year, as follows:[24]

> ...Mr. W. A. Beckwith, who runs the sawmill on Turtle Bayou, is kept busy supplying the local demand for lumber, for fencing, and building purposes. He recently sold an order for 40,000 feet from the Galveston Cycle Club, to be delivered next month...

Two sources reveal that Captain T. A. Kilgore was operating a sawmill independently on Turtle Bayou in December, 1897, although one source was quoting the other.[25] However, a Liberty editor's comment of November, 1897, certainly verifies that Beckwith and Kilgore became partners following their removal to Wallisville, as follows:[26]

> ...Messrs. Wm. Beckwith and T. A. Kilgore have purchased the Turtle Bayou sawmill and will remove it to the Kilgore estate on the Trinity, at Wallisville. Mr. Beckwith is a sawyer of experience. and Mr. Kilgore (is) an old resident of established reputation and well-liked. Mr. Beckwith will at once begin work on his residence, which will also be on the Kilgore estate...

No information survives about the size of the Beckwith and Kilgore mill or its ultimate fate. With the exception of the Wallisville Lumber Company and the two Cummings mills at Anahuac and Liberty, the writer believes that most of the remaining Chambers County sawmills were relatively small in size, each with a single circular saw and a cutting capacity between 10,000 and 20,000 feet daily.

Cove is a Chambers County community situated on Old River, a few miles west of Wallisville, at the western edge of Trinity flood plain. The surrounding country there includes mostly rice fields, but it is no longer known how extensive the surrounding forests may have been a century ago. However, the writer believes that that area may have contained a substantial amount of pine, cypress and hardwood before the timber was cut out and the land was converted to rice fields. The only surviving record of a Cove sawmill was that supplied by J. H. Harry in his thesis of Chambers County, a part of which is quoted as follows:[27]

> ...Captain William Icet and his two sons, W. S. and H. C. Icet, operated a sawmill at Cove...from 1893 to 1898...The Icet mill resumed operation in 1902 after the 1900 Galveston storm and continued sawmilling until 1915...

Another sawmill site at the "turn of the century" period was that of John W. Cook, located adjacent to Wallisville on the east side of the Trinity River. The Cook sawmill site also appeared on W. O. Work's 1930 "Map of Lost Lake in Chambers County," which also indicated the site of "Cook's Well," an artesian well located on the mill site.

Newspaper references to Cook's sawmill observed that its origin may have been as early as 1895 or 1896. Apparently, its mill capacity was larger than others in the county because one source described it as "a thriving operation and into full swing" by 1899. It appeared as well that Cook's mill, like the Cummings sawmill, was a hardwood mill, which catered primarily to furniture manufacturing interests in Germany, before being destroyed by the hurricane of 1915. An editor of 1899 recorded that:[28]

> ...The steam tug *Horatio* with two barges are at Cook's mill loading logs and lumber for Hamburg, Germany. Mr. W. Gartner is checking out the lumber...

THE CUMMINGS BROTHERS' FIRST LARGE MILL

The Cummings Brothers Lumber Company of Texas, organized in 1893, consisted of Charles R. and Jesse Cummings, who came to Texas from Michigan, with a stopover in Arkansas. Jesse Cummings was only with the company for four years, before he sold out his interests "to parties in Berlin, Germany," in October, 1897, after which Jesse Cummings moved to Louisiana and the milling firm was reorganized as the C. R. Cummings Lumber Company, with headquarters at Houston.[29]

Unlike the Neches and Sabine rivers, where the huge Beaumont and Orange sawmills of 1890 cut only pine and cypress logs, the Trinity River sawmills made abundant use of hardwoods. In addition to stands of cypress and pine, there were abundant forests of white and red oak, ash, blackjack, hickory, cottonwood, beech, red, black and Tupelo gum, and walnut. The wine barrel coopers of France and Germany sought barrel staves made of oak, whereas the furniture makers of England and Germany wanted several varieties of hardwood, including oak, ash, gum, walnut, hickory, and cottonwood. Although it is unknown which of the Trinity River sawmills began cutting hardwood first, by 1897 only two of the dozen or more Liberty County sawmills (Keno and Devers) were cutting pine lumber. By 1896 the two Cummings sawmills at Liberty and Anahuac, in addition to the Beckwith-Kilgore and Wallisville Lumber Company sawmills at Wallisville, were apparently engaged almost exclusively in the hardwood export trade.

There are some articles which told of Cummings Brothers' earliest days at their Anahuac sawmill, one of September, 1893, observing that:[30]

> ...V. R. McManus, postmaster at Lake Charlotte told the *Vindicator* that "work has already begun on the large sawmill to be erected at the mouth of the Trinity near Anahuac. Sailboats are already conveying the lumber from Beaumont for the purpose of building a wharf, mill shed, etc...

A Galveston reporter's comments of April, 1897, acknowledged the existence of the Anahuac mill and its class of lumber, noting that:[31]

> ...The large forests of oak, ash, beech, cottonwood, elm, which lie on the Trinity both north and south of Liberty, must come

> into market. Already the enterprising firm of Cummings Bros. has utilized this timber at their mill, situated near the mouth of the Trinity River in Chambers County. The large exporting trade they have built up speaks for itself...

An Anahuac letter published in Liberty *Vindicator* in 1904 revealed some of the details about the Anahuac sawmill before it moved and one of the reasons why it had to be moved, as follows:[32]

> ...In 1894 Cummings Brothers erected a large band sawmill here (Anahuac), with capacity of 150,000 feet (error) of lumber per day. All kinds of lumber were sawed and exported direct to Hamburg, Germany. Even walnut obtained from the Indian Territory (Oklahoma) was exported from here. But owing to their inability to hold logs in a boom here in bay, they moved their mill to Wallisville.

Immediately, errors are visible in the foregoing comments, because Cummings' Anahuac sawmill was not a band mill (although a band saw was later installed at Wallisville). It was probably a 3-block Allis circular mill, with a daily capacity of about 50,000 feet, and an 1896 Galveston article tended to confirm that supposition, as follows:[33]

> ...Now there is a wharf here (Anahuac) extending to the channel in the bay, a fine sawmill with a capacity of 40,000 feet of hardwood lumber per day (all of which is exported direct to Hamburg, Germany), two stores, two boarding houses, and twenty dwellings, all put here in the past two years. The mill gives employment to sixty hands here, and a large number of logmen and raftsmen up the Trinity River...

With sawmills at both Liberty and Anahuac, Cummings Brothers quickly learned that the firm needed only to maintain one commissary, that is if it were movable and mounted on a barge. Apparently the company commissary was towed to Liberty and docked there for about two or three days weekly before being towed back to Anahuac for the remainder of the week. A newspaper article of 1896 observed that: "...Cummings Bros.' tug, with commissary boat (barge) in tow, went up the river this morning."[34]

An article of January, 1896, revealed how the Trinity River was choked with saw logs following each river freshet, as follows:[35]

> ...Heavy rains caused a big rise in the Trinity River and white oak logs coming down on this rise for Cummings and Bro., who own the large sawmill at Anahuac. Thirteen hundred

> passed down on the 1st, and for twenty miles above, the river is full of ash logs for this mill. It also receives a great many walnut logs, and after being manufactured into lumber, is shipped direct to Europe...The mill has ash logs enough on hand to keep it busy for twelve months...

With a river choked with free-floating logs, log rafts, and 'circle booms,'logging on the lower Trinity was always a risky business, particularly the log rafting through Turtle Bay to the Anahuac log boom. In addition, it was a very costly operation, which apparently helped convince C. R. Cummings of the wisdom of consolidating all his milling operations at Wallisville.[36] Probably the deciding factor, however, was the severe hurricane of September 13, 1897, during which hundreds of Cummings logs were lost in Galveston Bay. Although the 1897 storm was dwarfed by the 1900 hurricane, it was still a monstrous storm, probably 110 miles an hour, that killed ten people at Port Arthur and completely destroyed the C. N. Fisher (Liberty Lumber Company) sawmill at Liberty and the S. D. Likens sawmill at Devers.[37] The storm came ashore at High Island, and although the writer has no record of damage at Anahuac and Wallisville, both towns and their sawmills must have endured much destruction as well.

After the Liberty mill was destroyed, C. R. Cummings bought the steamboat or tug *Horatio* from C. N. Fisher. By October, 1898, the dismantling of the Cummings mills at Liberty and Anahuac was well underway, as may be noted in one newspaper account, as follows:[38]

> ...The Cummings mill that was at Liberty is now at Wallisville—that is, the machinery is here. It was brought down by the steam tug *Horatio* and landed on the river bank just below the Stephens (Wallisville Lumber Co.) mill. It will be speedily put up and will probably be sawing by the first of next month...

THE CUMMINGS SAWMILL AT WALLISVILLE

The years 1898 until 1915 were more or less the golden age of Wallisville, with three sawmills in operation there during many of those years. The city was the county seat of Chambers County, but its golden image began to tarnish when in 1907 the seat of county government was transferred to Anahuac.

Apparently as early as 1896, Cummings Brothers were already planning the move of their two sawmills to Wallisville; otherwise, how could one account for the news article that was published in December, 1896, which follows:[39]

> ...Having purchased the right from J. J. Mayes, Cummings Brothers of Anahuac and the Wallisville Lumber Company are putting in a substantial (river log) boom at this place, to stop their logs which they raft down the Trinity...

In November, 1899, C. R. Cummings acquired a 20-year lease from J. J. Mayes (at a price of $100 a year) for a mill site of 25 acres on the west bank of the Trinity River. From that date, one can assume that the lease date was a full year or more after the Anahuac and Liberty sawmills were dismantled and the machinery had been moved to Wallisville.[40] Hence, when the new Cummings mill was erected in that city in 1898, the company had two small circular sawmills, each of about 60-inch diameter, to work with.

Apparently between the years 1898 and 1901, C. R. Cummings made no startling changes in his mill operations at Wallisville. All the logs that were milled were those that had been floated down the river from some point north.

In 1900 only 44 persons in the Chambers County census were connected with sawmills, and that at a time when William Beckwith's, William Stephens', and John Cook's mills were also in operation there. Since until 1901, Cummings' mill property was assessed on the tax rolls at only $4,000, it appeared that C. R. Cummings was still operating a very modest business, with a single circular probably sawing no more than 40,000 feet of hardwood daily. And with only 44 mill hands in the entire county in 1900, no mill could possibly have had sufficient manpower to conduct forest operations as well.[41]

By 1901, however, Cummings' mill property suddenly increased in its tax appraisal from $4,000 to $17,000. Hence, 1901 was certainly the year that the Cummings mill installed a band saw. A shingle mill was added at the same time, which probably utilized the second circular mill for sawing cypress blocks for the shingle machine. Also, the writer surmises that the first short tram road for hauling logs may have been built (requiring a locomotive and tram cars) since in 1900, Cummings bought 1,000 acres of timberlands in the B. M. Spinks survey, followed in 1901 by the purchase of 1,600 acres of timber from the T. Devers survey.[42]

One source noted that by 1903, the Cummings sawmill was known to possess a band saw and a circular saw in the main mill, a shingle mill (which would entail at least a circular saw and a shingle machine), a planing mill, three dry kilns, as well as dry sheds.[43] And certainly, a newspaper article of January, 1904, confirmed the foregoing statement, reading as follows:[44]

> ...There are two sawmills at Wallisville...One of these mills has a capacity of 80,000 feet. It has a band saw and circular saw, a shingle (mill) and planing mill, and three dry kilns. All the clear output of this mill is shipped direct to Germany...

Also late in 1904, C. R. Cummings incorporated his mill under a different name, as the following account reveals:[45]

> ...The C. R. Cummings Export Company of Houston is the very latest incorporated enterprise noted. Its capital is $500,000, and its purpose is to buy and sell timber and logs, manufacture, build and operate trams, tugs, and barges...

THE CUMMINGS MILL OPERATION AND MACHINERY

Surviving photographs of the Cummings mill show several members of the log pond crew "poling" logs in the direction of the endless chain log "haul-up." The Cummings log pond was the company log boom in the Trinity River. After entering the mill, logs were cut to the desired length by the drag saw, before arriving on the log deck. The deck was provided with two steam mechanical devices, a "log kicker," which kicked the log onto the log deck, and a "log stop-and-loader," which permitted the sawyer to allow only one log at a time to roll onto the carriage. The most valuable device on the carriage was the steam log turner (sawmill vernacular called it the "steam nigger"), which allowed the sawyer to move and lock the log into whatever cutting position the situation demanded.

The fact that the book *Cultural Resources Evaluation...of the Cummings Mill* records mostly generalities about sawmill machinery may suggest how very little about the Cummings mill machinery is actually known. The log carriage at the mill may have cut on both the forward and return motions of the carriage, or it may have idled on the return trip, depending on whether the Cummings head saw was a "double-cutting" or "single-cutting" band saw. An endless band saw was 33 feet long, or longer, twelve or fourteen inches in width, a quarter-inch in thickness, and if the saw had teeth on both edges, it could function as a double-cutting band by slicing the log on both the forward and return trips of the carriage. The saw was held taut in position and rotated on two 8-foot or 9-foot circular "dogs," the saw cutting on the downward stroke.

It is known that by 1913, the Cummings mill had five boilers, each about 60-inch by 16 feet, or larger, in size, and each capable of producing about 150 horsepower. Normally, a combination band and circular mill required a 300-horsepower or larger steam engine and at least a 16-foot diameter flywheel. In February, 1913, one boiler exploded, killing one Mexican and three Negroes, and injuring five more employees. Since no explicit information about planer mill machinery, dry kilns, or dry sheds is included in the book, the writer assumes that those details are unknown as well.[46]

The 1900 census furnished so little information about the Cummings employees that it is hardly worthy of mention. Of four sawmill managers, three were known to have been connected with other mills, leaving James Cleek as the probable Cummings mill superintendent. In the 1910 census, six administrative heads were believed to be Cummings employees, as follows: Robert Burns, mill superintendent; W. J. Bradley, steam engineer; Rufe Dutton, sawyer; Joseph Laflour, commissary clerk; R. L. Sproule, steam engineer; and William Robertson, bookkeeper. One source added that: "...Mr. C. J. Robertson was general manager for both the lumber mill and the commissary. After his death the younger Charlie Robertson became general manager..." In 1914, D. C. Hines was promoted to mill superintendent, and he was still on the job when the Cummings mill was totally destroyed in the August, 1915, hurricane.[47]

LOGGING THE CUMMINGS MILL

According to one photograph, the Cummings locomotive and its trams cars were known as the "Cummings Express."[48] Although the writer knows nothing about steam locomotives, he readily recognizes that the dinkey engine had four main drive wheels and was of a class used on tram roads throughout East Texas. In 1985, V. R. Hylton, Sr. recalled the Cummings tram road quite well, when the Cummings loggers were sawing trees on the McFaddin and Spinks leagues about 1904. He added that Cummings built a tram road east to Turtle Bayou, thence north to Spinks Creek and into south Liberty County. At a point north of the Chambers County line, the sawmiller erected a semi-permanent logging camp, a blacksmith shop, "houses for the men to live in," and corrals for the livestock. The log tram had a skidway into the Trinity River, where logs were skidded into the water and floated down to the log boom.[49]

One amazing fact was that in 1905-1906, Cummings made ten purchases, totalling more than 3,000 acres of Liberty County timberlands, from W. F. Blair. The amazing part was that Blair owned W. F. Blair and Company of Dayton, one of Liberty County's foremost hardwood mills from 1890 until after 1900, and apparently the lands had been a part of Blair's timber reserve for his own sawmill. However, the writer assumes that the timberlands bought from Blair were on the east side of the Trinity River, whereas Blair's sawmill at Dayton was on the west side and perhaps out of reach of the Blair tram road.[50]

No mention has ever been located about the Cummings loggers' use of a steam log loader, a steam log skidder, or some other logging devices. However, that does not prove that one or both of those were not in use, and that the loggers were wholly dependent on mule teams or oxen for dragging and loading logs on the tram cars.

Again, the writer concludes that the Cummings tram road must have extended to tracts on the west side of the river as well. According to a map in *Cultural Resources Evaluation...of the Cummings Mill Etc.*, several Dorsett and Dugat tracts of land were located on the river's west side near Lost Lake, and a 1904 news article read as follows:[51]

> ...Grading for the Cummings railway through the Williams, Dugat, and Dorsett leagues is about completed, and track-laying is now underway. The dinkey engine will be put on the track next week...

A news article of 1905 observed as well that:[52]

> ... C. R. Cummings, the sawmill man...closed a trade the past week (Feb. 1905) for 1,600 acres of fine pine land near Wallisville in Chambers County. It is the intention of Mr.Cummings to build a tram from his Wallisville mill to the timber and get out 30,000 railroad ties, which will be sent to Galveston and creosoted...

A subsequent account left it unclear whether the afore-mentioned tram extension was ever made. It read as follows:[53]

> ...It was announced some time ago that the Cummings Lumber Company would extend their tram from Hickory Island to Cedar Bayou in Chambers County. The statement is now made that the extension will not be made because there is not a sufficient stand of timber on Cedar Bayou to

justify the extension...

C. R. Cummings had other logging and milling operations than those around Wallisville. In 1909, Cummings logging and milling activities were being conducted near Rye, in extreme northeastern Liberty Couny. One news article stated that: "...Cummings' log camp at Clark (south of Rye) has had to shut down on account of low water in the Trinity..." Another reporter added that: "...The mill of the Cummings Export Company at LaSalle (Rye) will resume operation this week..." Hence, it appeared that the Cummings loggers at Clark were skidding logs into the Trinity for the Wallisville mill as well as logging a Cummings sawmill at Rye.[54]

In addition, there is evidence that C. R. Cummings owned three other sawmills during the years 1906 and 1907. In 1907, Cummings built a new band sawmill adjacent to Magnolia Cemetery, on Brake's Bayou at Beaumont. One editor observed that: "...The (new) Cummings sawmill on Brake's Bayou has been running in a small way for the past few days..."[55] Its daily capacity was 50,000 feet, and when the writer wandered through the mill in 1936, it had already been abandoned for many years.

Also, in a list of sawmills on the Texas and New Orleans Railroad published in 1906 in *Southern Industrial and Lumber Review*, there were two more C. R. Cummings' sawmills listed. One was a 35,000-foot mill at Houston, and the other was a C. R. Cummings sawmill (daily capacity not listed) at "Bayou Saline," Texas. The latter location or bayou is unknown to this writer, but was apparently somewhere on the Southern Pacific system.[56]

CUMMINGS MILL LIFE AT WALLISVILLE

In addition to the steam tug *Horatio*, the Cummings Export Company also owned the steam tugs *Helen* and *Dick*.[57] Hence, in addition to mill workers and loggers, Cummings also employed a number of lumberjacks (river raftsmen), boatmen, and shipbuilders at the company shipyard. Thus, between 1905 and 1915, C. R. Cummings probably employed in all categories somewhere between 150 and 200 workers. With no rail access to market, all Cummings products were barged to Galveston for reshipment abroad.

The sawmill also owned and operated the Wallisville Mercantile Company, the inventory of which was carried on tax rolls at from $2,000 in 1903 to $5,000 in 1910. However, for a mill of that size, the inventory must have been notoriously understated, for some East Texas mills smaller than Cummings carried commissary inventories that were much higher in value. In the decade from 1900 to 1910, such a mill commissary would have handled the greater part of its inventory in bulk, that is, in barrels, sacks of dry stock, and bolts of yard goods rather than the canned, pre-packaged, and ready-to-wear merchandise and apparel of a much later date.[58]

The Cummings mill employed workers in three racial categories of white, Negro, and Mexican. According to one person, most of the mill workers were white, except for perhaps twenty Negroes. Since in 1913, one Mexican was killed and three more were injured in a boiler explosion, there must have been a significant number of Mexicans employed also. Apparently only two (or at least a very few) families lived on the mill side of the river, whereas the remainder had to cross the river on a flat boat twice daily en route to and from Wallisville.

The company-owned housing west of Wallisville, except perhaps for about a dozen top administrative personnel, were probably quite primitive in appearance. Houses was all unpainted, without utilities, although running water may have been piped to some residences. All residential areas were segregated by race. Houses for Mexicans were described as "all alike," whereas Negro cottages were described as "shanties" (probably two-room "shotgun houses"). All housing was probably of "clapboard construction," which means one-inch by 12-inch upright boxing, with the cracks covered with 1"x4" battens. There were about seventy-five company-owned houses.[59]

One source noted that two boarding houses were operated on the west side of the river, one for Negroes and the other for white employees. No mention was made of company-owned public buildings in Wallisville other than the commissary, and of course, Wallisville already had an adequate number of churches and schools before the Cummings mill arrived.[60] In 1900, unskilled sawmill labor in East Texas earned $1.50 for a ten or eleven hour work day, whereas skilled workers earned from $2.50 to $3.50 daily. Wages were usually paid in mill checks, exchangeable only at the commissary. The writer found no references to a mill physician or of a company health insurance plan.[61]

The Cummings mill at Wallisville prospered until August, 1915, when the sawmill and commissary were totally destroyed by the 1915 hurricane which struck Galveston. Also, about 3,000,000 feet of sawed lumber floated away and was never recovered. The huge loss bankrupted C. R

Cummings, and the mill was never rebuilt. Two years later, he died of a heart attack. The salvageable machinery was sold to C. J. Robertson and Roy McDonald, but a greater amount was sold as junk to E. A. Dunman.[62]

THE LATER MILLS

The hurricane of 1915 ended the sawmilling era at Wallisville, and never again did sawmilling rise to its pre-1915 zenith, although a number of small mills continued to operate in the county for another forty years. The hurricane was equally destructive to another long-time Wallisville landmark. John W. Cook's sawmill had cut hardwood for export to Germany for close to twenty years, and according to W. O. Work's 1930 "Map of The Lost Lake Area," the Cook mill property was located adjacent to Wallisville on the east side of the Trinity River. A photograph on page 16 of *The Photogaphic Heritage of Wallisville* depicts Woodson Mayes and the owner, John Cook, surveying the remnants of the destroyed sawmill after the hurricane winds had reduced it to rubble. A caption beneath the photograph reads: "The artesian well at Cook's sawmill was the only thing operational after the 1915 hurricane."

After purchasing the salvageable remains of the old Cummings mill, C. J. Robertson and Roy McDonald moved it to Turtle Bayou, where they erected a new sawmill capable of cutting 60,000 feet daily. After the storm, Art J. Long built a sawmill on property west of and adjacent to the old Cook mill site. How many years the Long sawmill operated is unknown, but it appeared on W. O. Work's map of 1930 as "A. J. Long's sawmill (although of course the mill may have been shut down at that time). No record of types of machinery or daily cutting capacity for either Cook's or Long's mill has been located to date.[63]

J. H. Harry's MA thesis recorded that many sawmills sliced logs in Chambers County between the two World Wars. In 1940, there were still five sawmills cutting timber there, the largest of which was the Sartain-Williams Lumber Company at Anahuac. The mill, which was founded about 1935, cut about 20,000 feet daily. The company employed 45 men and kept a finished stock of lumber on hand worth $20,000.[64]

Four other sawmills in Chambers County in 1940 included the R. H. Harlow sawmill at Hankamer, the Simon sawmill at Turtle Bayou, the J. C. Storms sawmill at Anahuac, and the J. E. Cline sawmill at Eminence. All of the latter were small sawmills of perhaps 10,000 to 15,000 feet capacity daily. The first four mills appeared on the tax rolls at only $2,160 in value. The Cline mill began cutting in January, 1940, and in five months time, it had cut 400,000 feet, or about 80,000 feet monthly. The last four mills employed "fifty men per month," and the 1940 sawmills added about $300,000 annually to the Chambers County income coffers.[65]

According to one record, lightning struck the R. H. Harlow sawmill on July 22, 1942, killing two men and injuring one other man. About eight men had taken refuge under a large tree when the thunderstorm began, and lightning struck the tree.[66] In January, 1949, a fire destroyed a dry kiln at the Curry-Jackson Lumber Company sawmill on Jane Brown Road in Anahuac, doing $6,000 worth of damage. Rain was falling when firemen arrived, and it helped save the remainder of the mill. The Curry-Jackson mill fire is the last record the writer has of sawmilling in Chambers County.[67]

A SUMMARY OF CHAMBERS COUNT SAWMILLING

In summary, sawmilling continued in Chambers County for at least 125 years. And it is probably continued to some degree still, by a few small circular saws run by tractor engines. Chambers County still has perhaps a 100 square miles of heavily-forested, second growth timber, that needs to be harvested at intervals. At sawmilling's peak between 1895 and 1915, it seems probable that 200 men or more received their sustenance from either mill or logging operations.

Certainly the 1915 storm and the loss of its sawmills was especially destructive to and the beginning of the end for the old community of Wallisville. Situated as it was near the mouth of the river, Wallisville was for years the end of the line for the cotton floated down the Trinity River on keelboats before 1880. Keelboats, being without engines, sails, or steerage equipment, could not navigate in Galveston Bay, and Wallisville's commission merchants, such as A. G. Van Pradelles were on hand to buy their cotton and sell them the necessities for frontier living to take back up the river. In fact, Van Pradelles was one of the ten largest cotton factors of all time on that river. After cotton-factoring on the Trinity died out about 1885, lumber manufacturing became the mainstay of Wallisville's sustenance for the next 35 years. Already, Wallisville had lost its county seat and several county office positions in 1907, and when the sawmilling industry died in August, 1915, it might as well be said that Wallisville as a commercial entity died with it.

Since Chambers County lacked the huge stands of timber that many

East Texas counties had, one would hardly expect to find the huge pine mills there that were equal to Wiergate, Bessmay, or Diboll. However, considering the size of the forests within the county's boundaries, the few sawmillers there made adequate use of their timber resources in days gone by and will probably continue to do so in the future.

Jefferson County

Timber stands such as these greeted early sawmillers in Jefferson County.

JEFFERSON COUNTY

THE GEOLOGICAL OUTLOOK

Jefferson County has a most unusual sawmill history, not because of any great forests which stand within its boundaries, but due to its proximity to the Neches River and its tributaries, Sabine Lake, and the Gulf of Mexico. Throughout sixty years of its early history, its sawmills would depend on logs floated down the Neches River, and even across Sabine Lake. Of the county's more than 900 square miles within its borders, fully 750- square miles are coastal prairie, known geologically as the "Beaumont Clay," which because of its unfavorable chemical structure, did not support the existence of the virgin long leaf pine forests. There were basically only two forested areas within the county, one being the Gilbert Woods near Fannett, perhaps eight by twenty miles in size, and the other being a pine belt running along the south side of Pine Island Bayou, containing perhaps an additional fifty square miles of forests. So little was thought of the timber resources of Jefferson County that in a list of "stumpage" (log measure of uncut trees or logs) for 38 East Texas counties, published in 1900, Jefferson County was not even listed.[1]

Nevertheless, prior to 1880, the marshes of the Neches River and Sabine River, their tributaries, and Sabine Lake contained a huge reserve of large cypress trees, a wood very noted for its ability to absorb moisture and for shingle-making. The existence of mills that manufactured cypress shingles by hand and exported them by water is recorded in several of the early records. Even the British took note of the potential value of the county's cypress forests and reacted accordingly. In 1838, an English brig (sailing ship) was anchored in Sabine Lake, while its crew cut cypress timber along the shores, floated the logs out to their ship, where they were loaded aboard and carried to England.[2] As early as 1839, the schooners *Santa Anna* and *Waterwitch* carried loads of hand-hewn cypress shingles to Galveston,[3] and by 1857, 6,120,000 shingles were exported to Galveston from Sabine Lake, most of them probably made at Orange.[4]

Very early, the residents of Beaumont recognized the potential for industry and profits that lumbering offered, and in 1838, the Texas Congress chartered the Neches Steam Milling Company at Beaumont, However, nothing materialized from that early effort.[5] In 1837, the Beaumont townsite proprietors reserved the "steam mill square" on Brake's Bayou as the site for a future sawmill, and conveyed the site to W. H. Irion upon his agreement to build a steam sawmill there, but again nothing materialized on that site for two more decades.[6]

SAWMILL AT SABINE PASS

Perhaps the most unlikely spot in Jefferson County for a lumber industry, Sabine Pass was chosen by Sidney A. Sweet in 1846 to become the first location of a steam sawmill in the county. Since steamboats and schooners carried only cotton in preference to lumber during the six-months of the cotton-shipping seasons, Sweet was convinced he could surmount some of his transportation problems by floating log rafts down the Sabine River, towing them across Sabine Lake, and milling them at Sabine Pass. In furtherance of his plans, he built a steam sash sawmill, a sash and door millwork plant, and a shipyard at Sabine Pass in 1846.[7]

In January, 1848, Sweet sold the sawmill to Henry Hubbell, a Galveston lumber dealer, who soon increased the mill's cutting capacity. In January, 1849, he added three Page single-circular saws to the Spartan Mill Company at a total cost of $10,000.[8] During the same month, Hubbell sold Spartan Mill Company for $12,000 to David Bradbury, Orrin Brown, Isaiah Ketchum, and Benjamin Granger, all of whom were Sabine Pass ship carpenters, who formed a partnership to saw cypress timbers. The mill's lien was payable either in cash or in "good merchantable cypress lumber at $18 a thousand," and John Sealy of Galveston was appointed trustee to auction the sawmill in case of lien default.[9]

In 1850 the Spartan Mill Company cut 4,000 saw logs, worth $5,500, into 1,200,000 feet of lumber, valued at $23,000. The plant employed fifteen men paid a total of $637 in wages each month.[10] By 1857, however, hard times had already befallen the Spartan sawmill, and the equipment lay abandoned and rusting.[11] To protect the lien balance of $2,000, Sealy sold the sawmill in March, 1858, to David R. Wingate, a Newton County resident who, intent on building the sawmill into the largest in Texas, immediately moved to Sabine Pass.[12]

In 1859, D. R. Wingate and Company, capitalized at $30,000, cut 7,488 saw logs, worth $11,980, into 2,496,000 feet of lumber, valued at $43,680. Wingate employed ten men, who were paid a total of $300 monthly in wages.[13] Some of Wingate's thirteen slaves at Sabine Pass also worked in the mill. (Wingate also owned about 75 slaves on his plantation in Newton County.)

In 1860, Wingate installed the latest saws, engines, and boilers until his mill was cutting 30,000 feet daily, the most of any mill in Texas. However, in August, 1860, a boiler explosion, which killed and injured four people, disabled the saw and planing mills for several weeks.[14]

A biography of Wingate explained in detail the early Sabine Pass lumber industry, recording that the business included:[15]

> ...A sash and door and blind factory that turned out windows and door frames. He manufactured cisterns and tanks. The window and door frames were made from pine; the other stock turned out was from cypress...He found ready sale for his goods all along the coast. He also turned out special patterns of some of them that were sent to Mexico and Cuba. He was doing a fine business in lumber tanks and sugar vats with the West Indies when the War Between The States came on and put a stop to shipping. In 1861 he finished sawing up logs that had been run down in the winter and spring of that year, and closed down with at least 1,000,000 feet of choice shop cypress and several million feet of lower grade cypress and pine stacked in his yard...

By 1860, Wingate had built several lumber schooners, which carried his products throughout the Gulf of Mexico. Later, he used them to carry cotton through the blockade, eventually losing one schooner and its cargo to the offshore blockade fleet. On October 21, 1862, a naval detachment from the Union blockaders came ashore and burned Wingate's palatial residence, sawmill, planing mill and sash and door factory, thus ending the history of Jefferson County's first sawmill and closing the chapter of steam sawmilling at Sabine Pass.[15a]

THE SAWMILL AT PORT NECHES

Another sawmill was built on Jefferson County's treeless coastal plains, on the Neches River at Port Neches. In 1856, Samuel Remley and John T. Johnson, who were brothers-in-law, built a steam sawmill adjacent to Port Neches Park and depended on logs rafted down to them from Beaumont. In 1857, the Remley and Johnson mill was one of seven such steam sawmills in the Sabine-Neches localities, each cutting from 3,000 to 4,000 feet daily.[16] In 1859, the sawmillers cut 5,000 logs, worth $5,000, into 1,000,000 board feet of "planks and scantlings," valued at $12,500. The sawmillers employed six men paid a total of $200 monthly, in addition to the labor of their slaves.[17] In 1861, Johnson entered the Confederate Army as captain and adjutant of Spaight's 11th Texas Battalion, and during the war, Remley used only the boiler to drive his grist mill as he ground corn for the Confederate Army.[18]

(The writer was born in Port Neches in 1920 in an antebellum home, built in 1858 on the banks of the Neches River, and constructed of lumber cut in Remley's mill. When the old house was torn down, it revealed rough (unplaned) boards, 1 1/4 inches thick by up to thirty inches in width, surfaced only on two sides, the edges still bearing the out surface of the logs, less the bark. The boards were nailed with square nails, and the "building blocks" were round, wooden blocks, three feet wide and three feet tall, each cut from the trunk of a cypress tree.)

In 1866, Johnson and Remley borrowed $800 to buy logs with, so that they could reopen their mill. Johnson soon left the business when he moved to Cow Bayou, Orange County, and opened a blacksmith shop, and it appears that Remley was left in financial straits for several years thereafter. Remley's mill burned in 1870, and in June, 1870, he bought another steam sawmill from H. C. Pedigo that had formerly been in use at Wiess Bluff, Jasper County. To finance his purchase, Remley had to sell a half-interest in the "Grigsby's Bluff Sawmill Company" to C. H. Alexander, a Sabine Pass merchant. In July, 1876, the Grigsby's Bluff Sawmill Company burned a second time, thus ending the lumbering epoch at Port Neches.[19]

EARLY SAWMILLING IN BEAUMONT

After a twenty-year wait, steam sawmilling finally came to Beaumont in 1856, and by 1858, three single-circular mills had been completed there. In January, 1856, the townsite proprietors transferred the "steam mill square" to William Phillips and Loving G. Clark and within six months the Phillips mill was in operation.[20] In 1858, William Lewis, a Beaumont attorney, bought out Clark's interest in the mill for $2,500 and acquired complete ownership a year later.[21]

In the census year of 1859-60, the Lewis mill cut, 2,000 cypress and pine logs, worth $2,000, into 300,000 board feet of lumber, worth $5,400. He paid five employees $95 monthly in wages.[22] Nothing is known of William Lewis' mill during the Civil War. He lost heavily on his mill investment, and in 1866 sold out to J. D. Bullock for $700. The latter lost heavily as well, selling out to Rev. John F. Pipkin and his son-in-law, Dr. N. G. Haltom, for $300 in 1869.[23] The Pipkin and Haltom sawmill was listed in the county's Products of Industry census as one of two small steam sawmills left in Beaumont in 1870, where it cut 150,000 feet of lumber during the only two months of the census year that it was in operation. The sawmill burned on September 18, 1873, a $7,000 loss, and was never rebuilt. The fire also consumed 150,000 feet of lumber, although several thousand feet more were saved by tossing it into the river.[24]

Early in 1857, John R. Ross and James R. Alexander freighted a single-circular steam sawmill overland from the Trinity River by ox wagon and erected it on Brake's Bayou on present-day Pine Street.[25] The following December, H. R. Green, a Galveston newspaper correspondent who also taught a school on Pine Street, observed:[26]

> ... I was shown yesterday a novel invention, a self-starter at Ross and Alexander's mill, with which the logs are set to the saw, and which reduces the boards to an exact precision in width and thickness at both ends...Moreover all you have to do is roll the logs on the carriage, drive in your mainstays. . .put on steam, and then go to dinner, and by the time you get back, the log is slabbed, sawn up, shoved out of the mill, stacked up, and the price marked on the top most plank - and all of it done by steam...

The Ross and Alexander mill soon began exporting lumber to Galveston, but the ill-fated partners were soon in financial straits. In November, 1858, Michael Alexander was appointed receiver to pay off the sawmillers' indebtedness.[27] In February, 1859, H. R. Green, the school teacher-correspondent, left a long account of a midnight conflagration that robbed Beaumont of one of its best early manufacturing facilities, as follows:[28]

> ...Messrs. Ross and Alexander...have lost their grist and sawmill, a large lot of the very best quality of cypress lumber, amounting to. . .fifteen or sixteen thousand dollars. ...Flaming shingles, streams of fire, cinders, ashes, and volumed smoke swept southward for several miles...The county was illumined so brilliantly that persons and objects were recognizable at a great distance...

In March, 1860, Michael Alexander sold the mill site and burned-out machinery to James Long and his brother-in-law, Frank L. Carroll.[29] After Long and Carroll entered the Confederate Army, the mill site and burned-out machinery remained vacant for several years until the owners returned from the war. They then built their saw and shingle mill, known at first as Long and Company, into one of the largest sawmills of its type in the South.

THE LONG AND COMPANY SAW AND SHINGLE MILL

When James Long built the Long and Long sawmill in 1867, it cut only lumber, and its shingle-making activities were still many years in the future. In 1866, his father, Davis Long of Leesville, Louisiana, moved to Beaumont with several of his daughters, and he acquired a half-interest in his son's sawmill. In 1870 Long and Long sawmill was the second of two steam sawmills at Beaumont. The mill had a single-circular, which with related equipment was valued at $10,000. The sawmill operated a full twelve months of the census year ending July 1, 1870, and paid a total of $3,300 in wages to its nine employees. In 1870, the mill cut 5,000 saw logs, valued at $6,000, into 1,200,000 feet of lumber, worth $14,000.[30] James Long died on June 3, 1873, robbing Beaumont of one of its most promising young lumbermen.[31]

A few facts about the Long and related families are so pertinent to Jefferson County's sawmill history that they must be clarified before this story can proceed. Davis Long had four daughters, each of whom married some one connected with the Long sawmill interests. Sarah Long had already married Frank L. Carroll in Louisiana; the latter would eventually own the Beaumont Lumber Company, along with its affiliate firms the Nona Mills Company of Nona, Hardin County, and Odelia, Jefferson County, and the Nona Mills Company of Louisiana, with its sawmill at Leesville. Julia Long married William A. Fletcher, who would eventually own the Texas Tram and Lumber Company of Beaumont and the Village

Mills Company of Hardin County. Hasseltine (Hassie) Long married John W. Keith, who would remain the active manager of Long Manufacturing Company, as well as an officer of W. A. Fletcher's mills. The fourth daughter, Martha Long, married Joseph A. Carroll (Frank's brother), who for many years would supervise the Beaumont Lumber Company's Yellow Bluff Tram Railroad and the logging camp of Buna in Jasper County before founding his own sawmill at Rockland, Texas.

When James Long died in 1873, his widow signed an indenture, authorizing her brothers-in-law to operate the Long sawmill as they saw fit; and they quickly converted the sawmill into a shingle mill known as the Long Manufacturing Company. In time, the brothers-in-law formed an interlocking directorate over most of the Beaumont lumber facilities, and accounting over the next quarter century for four-fifths of Beaumont's lumber production.[32]

Also the railroads became very important to the Jefferson County sawmillers after 1876, and very quickly a number of other sawmills and planing mills were built at Beaumont. Nevertheless, whenever freight rates remained too high or the perennial rail car shortages lingered, the lumbermen always had the option of relying upon water transportation to counterbalance the railroad situation. However, the great market for East Texas lumber during the 1870s-1880s was in South, Central and West Texas, and if lumber were shipped by water, it still had to be unloaded at a seaport and shipped inland by some means.

The Texas and New Orleans Railroad built from Houston to Beaumont and Orange the first time in 1860, but the line was so shoddily constructed that the rails were completely inoperative by 1867, the bridges, trestles, and crossties already rotted out. The line was completely rebuilt in 1875-1876, and for twenty years would remain Beaumont's sole rail route to the West Texas markets.[33] In 1881 the Sabine and East Texas Railroad laid its trackage from Beaumont to Rockland in Tyler County, and some Beaumonters built new sawmills along that route.[34] In 1895 John H. Kirby built the Gulf, Kansas City and Beaumont Railroad from Kirbyville to Beaumont, and the Gulf and Interstate built a railroad from Galveston to Beaumont. In 1897, the Kansas City Southern was completed from Port Arthur to Beaumont and on to Kansas City, thus opening up the Midwest for Beaumont lumber. Within a brief time, 1900 to 1905, three Sante Fe routes passed through Beaumont as follows: Galveston to Beaumont to Longview; Beaumont to Sour Lake, Conroe and on to Somerville; and from Baton Rouge to Houston, making of Beaumont a major rail junction.

After James Long's death in 1873, his brothers-in-law quickly converted the old sawmill to a shingle mill. Long Manufacturing Company soon became the largest employer in Beaumont, and many successful Beaumont businessmen of a later era would recall the beginnings of their business careers there, as 15-year-old shingle bundlers earning fifty cents for an eleven hour work day.

By 1875, loggers in Hardin, Tyler and Jasper counties were sending down thousands of logs on each new river freshet, and during the week of October 25, 1877, from 8,000 to 10,000 saw logs were expected to arrive at Beaumont, where they had to be sorted out and corraled in the river log booms of the Beaumont Log Boom Company and the Neches River Boom Company.[35] A log boom was made of floating logs chained together and served much the same purpose as a barbed-wire pen for cattle. Each independent log company in the woods and each Beaumont sawmill had its own mark or log brand, which was burned into the log to identify it. During the week of November 10, 1877, about 5,000 pine logs and 1,000 cypress logs were due to arrive in Beaumont, and it was from the latter type of logs that the Long firm found the raw materials for its shingles.[36] Each firm sought to keep enough logs on hand to stay busy for months ahead, but during dry summer months when the Neches River water levels fell much too low to float logs, there were sometimes log shortages sufficient to force the sawmills to shut down for months at a time.

In 1877, Long Manufacturing Company shipped 2,520,000 shingles between November 2 and December 25th. During the week ending January 4, 1878, the company shipped six carloads of shingles. The same article noted that "Long and Co. shingle mill, with an average capacity of 140,000 per day, gives employment to sixty persons."[37]

Jefferson County's Schedule IV, Products of Industry, for 1880 carried the following mill statistics for the Long shingle mill, as follows:[38]

> ...Capitalization: $50,000; employees: maximum, 60, average 35 men, 12 boys under 16 (as shingle bundlers); daily work hours: 11 winter and summer; daily wages: skilled, $3.00; unskilled, $1.50; annual wages paid: $18,000; months in operation: 10, idle two months; equipment: 3 circular saws, one band saw, 4 boilers, three 75-horsepower steam engines; raw material and value:

> cypress logs worth $20,000...; product and value: 24,000,000 shingles, worth $50,000; origin of logs; Jasper, Hardin, Jefferson counties; mill did one-tenth of its own logging.

One article of 1881 revealed that Long Manufacturing Company had the first telephone in Beaumont, a private line to the Beaumont Lumber Company office, a line that was installed years before the first telephone exchange was installed in Beaumont, and only five years after Alexander Graham Bell first demonstrated his marvelous invention in 1876. An early Beaumont visitor was responsible for the article of 1881 that reported:[39]

> ...The great shingle mill of Long and Company is not in operation at present, owing to the extreme high water (of Brake's Bayou). The mill has on the yard now 3,000,000 shingles, and the company is buying timber, having just paid...$404.88 cash for fifty logs, which was at the rate of about $7.00 per thousand (stumpage)...We had the pleasure of meeting Mr. F. L. Carroll, also of conversing through a telephone with a gentleman in the office of Beaumont Lumber Company a mile distant...

In 1891 the Long Manufacturing Company installed the first band saw in Beaumont. Two almost identical articles about Long Manufacturing Company appeared in Galveston and New Orleans papers in 1888 and 1889, as follows:[40]

> ...James Long, deceased, laid the foundation of the colossal shingle mill of the Long Manufacturing Company of this place almost twenty years ago...To do this required expensive machinery and...imported skilled labor from far away Michigan...He put in the machinery...and commenced the manufacture of his "O. K." brand of cypress shingles...
>
> The old firm of Long and Company was merged into the present firm of Long Manufacturing Company, but the stockholders were the same...W. A. Fletcher, president; and John W. Keith, vice president...Mr. John L. Keith, the secretary-treasurer...They claim to have made and sold during the past twenty years more shingles then any other manufacturer...The annual cut has for years averaged 36,000,000 a year...A feature of their business that should be imitated...is that any of their employees can put all his

> savings into the stock of the company, but no one else can buy it at any price.

Long Manufacturing Company enjoyed immense prosperity during the decade of the 1880s, but after 1890, the firm experienced a severe shortage of cypress logs, a depression that killed off home-building, and over-production of shingles, which forced it into bankruptcy in 1898. Its assets were liquidated to pay off creditors.

THE RELIANCE LUMBER COMPANY SAWMILL

Another Beaumont sawmill, the historical continuity of which spanned more than sixty years, was the great Reliance mill of Valentine, William, and Mark Wiess, three of Beaumont's earliest and most successful entrepreneurs. Early in 1859, an early Beaumont merchant named Otto Ruff, an immigrant German, purchased a new, single-circular sawmill from Steadham Foundry of Indiana. He shipped it via steamboat to New Orleans and via schooner to Beaumont. The mill was powered by a double-flue, 42-inch by 32-foot boiler and a 12-inch by 24-inch steam engine, with a fly-wheel ten feet in diameter.[41] During the census year ending July 1, 1860, the Ruff sawmill cut 8,000 saw logs, worth $8,000, into 1,250,000 feet of lumber, valued at $18,750.[42]

In December, 1860, Ruff sold his mill to A. J. Ward, who reputedly used it to cut timbers for the Confederate States government. It also was used to provide shelter for Confederate soldiers passing through Beaumont. In October, 1865, Ward sold the mill to E. L. Goldsmith and M. W. Reagan, both of whom died during the yellow fever epidemic at Houston in 1867.[43] Other sales of the mill took place between 1867 and 1878, including one sale to Mark Wiess and H. W. Potter (by which name it came to be known-the "Wiess and Potter" mill), but for much of a decade, it often lay unused and abandoned until 1878.

In 1878, Mark Wiess formed a partnership with James F. Ward, H. W. Potter, and W. P. H. McFaddin to buy, rebuild, and operate the mill under the name of the Reliance Lumber Company. At the same moment they bought all new equipment, $6,600 worth of sawmill machinery from E. P. Allis and Company of Milwaukee, to include a double-circular saw, gang saw, and related equipment.[44]

In 1880, mill statistics of the Reliance Lumber Company appeared in the industrial census as follows:[45]

> ...Capitalization: $50,000; employees: maximum 60, average

> 50 men, 3 boys under 16; daily hours operated, 11 winter and summer; daily wages: skilled, $3.50; unskilled, $1.50; annual wages paid: $20,000; months operated in year: 9, three months idle; equipment: one 5-gang saw, one double-circular saw, 3 boilers, one 80-horsepower steam engine; raw materials and value: saw logs, worth $50,000, mill supplies worth $5,000; products: 9,000,000 BF lumber, 200,000 lathes; value: $90,000; origin of logs: Jasper, Hardin, Tyler counties; mill did no logging of its own.

In September, 1881, Valentine Wiess bought out the interest of James F. Ward in the company, and William Wiess purchased W. P. H. McFaddin's interest. For the next twenty years, William Wiess was the Reliance's active manager; Mark Wiess served as sales manager and traveling emissary; and Valentine Wiess as president and financier. Valentine Wiess was the owner of Beaumont's first private bank, operated in conjunction with V. Wiess and Company, and he organized and served as president of the East Texas and Louisiana Lumbermen's Association, with its headquarters on the second floor of Wiess' store. The mighty Reliance maintained sales offices as far away as London, England, and because Mark Wiess spent his every traveling hour as Beaumont's unofficial Chamber of Commerce, forever promoting the advantages of his hometown, he was known far and wide as the "Bishop of Beaumont." The 'three Wiess men' were Confederate cavalrymen in Company A of Spaight's Battalion.[46]

Mark Wiess was also an able mechanic and innovator, who in 1878 realized that lumber production at the Reliance would always remain about 8,000 to 10,000 feet daily, a captive of mechanization as long as the inefficient friction-feed log carriage remained in use. He began experimenting with steam as a means of moving the log carriage and he perfected a crude steam cylinder which, when attached to the track of the log carriage and energized by steam, would move the carriage forward on the track, exhaust, and then reverse the direction of the carriage. He soon sold his first working model to a Wisconsin foundry, which in turn provided the Reliance sawmill with the first experimental model of the "shotgun feed" log carriage. For a while, the Reliance was the only mill in the South that was so equipped. "Shotgun feed" revolutionized sawmilling everywhere, and soon enabled the Reliance to raise its daily output to 50,000 feet with only one set of double-circular saws.[47]

One of the Wiess brothers' first actions was to add a new steam engine and a second set of double-cutting circular saws.[48] In November, 1880, the mill owners completed a 1,000-yard canal to connect the head of Brake's Bayou with the Neches River, thus eliminating a 2 1/2 mile pull for the log raftsmen and creating an automatic log boom and log pond for the Reliance Company's logs.[49]

By 1885, the Reliance mill could boast of the following equipment, some of which was the first to be used by any Texas sawmill, as follows: automatic slab carriers, a set of line rollers, an Orescott direct-action steam "shotgun feed" log carriage, a log trip on the log dock, elevated tank and overhead sprinkler system throughout the mill, Curran and Wolf steam dry kilns, separate boiler houses (across Brake's Bayou from the mill), steam-powered log turner, Corliss steam engines, a boom that held 20,000 logs, and a carriage capacity for cutting logs 50 feet long.[50]

At the peak of its production, the Reliance mill provided employment for 125 people. Since the mill had no tram road, it employed no loggers in the forests.[51] By 1887, the mill was producing many types of artistic millwork, including all types of mouldings, lathe work, pickets, banisters for stairways, even pulpits. One article noted how the mill outdid its competitors in Tennessee as follows: "The Reliance sawmill of Beaumont (has) just finished a pulpit for a church in Murfreesboro, Tennessee, which is a piece of really artistic work. The design...is octagonal, with gothic panels and brackets supported by turned columns..."[52] In September, 1891, the Reliance mill signed a contract with the Omaha, Kansas Central, and Galveston Railroad to furnish 100,000,000 feet of crossties, bridge and trestle timbers, and depot materials over the next two years, the largest lumber contract ever recorded in Southern sawmill history. An order of such magnitude was unheard of, and many people would not believe it until it had been confirmed by several sources. In fact, the order was so large that it would have consumed the total output of the Reliance mill for three or four years, and the lumber company had to buy much of the output of other mills to fill it. It was a Godsend in many ways, for the area mills were already scrambling for markets due to lumber overproduction.[53]

Again the best description of the Reliance Lumber Company came in the Galveston article of 1888, as follows:[54]

> ...This mill is owned by the Wiess brothers and H. W. Potter. ...Its capacity is 20,000,000 feet a year. They have a Reynolds-Corliss engine...and a sixteen-foot driving (fly) wheel, thirty-one inch face...The mill is a double-rotary, with improved steam feed, powerful steam log turner, line rollers, edgers,

> trimmers, sizers, and automatic slab chain...E. P. Allis of Milwaukee constructed all except the slab chain...They also run a manufactory, where they manufacture doors, (window) sash, stair rails, balusters, handsome office desks, tables, etc...They also make a specialty of Bagdad (picket) fence... Mr. William Wiess is general superintendent and Mr. V. Wiess financier...

Despite the Reliance mill's huge railroad order, the decade of the 1890s was especially noted for the vast overproduction of lumber in East Texas, which sent mills scurrying for foreign markets, and a national depression which killed off home-building. In fact, had it not been for the continued expansion of the railroads, probably more East Texas sawmillers would have faced bankruptcy than actually did. By 1902, the Wiess brothers had grown tired of the lumber business, had become heavy investors in oil production, and in fact, they brought in the Batson oil field a year later. In addition, an end to river logging was already in sight. The brothers soon sold the big Reliance mill to the Kirby Lumber Company, who gave it the designation as "Mill A" in the new Kirby system of mill-naming. The new owner soon shut down the Reliance for a year while it was completely rebuilt and ovehauled. Old saws were torn out, and large, double-cutting band saws replaced them.[55] For the decade and a half following 1905, the Reliance sawmill (Mill A) received its logs steadily, but at much greater expense via train loads from Jasper County. Finally, its heyday long past and its value as a lumbering facility ended, the old Reliance bandsaws screeched to a halt in 1920, and the mill was dismantled and moved away.

OLIVE AND STERNENBERG'S CENTENNIAL SAWMILL

In 1875, John Abraham Sternenberg, a Houston lumberman, teamed up with Sidney C. Olive of Waco to found one of Beaumont's earliest large sawmill firms (some say the earliest), Olive, Sternenberg and Company. Olive, a former Confederate soldier, moved his wife Amerika and two children to Beaumont, but Sternenberg, also a Confederate veteran, boarded at the old Telegraph Hotel, but visited his family in Houston whenever he could. Early in 1876, Olive took his family to the Centennial Exposition in Philadelphia, where he "purchased the prize engine of E. P. Allis and Co. (Milwaukee sawmill manufacturer), shipped it to Beaumont, where it was the first (sawmill) engine south of the Mason-Dixon line that had a capacity of 50,000 feet a day."[56]

On October 10, 1876, Olive and Sternenberg purchased the "steam mill square" from the Nancy Tevis Hutchison estate, a site that had lain vacant since the Pipkin-Haltom sawmill burned there in 1873.[57] They installed one steam-driven shingle machine and three lumber saws, an installation the size of which Beaumont had not previously witnessed, and they named their new mill the Centennial Sawmill after the centennial birthday of America which Beaumonters were still celebrating. Despite the sawmill's capacity, it was hampered by its old friction-feed log carriage, which reduced its output to about 33,000 feet daily. Nevertheless, Beaumont had never witnessed such production from one mill prior to 1878, a year when three other sawmills would also install complete new equipment. In December, 1877, one editor observed that the "Centennial Mill of Messrs. Olive and Sternenberg cut 805,000 feet of lumber last month in 26 days."[58]

A month later, a newspaper article reported that "the Centennial mill of Olive and Sternenberg 'chaws' up logs at the rate of 40,000 feet a day and employs twenty hands."[59] For the year ending in August 1878, the six Beaumont sawmills shipped a total of 21.1 million feet of lumber, of which two-fifths (8.85 million feet) was shipped by the Centennial mill.[60]

In September, 1878, a Galveston editor observed that:[61]

> ...Another firm, Olive and Sternenberg of the Centennial Mills, are among the prominent and reliable lumber manufacturers of Beaumont. They turn out nearly 10,000,000 feet annually. Their planing mills are located in Houston, and the dealings with this firm are always prompt. Their mills in Beaumont turn out 34,000 feet a day...

The 1880 Products of Industry census Schedule IV for Beaumont added a great deal of information about the town's first major lumber producer, as follows:[62]

> ...Olive and Sternenberg's Centennial Sawmill, Beaumont, Texas. Capitalization: $56,000; employees: maximum, 160; average, 60 men with 6 boys as shingle bundlers; work hours: 11 winter and summer; daily wages: skilled, $3.00 daily; unskilled, $1.50; annual wages paid: $20,000; months mill in operation: 10; shut down for logging, 2; equipment: one 5-gang saw, 2 double-circular saws, two 75-horsepower steam engines, 3 boilers; raw materials: saw logs, worth $50,000; mill supplies worth $3,400; products: 9,000,000 feet of lumber, 4,000,000 shingles; value of products: $88,000; origin of logs: Neches River and its tributaries-mill did 80% of its own

logging.

Thus, when the sawmill closed down for two months, Olive and Sternenberg employees were not laid off. Previously, the owners had purchased several thousand acres in eastern Hardin County, and the employees left for the forests for two months, to work either as loggers or as raftsmen, rafting logs downriver to Beaumont.

Jefferson County's Log Brand Book of that era reveals that Olive and Sternenberg registered its log brand, the letter 'S,' on August 4, 1879.[63] There was a code of honor among Beaumont's sawmillers that in case a log were delivered to the wrong mill, the log would be sawn up, but first it would be measured, after which proper payment for it would be made to the rightful owner. This often occurred because the logs in the river belonged to many different owners. Each sawmill or independent logging firm branded its logs, in much the same manner that ranchers branded cattle, and upon reaching Beaumont, the lumberjacks had to sort out and deliver the logs to the different log booms of the rightful owners.

By 1881, Olive and Sternenberg had moved their Houston planing mill to Beaumont.[64] During 1880, Olive and Sternenberg had suffered greatly from two causes: a shortage of Texas and New Orleans rail cars, often receiving for their shipments only half as many as were needed, and also severe log shortages due to water levels in the river too low to float logs. It had been a severely dry season in those counties tributary to the Neches River. Hence, when the Kountze Brothers, investment bankers of New York and Sabine Pass, announced that they would extend the Sabine and East Texas Railroad from Beaumont to Rockland, Texas, the sawmillers greeted the announcement with eagerness, especially Olive and Sternenberg, since the railroad would have to cross their property three miles north of present-day Kountze. The sawmillers hoped that by switching their base of operations to the East Texas Railroad, they could avoid the perennial box car shortages experienced by all the Beaumont sawmillers. At worst, the Centennial Mill could be logged by rail from Hardin County.

By January, 1881, the first rails of the East Texas Railroad crossed Pine Island Bayou en route north, and by July the railroad had reached Kountze, which at that time was only a work camp for rail workers. Olive and Sternenberg began delivering lumber and new sawmill machinery at Kountze and freighting it the last three miles by ox wagon to their property, where they built the new Sunset Sawmill and founded the mill town of Olive. By August, 1881, the sawmillers were rushing their new mill to completion. Other Beaumont millers did likewise, the Carroll families rushing to completion their new mill at Nona and the Long and Company owners, W. A. Fletcher and John W. Keith, building the Village Mills, all of which will be dealt with in a separate chapter on Hardin County. In October, 1881, a Beaumont editor observed that: "A few miles farther (north of Kountze) the Sunset Sawmill of Olive and Sternenberg will cut its first lumber next Monday morning..."[65]

Nevertheless, the owners did not close the Centennial mill immediately. In Hardin County, they purchased a locomotive and flat cars, built five miles of tram railroad into their pine forests, and thus were guaranteed an ample supply of logs.[66] After waiting two years to determine if their new Hardin County sawmill at Olive would succeed, Olive and Sternenberg closed the Centennial sawmill in 1883 and dismantled it.

THE ADAMS AND MILMO SAWMILL

Another early Beaumont sawmill was the Bernard Milmo mill. Like the Centennial mill, it was not destined to play a very important role in the county's sawmill history except for a very few short years. In March, 1878, Milmo, an immigrant Irishman,[67] formed a partnership with Columbus C. Caswell, a Beaumont merchant, to mill lumber, Milmo to supply the sawmill, and Caswell to furnish logs and operating capital. Caswell dissolved the partnership, and in 1882 Milmo sold his sawmill to Francis Adams for $6,900.

Adams soon sold a one-third interest in the sawmill to W. B. and S. A. McCarty and another one-third interest to Henry Solinsky, a Beaumont merchant, and after 1882, the business was operated as Adams, Solinsky and Company.[68] The 25,000-foot per day sawmill was located on Brake's Bayou, adjacent to the Reliance mill.

The following statistics for the Adams and Milmo sawmill appeared in Beaumont's Products of Industry census for 1880, as follows:[69]

> ... Capitalization: $9,000; employees: maximum, 22; average, 18; daily hours: 11 winter and summer; daily wages: skilled, $2.50; unskilled $1.40; annual wages paid: $2,500; months mill in operation: 4, mill idle for 8 months: equipment: one 3-gang saw, one circular saw, 2 boilers, one 45-horsepower steam engine; raw material and value: saw logs, $6,000; mill supplies, $300; product and value: 1,200,000 BF lumber worth $12,000; origin of logs: Jasper, Hardin and Tyler

counties; mill did one-fourth of its logging.

In 1881, an Orange editor visited the Adams and Milmo sawmill. He wrote that: "...It is the smallest in town, but runs all the time, does splendid work, and turns out about 20,000 feet of lumber a day. Its yards are well-stocked and its conveniences many..."[70]

In May, 1881, Adams and Milmo sawmill shipped 84 car loads of lumber, which was certainly an impressive amount for such a small mill. (The Texas and New Orleans rail cars were very small and carried insignificant amounts of lumber, about 7,000 feet.) During the same time span all of the Beaumont mills shipped a total of 676 cars.[71] In 1883, Henry Solinsky converted the sawmill to a cypress shingle mill, and during a Neches River flood of June, 1884, one editor observed: "...Mr. Solinsky had almost 12,000 feet of cypress timber stacked upon the banks near his mill, and it was washed away."[72] No further record of the Solinsky mill was ever located. The writer believes that it was soon afterward sold, was dismantled, and moved to the East Texas Railroad.

THE INDEPENDENT PLANING MILLS

Although it was not the writer's intent to deal with the independent planing mills at Beaumont, not otherwise connected with a sawmill, two such independent planing mills were established there about 1880. Both of them can be dealt with quickly, and each of them apparently performed a much needed service. The first was the Globe Planing Mill Company, founded in 1878 by Alexander Young and J. L. Williams. In 1882 Young left to found the two large sawmills at Warren, Texas, and eventually J. L. Williams became well-known as a Texas and Louisiana sawmiller.

Beaumont's Products of Industry census of 1880 for Jefferson County recorded the following statistical data for the census year ending July 1, 1880, as follows:[73]

> ...Globe Planing Mill, Beaumont, Texas. Capitalization: $15,000; employees: maximum, 75; average, 40; work hours daily: 11 winter and summer; daily wages: skilled, $2.50; unskilled, $1.50; annual wages paid: $20,000; months in operation: all year; equipment: planers, 2 boilers, one 75-horsepower steam engine; value rough lumber and mill supplies: $55,000; value finished products: $85,000.

From the language used by a visiting journalist in March, 1881, he was apparently more impressed by the planing mill than by any other lumber industry in Beaumont, as follows:[74]

> ...What we perceived to be the best institution in Beaumont for Beaumont is what is known as the Williams Planing Mills (Young and Williams' Globe Planing Mill). Here is found a perfect mass of machinery, planers, picket, moulding, and various other machines - driving away at full tilt. This mill does not come in as a competitor with the other mills; it cuts no lumber, but draws its supplies from all the various mills - millions of feet of lumber, even from Orange. And by working it into finer material, it enhances its value, and thus retains and turns loose in Beaumont an average of $100 daily, which without it, would go elsewhere, while there would be no diminution in the amount of lumber...

The last mention of the Globe mill, also known as J. L. Williams and Company, occurred in August, 1885, when lightning struck the sawdust blower of the firm, creating a fire which was confined to the destruction of one small building.[75] The Globe Planing Mill Company was not mentioned in either of two very long articles about the Beaumont mills in 1888 and again in 1889. The company could have been absorbed by the larger mills. Since both Young and Williams became involved in sawmills elsewhere, the writer believes the planer mill was moved to some other sawmill that one of them owned at some remote point.

In March, 1882, Mark Wiess, S. B. Bacon and James F. Ward organized a planing mill company at Beaumont and commenced operations immediately.[76] The following September, Wiess, Bacon and Ward sold out their business to the Beaumont Planing Mill Company.[77] In January, 1883, an editor observed that "the Beaumont Planing Mill Company is adding numerous improvements to their mills."[78] Apparently the owners were soon in financial straits, for in June, 1884, the planing mill company entered receivership, and a receiver was appointed to administer its affairs.[79] Since no further mention of Beaumont Planing Mill Company was ever found, its assets were probably auctioned to satisfy creditors or else it was absorbed by one of the larger mills.

THE BEAUMONT LUMBER COMPANY SAWMILL

One of the first spin-offs of the Long Manufacturing Company was the Beaumont Lumber Company, founded in March, 1877, by Long and Company, Frank L. Carroll, George W. Carroll, John C. Ward, and John

N. Gilbert. As time passed, that company came under the exclusive control of the Carroll families and Gilbert. The partners soon afterward bought fifty acres in "the bend of the Neches River" - the present-day Beaumont dock area - where they soon built two sawmills.[80] The Beaumont Lumber Company mills, which were also near the court house, were the southern-most sawmills of the many that lined the Neches River. Farther north along the river, opposite the present-day (1992) city hall and public library, was the Eagle Mill, which in 1883 became the Texas Tram and Lumber Company. On the "steam mill square," where Brakes's Bayou flowed into the Neches River, was the Centennial Mill of Olive, Sternenberg and Company. Going north on Pine Street, which ran parallel to Brake's Bayou, was the Long Manufacturing Company and its company store. The last two mills on Pine Street were the Adams and Milmo mill and the Reliance Lumber Company sawmill. Also on Pine Street, between the Reliance and Magnolia Cemetery, were several rows of identical "shotgun" houses, built by Mark Wiess for the Reliance's Negro employees. Each house was painted a bright red with red lead paint, and the area soon became known as "Red Town."

Beaumont Lumber Company operated two sawmills simultaneously. The company bought as its first mill the old W. B. Black sawmill of Orange, a small mill of 25,000 feet daily capacity, dismantled it, and erected it again on the river bend that they had just purchased. The second mill, which had a daily capacity of 40,000 feet, was a new double-circular sawmill, bought from E. P. Allis and Company of Milwaukee and was erected adjacent to the old Black mill.[81]

During the late 1870s, several Beaumont lumber firms began buying up large tracts of timberlands in Hardin and Jasper counties, realizing that their mills would be dependent upon logs from those counties for years to come. By 1893, Beaumont Lumber Company (the Carroll and Gilbert families) owned 70,000 acres there, mostly in Jasper County; Texas Tram and Lumber Company (Fletcher and Keith families) owned 110,000 acres; and the Reliance Lumber Company owned 90,000 acres. Two other large timberland owners included Kountze Brothers' East Texas Land and Improvement Company, whose 260,000 acres in Hardin and Tyler counties was their principal reason for extending their East Texas Railroad from Beaumont to Rockland; and John H. Kirby's Texas Pine Land Association, which owned 140,000 acres.[82]

During the census year ending July 1, 1880, the following mill statistics were compiled for Beaumont Lumber Company, as follows:[83]

> ...capitalization: $35,000; employees: maximum, 50; average 35 men and 6 boys under 16; daily work hours: 11 summer and winter; daily wages: skilled, $3.00; unskilled, $1.50; annual wages: $18,000; months in operation: 11, idle for one month; equipment: one 5-gang saw, one double-circular saw, 2 boilers, one 100-horsepower steam engine; raw materials and value: saw logs worth $50,000, mill supplies worth $2,200; products 10,000,000 BF lumber, 2,000,000 shingles; total value of products: $78,000; origin of logs: Neches River and tributaries; mill did nine-tenths of its logging.

Early in 1877 Long Manufacturing Company and Beaumont Lumber Company organized the Yellow Bluff Tram Company to harvest pine timber on their lands in Jasper County. The company laid cypress rails from Yellow Bluff (ten miles north of Evadale) to Cairo Springs, five miles to the east, where they built a logging camp and from whence mule teams pulled the log cars to the river, where logs were rolled down the spillway and floated south to Beaumont. B. F. Van Meter was the first superintendent of the Yellow Bluff tramway, but was replaced a year later by Joseph A. Carroll, a brother of Frank Carroll.[84]

By April, 1878, iron rails had replaced the cypress rails of the tram road, and a locomotive, the first one ever in Jasper County, had replaced the mule teams, guaranteeing the Beaumont Lumber Company a sufficiency of logs whenever the water level of the Neches River permitted. The tram road was described as follows:[85]

> ...The tramway at Yellow Bluff is in full blast, and the screech of the locomotive, the 'Colonel No. 2,' is heard across the tall pines of Jasper County, echoing and re-echoing along the banks of the turbid Neches (River). As she goes tearing throught the forest after the mighty monarchs..., it gives new impetus and new life to everything in that section. From 150 to 200 logs are put into the Neches daily, giving employment to about 60 men...

Eventually, the Yellow Bluff tram road reached Buna, a logging camp which Joseph Carroll named for his first cousin. Another branch of the tram road was built to Ford's Bluff (Evadale), which shortened by ten miles the distance each log traveled to the mill. As evidence of the value of the Yellow Bluff Tram Company to the sawmills at Beaumont, 18,000 logs ran down the Yellow Bluff spillway into the water during one river freshet of November, 1878.[86]

A long article of 1888, entitled "Beaumont's Big Business," gave one of the most comprehensive accounts of the Beaumont Lumber Company that survives in print, as follows:[87]

> ...The Beaumont Lumber Company is composed of fresh young...blood...F. L. Carroll, president; G. W. Carroll, vice president; and John N. Gilbert, secretary-treasurer. . .This mill has a capacity of 60,000 feet per day, and their planer can cut through 50,000 feet per day...They sold 35,000,000 feet last year...With never less than 5,000,000 feet of dry stock on hand, dealers may safely rely upon prompt dispatch of orders...
>
> Owning a large line of railroad fully-equipped and landing their logs in tidewater daily, they are prepared to take orders for dimension lengths, cut and ship same the year round, on shorter notice than any other mill. They have a paid-up capital of $239,000. Two of Prescott's steam feeders are en route from the factory for their mills which ...will largely increase their facilities...With fair dealing, liberal wages, and prompt pay, no strike has ever been contemplated by their men...

In 1884, the Beaumont Lumber Company chartered the Nona Mills Company, Ltd., headquartered in Beaumont. The Nona Mills Company of Texas operated sawmills at Nona in Hardin County and at Odelia in Jefferson County, and Nona Mills Company of Louisiana operated one sawmill at Leesville. Over a long span of years, F. L. and G. W. Carroll and John N. Gilbert served as the principal officers of both companies.[88]

Through the constant improvement of their equipment, Beaumont's three principal sawmills had increased their daily cutting capacities to 275,000 board feet by 1893, as follows: the Reliance sawmill, 90,000 feet; Texas Tram and Lumber Company, 100,000 feet; and Beaumont Lumber Company sawmill, 85,000 feet.[89] In March, 1896, the two double-cutting circular saws at the latter's mill were torn out and replaced by a single-cutting band saw and a pony saw. Two newspaper accounts of that year revealed that:

> ...The Beaumont Lumber Company's band sawmill will be under headway in a few weeks more, when they will be enabled to turn out lumber that will make the retailer's mouths water...[90]

> ...The Beaumont Lumber Company has been shut down... A good deal of general repairing will be done, but the item of most interest is the band saw. A conservative estimate places the capacity of the band mill at 65,000 feet per day, while the pony circular saw will manufacture 20,000 feet ...The band saw works more accurately than a circular, and with the saving of the sawdust made in excess of the circular, a handsome residence could be the result each month...[91]

{Note: In order to be a wholly flexible, endless bandsaw blade, its thickness had to be about 3/16-inch, less than one-half that of a circular blade. Band saw blades required a specially-hardened or tempered steel.}

Much like the owners of the Reliance sawmill in 1902, the owners of the Beaumont Lumber Company, Frank L. and George W. Carroll and John N. Gilbert, were tiring of the sawmill operations in the town in which they lived, and too their eyes had become focused on oil production. Their property occupied a critical area of river frontage that would soon be needed for a dock area; also the era of river logging was rapidly growing to a close. The company still had three sawmills elsewhere in Texas and Louisiana. Like the Wiess brothers, the owners might well have entertained an offer to sell out to the Kirby Lumber Corporation in 1902 except for one factor. The Beaumont Lumber Company burned to the ground on June 2, 1902, with the sawmill, planing mill, dry kilns, dry sheds, and stacked lumber a total loss.[92]

THE EAGLE MILL

The small Beaumont sawmill that became known as the "Eagle Mill" came into existence in April, 1873, under the firm name of "J. G. Smyth and Brother." The new company was composed of J. G. Smyth, George W. Smyth, Jr., and John C. Ward. The brothers were sons of George. W. Smyth, Sr., of Jasper County, who had signed the Texas Declaration of Independence. In April, 1875, Ward sold back to the Smyth brothers his one-third interest in the firm for $1,120.[93] The writer is uncertain where the original Smyth sawmill of 1873 originated. After George W. Smyth's return from the Confederate army, he operated a sawmill at Smith's Bluff (where Unocal refinery is located), two miles north of Nederland, from 1867 until 1869. It is possible that he dismantled the Smith's Bluff sawmill in 1869 and kept it stored in Beaumont until 1873. The first Eagle sawmill was erected at the intersection of Hickory and Cypress Streets, on lots 600-608, Block 24, of the Beaumont townsite, which the brothers had purchased from Gilbert Stephenson.[94]

In 1877 Elias T. Seale of Jasper County bought out J. G. Smyth's interest in the Eagle Mill, and the new company became known as Smyth and Seale.[95] In 1878 Smyth and Seale decided to build a much larger sawmill, and lacking the land needed to build it on, they purchased 6.8 acres of land, fronting on the Neches River, from several grantors and "beginning at the northeast corner of Block 3 at the east end of Washington Street."[96]

In July, 1878, the Eagle Mill owners bought $20,600 worth of sawmill machinery from E. P. Allis and Company of Milwaukee, to be installed on Block 7 of the townsite, as follows: a 20-inch x 40-inch steam engine at $2,500; three 12-feet x 42-inch boilers at $2,500 each; one 5-gang saw edger at $500; one log turner at $300; and one "large size double-circular sawmill at $2,100." The sawmillers disbursed cash and notes to the manufacturer for the balance of their indebtedness.[97] A few months later, Smyth and Seale sold their old sawmill adjacent to the steam mill square to Olive, Sternenberg and Company for $3,500.[98]

The industrial census for the year ending July 1, 1880, recorded the following statistical information for the Eagle Mill as follows:[99]

> . . . Eagle Mill, Beaumont, Texas. Capitalization: $55,000; employees: maximum, 115; average 45; daily work hours; 10 winter and summer; daily wages: skilled, $3,00; unskilled, $1.50; annual wages paid: $12,500; months in operation: 7, idle for 5 months; equipment: one 5-gang saw, one double-circular saw, 3 boilers, one 139-horsepower steam engine; raw materials and value: saw logs worth $31,250, mill supplies worth $3,500; product and value: 6,250,000 BF lumber worth $62,500; origin of logs: Jasper, Hardin, Tyler counties; mill did no logging of its own.

By late 1879 and early 1880, deed records indicate that Smyth and Seale were in dire financial straits. The owners sold three small tracts of land, which included the company store, in a desperate effort to meet obligations, and the sheriff sold three other tracts of their land to satisfy court judgments.[100]

Apparently, in a last-ditch effort to avoid bankruptcy, the owners mortgaged for $11,500 all of their property on the Neches River and all sawmill equipment installed on it to C. R. Beatty of Galveston.[101] Still fearful of losing his investment and anxious to recover what he could, Elias Seale, only three weeks later, sold his "one-half interest in the sawmill, mills, engines, boilers," along with 350,000 feet of lumber and his interest in the Neches River Log Boom, to Beatty, the lienholder. The transaction gave Beatty a three-fifths interest in the Eagle Mill, which he promptly sold to Columbus C. Caswell, who was mayor of Beaumont.[102]

On June 5, 1883, G. W. Smyth and C. C. Caswell sold all their interests in the Eagle sawmill to W. A. Fletcher, John W. Keith, and S. F. Carter, who were also the owners of the Village Mills Company in Hardin County and of the Texas Tram and Lumber Company crosstie mill at Cairo Springs logging camp in Jasper County. The sale was for $17,500, but no cash changed hands. The grantors accepted as payment 175 shares of stock (Caswell, 105 shares; Smyth, 70 shares at $100 a share), which bought for Caswell a ten percent interest in the new Texas Tram and Lumber Company of Beaumont.[103]

Caswell's decision to sell was probably hurried by his own terminal illness, although there was also a moderate small pox epidemic in Beaumont at the time of the sale. Mayor Caswell died on August 6, 1883, and the sale did guarantee that his widow, Mrs. Elizabeth Caswell, would enjoy a sizeable income for the remainder of her life.[104]

Immediately after the sale of the Eagle Mill, George W. and J. G. Smyth founded the Wiess Bluff Tram Company, sixteen miles north of Beaumont, and began floating logs down the Neches River to the Beaumont mills. Had they taken that action ten years earlier to insure that the Eagle Mill would always have an adequate supply of logs, the sawmill probably would have prospered. Their Wiess Bluff tram owned 13,000 acres of virgin timberlands, a locomotive, thirteen log cars and five miles of trackage. In March, 1888, the Smyth brothers sold the Wiess Bluff tram road to the Beaumont Lumber Company. In 1889 the Smyth brothers used the proceeds of their sale to buy 25,000 acres of long leaf forests on the Sabine River in extreme south Newton County, where they immediately founded the Sabine Tram Company.[105]

THE TEXAS TRAM AND LUMBER COMPANY SAWMILL

In 1888, an article related the advances and improvements that the old Eagle Mill had accomplished under its new owners, as follows:[106]

> ...The Texas Tram and Lumber Company of this city is one

of the solid and enterprising institutions of Southeast Texas, having over a half million in hard cash invested in its business...The officers of the company are W. A. Fletcher, president...John W. Keith, vice president and treasurer; and S. F. Carter, secretary and business manager. The company has a sawmill with a capacity of 40,000,000 feet (annually) ...while is planing mill capacity is 25,000,000. . .

They own between 75,000 and 100,000 acres of...pine lands...They have equipped their tram roads with four first-class locomotives and an ample supply of cars...They have built eighteen miles of narrow-gauge railroad...They also own and operate the famous Village Mills (in Hardin County where the foregoing tram road was located)...

The shipments of the Texas Tram and Lumber Company for 1887 amounted to over 35,000,000 feet and had it not been for the great scarcity of cars...they would have shipped over 40,000,000 feet. Their shipments...have reached out of Texas into Old Mexico, Arizona, New Mexico, Colorado, Kansas, Missouri, Nebraska...They give employment to some 500 or 600 men at fair wages, and there has never been a strike among their employees...They also deal extensively in merchandise, handling close to a quarter million dollars' worth at their different stores during the year...

A year later, a New Orleans newspaper added some interesting statistics, which verify that the Texas Tram sawmill of Beaumont was second to none, as follows:[107]

...They (the Texas Tram sawmill) also have the best record for a single day's cut on one circular saw of 137,641 feet log measure {Note: Undoubtedly the cut was on large railroad bridge timbers.}...Their new planing mill, which is nearing completion is a mammoth concern...They do an immense merchandise business, aggregating hundreds of thousands yearly. Their receipts of grain and feedstuffs amount to over 1,500,000 pounds. A large amount of this is shipped to their extensive logging railway in Jasper County from which they draw their supply of logs...In fact there is no concern that is more thoroughly equipped...than this firm with $800,000 invested...

Another article observed that the new Texas Tram planing mill would cost $50,000 and would "give employment to from seventy-five to one hundred men."[108] The Texas Tram and Lumber Company owned the Village Mills tram road outright, but it shared ownership in the Yellow Bluff Tram Company and the Wiess Bluff Tram Company, both in Jasper County, with Long Manufacturing Company and Beaumont Lumber Company.

By 1893 the Texas Tram sawmill's cutting capacity was listed as 100,000 feet daily. It doesn't appear that that company suffered greatly from the economic depression and erratic lumber markets of the 1890's, most likely because that company catered to large railroad contracts of 20,000,000 or more feet in size, and which they could deliver with ease without neglecting their regular customers.

In 1902 W. A. Fletcher and John W. Keith sold the Texas Tram and Lumber Company (along with the Village Mills Company) to Kirby Lumber Company, and the Beaumont sawmill became "Mill B" in the new Kirby "alphabet soup" or system of identifying each of their mills by a single letter. For the next fifteen years, Kirby's Mill B depended mostly upon train loads of logs shipped into Beaumont over Kirby's Gulf, Beaumont and Kansas City (the "Kaycee") Railroad from Jasper County, and in 1918 Kirby Lumber Company dismantled Mill B and moved its machinery away. At the same time the Kirby Company had just purchased the Keith sawmill at Aldridge in Jasper County, which burned down in July, 1918. Since that sawmill was soon given the designation as "Mill B," it seems likely that the Texas Tram machinery was moved to Aldridge to rebuild the mill that was lost to the flames.

THE MILLER-VIDOR LUMBER COMPANY SAWMILL

The Miller-Vidor Lumber Company was one of the last lumber firms to move into Jefferson County, at a time when sawmilling at Beaumont was fast declining, and the age of river logging was ebbing. In 1905 the firm purchased the mill of the Beaumont Sawmilling Company, which had been built on the Neches River, in the refinery area about two miles south of town, some five years earlier, as the following article observed.[109]

> ...The Beaumont Sawmill Company...is pushing work along all lines and will soon be in a position to assume its place as one of the foremost producers in this locality. With the reorganization came new blood and new money. The latter was sadly needed...The concern has a large

quantity of logs in the booms and more are coming via tram and river...

The Miller-Vidor firm was organized in Galveston, and its two principal owners lived there. Charles H. Moore, chairman of the company, was an old retail lumber dealer of Galveston, who had owned lumber and shingle mills at Orange as early as 1875. A. W. Miller, also of Galveston, was the firm's president, and C. S. Vidor was vice president.[110]

Miller-Vidor moved rapidly to build or buy up four saw and shingle mills, as well as a vast acreages of pine lands in western Orange County. By 1907, a large logging camp, to which C. S. Vidor was to give his name, was established a few miles east of Beaumont on the Texas and New Orleans Railroad. The log camp had a capacity of 600,000 feet of stumpage daily, and gave employment in the forest to 175 men.[111] After 1920 Miller-Vidor logging operations switched to the Lakeview or Wiess Bluff area. Also the company operated about twenty miles of tram road in Orange and Jasper counties which ended either at the railhead at Vidor or at the Miller-Vidor spillway into the Neches River, from whence logs were floated down to the company's log boom. After deepsea shipping reached Beaumont in 1906, the Miller-Vidor firm's use of river logging dwindled accordingly, and eventually their Beaumont mill was logged entirely by log train from Vidor.

The Beaumont sawmill had a daily capacity of 90,000 board feet on its circular and gang saws, and was fully equipped with planing mills, dry kilns, and dry sheds. By 1910 Miller-Vidor Lumber Company showed phenomenal growth. Its four sawmills, at Beaumont, Orange; Milvid (in extreme northeast Libery County; and Timber in Montgomery County, had a total capacity of 100,000,000 feet of lumber annually (a daily cut of 330,000 feet), and the Beaumont and Orange mills also cut 50,000,000 shingles annually.[112]

E.H. Green, Jr. was manager of the Beaumont facility, and W. H. Harrington was mill foreman. The Beaumont sawmill of Miller-Vidor Lumber Company burned down on January 6, 1918; it was never rebuilt, but the company operations elsewhere continued for a few more years until all of their marketable timber was exhausted.[113]

THE DECLINE OF SAWMILLING IN JEFFERSON COUNTY

If one uses the year 1900, when river logging began to decline, as a yardstick to mark the decline of Jefferson County sawmilling in general, then it took another forty years for the industry's rigor mortis to set in. Several sawmills were to arrive on the local scene before their demise during World War II. Other firms, such as Industrial Lumber Company of Wichita, Kansas, maintained their sawmill headquarters and sales offices in Beaumont even though their sawmills were at Vinton, Sulphur, Seale, and Oakdale, Louisiana. For a brief period, however, Sam Park, the local manager for Industrial, operated one sawmill in Jefferson County.[114]

The town of Helbig in extreme north Jefferson County, on the Gulf, Colorado, and Santa Fe Railroad, grew up around a sawmill erected there in 1900 by the Foster Lumber Company, and a post office was established there during the same year. In 1903 Industrial Lumber Company bought the plant and continued to operate the sawmill until 1905, when all available timber was exhausted. In 1905, Industrial dismantled the mill and moved it to Seale, Louisiana.[115]

Another of the long-forgotten ghost towns of Jefferson County, where once a thriving sawmill town also existed, is Odelia. Located in the Gilbert Woods, fifteen miles southwest of Beaumont near Fannett, the sawmill and townsite of Odelia were built by the Interstate Lumber Company in 1905. Machinery for the new firm was manufactured in Fon Du Lac, Wisconsin. The followingnews release indicated that the mill's machinery arrived on the ground in January, 1905, as follows:[116]

> ...The machinery for the new mill of the Interstate Lumber Company at Odelia, Texas has arrived on the ground and is being rapidly installed...It is expected that the capacity of the mill will reach 50,000 or 60,000 feet a day, and sufficient orders have already been booked to keep the company busy for some time. Principally yellow pine for the general trade will be manufactured, and the management hopes to begin shipping within the next two or three weeks. The president of the company, Mr. P. E. Parminter has been at the mill personally, directing the installing of the machinery....

The mill, located on the Gulf and Interstate Railroad to Galveston, had a capacity of 60,000 feet daily, and was the first firm to cut logs from the Gilbert Woods. P. E. Parmenter was the mill superintendent.[117]

A history of the Nona Mills Company indicates that that Beaumont firm bought the Odelia sawmill early in 1906 and operated it until about 1915 when all the marketable timber was exhausted. Owners of Nona Mills Company of both Texas and Louisiana included Frank L. Carroll, president; George W. Carroll, vice president; John N. Gilbert, secretary-treasurer; C. L. Wallis, director; and G. R. Ferguson, director.[118] Charlsie Berly of Beaumont, retired Professor of English at Lamar University, recalled that her father, C. E. Berly, while paymaster and accountant for the Nona Mills Copmpany, often entrained on the Gulf and Interstate train to Odelia when he carried the sawmill's payroll.[119] To the best of the writer's knowledge, no further milling or logging was carried on in the Gilbert Woods until about 1935-1936 when the Peavy-Moore Lumber Company cut all the large second-growth timber and shipped it by log train to Deweyville, Texas, for milling.

Another large Jefferson County sawmill was at Voth, on the south bank of Pine Island Bayou. J. Frank Keith of Beaumont founded the 50,000-foot mill in 1902, and the town and post office that built up around it - Voth - was named for a logging contractor in Hardin County.[120] At first the mill was logged from the Saratoga area, but by November, 1904, a newspaper recorded that:[121]

> ...The mill is logged from Stutts on the Warren, Corsicana, and Pacific Railroad in Tyler County. The logs are hauled from Stutts to Warren, 22 miles, and thence to Voth, 32 miles distant, a total distance of 54 miles...A new dry shed with a capacity of 1,000,000 feet of lumber has just been put up. The logs at Voth are dumped into Pine Island Bayou and are used when needed...

In June, 1905, the Voth sawmill added two National dry kilns to its equipment.[122]

In 1922 the Kirby Lumber Corporation bought the Keith sawmill at Voth, and added a hardwood mill in 1924, which was dismantled in 1930. In 1948 the original sawmill of 75,000 feet capacity was still cutting pine lumber. In 1952, whenever the Kirby Corporation consolidated all of its milling operations at Silsbee, the Kirby mill at Voth became the last big sawmill in Jefferson County to close its doors.

The *Directory of American Sawmills* listed another large sawmill at Voth in 1915. In that year, the W. F. Ryder Lumber Company operated an 80,000 foot yellow pine mill, which manufactured dimension and yard stock and could saw, dress, and kiln dry timbers up to 32 feet long.[123]

SAWMILLS THAT BUILT THEIR OWN COFFINS

In 1907 C. R. Cumming of the Cummings Export Lumber Company of Houston organized the Neches Lumber Company, and built a 75,000 feet, double-cutting band sawmill on property where the old Adams and Milmo mill had once stood and adjacent to Kirby's 'Mill A' (formerly the Reliance Lumber Company).[124] While Cummings was the major stockholder in the venture, H. C. Huff served as president of Neches Lumber Company; Cumming as vice-president; and Sam Powell was secretary-treasurer.[125]

C. R. Cummings operated mills at Anahuac and Liberty until 1900, but as a result of losses suffered during the hurricane of 1900, he moved all of his millings operations to Wallisville. He cut only hardwood under contract to a German firm. He specialized in cottonwood lumber, and his products were delivered overseas to Hamburg.[126] Since Cummings cut no pine lumber at his Trinity River mills, the writer assumes the Neches mill was also a hardwood mill.

An unusual aspect of the Neches Lumber Company was its trade-off with the City of Beaumont waterworks of steam for lumber waste fuel, as follows:[127]

> ...The Cummings mill on Brake's Bayou has been runnig in a small way for the past few days....All of the machinery is not in place, nor are yard tracks laid, but some lumber is being cut for the mill's use...The dry kiln...will be supplied with steam from the waterworks powerhouse in exchange for fuel from the mill waste...The Cummings people have an immense supply of logs in Brake's Bayou. ...The mill will be logged by both river and rail...

It is not known exactly when the Neches Lumber Company ceased sawing timber, but apparently it closed down about 1930, soon after Wall Street crashed. For many years afterward, the mill lay untouched and abandoned, for during the Great Depression there was no market for second-hand sawmills. In 1936 the Neches Mill property, because of a rail switch located there, was leased as a site to unload and store redwood shingles, and as a 16-year-old lumber truck driver, the writer visited there many times while hauling shingles during that year. Once or twice the writer walked through the old mill which, except for the dust it had collected, looked as if it were ready to cut logs if it had a supply of steam. Instead, the big steam engine with its huge fly-wheel; the big double-cutting bandsaw, with its double rows of sharp teeth; the log carriage on its track, and the gang and circular saws all stood ghostly-silent and still, relics of an extinct lumber era that the sawmill had long outlived.

As late as 1941, three other big Beaumont bandsaws across town were still screeching away at full tilt. The three sawmills, Burris Lumber Company, on Crockett Street between 4th and Railroad (M. L. K.); the Roy Cloud Lumber Company at 11th Street and Hollywood; and the Southern Land and Lumber Company at Hollywood and 4th Street, were each cutting 50,000 feet daily and were logged by 18-wheel-trucks from the counties north of Beaumont. The mills had lumbered along at a turtle's pace during the depression years, eeking out only minute profits, but following the military draft of 1941, each of them received multi-million foot contracts to supply lumber for barracks at numerous army camps and air bases in Texas and Louisiana. Their profits quickly multiplied once more, but in a way, each contract was a nail in the sawmill's coffin. The barracks that they cut lumber for in 1941 were housing their former employees as soldiers by 1943, or else the former employees had found better paying jobs in the shipyards. By the time this writer returned home from World War II, the three sawmills had been dismantled and sold for scrap.

THE WRITER'S CONCLUSIONS

As surely as oil was the "mother" of Beaumont and Spindletop was the "father," old "King Lumber" and the sawmills were the Godparents. For forty-five years, sawmills offered the only industrial employment in the town. They nurtured the "sawdust city" from almost zero population in 1850 to a census enumeration of 8,000 in 1900. Beaumont's sawmills paid a "lion's share" of the taxes and provided the payrolls from whence taxes, groceries, housing, and all family necessities would be derived. And in four decades of time, sawmilling enriched a host of Beaumont families, whose palatial residences lined Calder Avenue and neighboring thoroughfares.

It was the several Beaumont lumber fortunes that provided the working capital with which to drill oil wells. For instance, in the summer of 1901 William Wiess bemoaned the arrival of the boom town, with its accompanying social problems and crime, and he longed for the return of the sleepy lumber town that was his choice for a home town. Nevertheless, by 1903 Wiess was so caught up in the new petroleum mania that it was one of his drilling companies that brought in the Batson, Texas oil field. In effect, William Wiess had only traded in his old Reliance sawmill to the Kirby Lumber Corporation for the new Reliance Oil Company and Paraffine Oil Company that he quickly organized. And each of those companies would be carried into the founding of giant Humble Oil and Refining Company (now Exxon) when that corporation was chartered in 1917.

The old sawmillers, however, must bear some stigma of guilt for transforming the beautiful East Texas forests into 10,000,000 acres of ugly, cutover stump lands, although reforestration was an unheard-of concept in their day. And the writer's hasty criticism is in essence an "ex-post facto" judgment, for to the old sawmillers' peers, cutover lands also meant jobs and payrolls, two items that make civilization bloom in a capitalist society. Another sad factor was that in sixty years' time Beaumonters witnessed a city where local dock yards contained millions of feet of lumber in 1900, awaiting shipment to foreign buyers, to dock yards containing millions of feet of lumber in 1960, just unloaded from ships arriving from the states of Washington and Oregon, for use in East Texas.

However, if one faults the earliest sawmillers for despoiling the East Texas forests, he should at least acknowledge the despoliation by the East Texas farmers who turned countless acres of forest into cleared lands for the plow and the cotton gin. Whatever their assets or failures, logging and lumbering provided an early means of livelihood for thousands of East Texans who might otherwise have moved on to California or some other "end of the rainbow."

Nacogdoches County

Giant pines such as these were characteristic arrivals at early sawmills in Nacogdoches County.

NACOGDOCHES COUNTY

With special gratitude to Jonathan Gerland for research graciously extended.

THE GEOGRAPHIC AND TOPOGRAPHIC DESCRIPTION

Nacogdoches County, perhaps the most historic spot on the East Texas map, is also one of nature's showplaces, most of its 963 square miles being quite scenic and neatly positioned between the Angelina River on the west and the Attoyac River on the east. Its neighbors include Angelina and Cherokee counties on the south and west, Rusk County on the north, and Shelby and San Augustine counties on the east. Nacogdoches County is also blessed with an abundance of spring-fed and clear-running bayous and creeks, a few of them being Lanana Creek, Dorr Creek, Bonita Bayou, Naconiche Bayou, and Nacoste Creek. The county's average altitude above sea level is about 400 feet, and its rainfall is quite adequate. Nacogdoches County's terrain might be described as rolling or gently undulating, with its lowlands being marked by river or creek beds. A fertile red land belt between four and six miles wide extends across the county from east to west. Otherwise, the soils vary from brown and gray sandy loams in the uplands to red and alluvial soils in the creek and lowland areas.

From its earliest history, the area now known as Nacogdoches County was one of East Texas' largest cotton-producing areas, growing 12,000 bales annually before the Civil War. Although the county was well-supplied with meadow lands suitable for agriculture, large areas of timberlands were most certainly cleared for cotton plantations as well. It is known, for instance, that as early as 1830, keelboats left Pattonia loaded with cotton and were floated along the Neches-Angelina water course, bound for Sabine Lake. By 1846, the first Neches River steamboat, the *Angelina*, under Capt. Moses Patton, made monthly voyages from Pattonia to the coast with cotton before it sank near Evadale, Jasper County, in 1850. As of 1947, forest lands still occupied 63 percent of Nacogdoches County's land area, but before the old growth forests were cut, the county's timberlands must have totaled closer to 85 percent. One government estimate of the volume of long and short leaf pine available for harvesting in 1900 amounted to 1.9 billion feet.[1]

The long leaf pine forests covered only the southern-most sector of the county, whereas short leaf and loblolly pines predominated in the central and northern regions of Nacogdoches County. A great variety of hardwoods also existed, principally in the Angelina and Attoyac river lowlands and the creek beds. The commonest varieties of hardwoods included several kinds of gum trees, red oak, white oak, pin oak, cottonwood, ash, elm, beech, magnolia, walnut, hickory, cypress, dogwood, cherry, birch, locust, hackberry, and others. A part of the Angelina National Forest covers the southern tip of Nacogdoches County.[2]

A tale published ninety years ago (Beaumont *Journal*, Nov. 13, 1904) portrays how an old-growth Nacogdoches County short leaf pine may have appeared at the turn of the century, as follows:

> ...A few years ago, some Mexican shingle makers in Nacogdoches County cut down a pine tree for the purpose of converting it into shingles. When the butt cut was sawed off, six inches diameter of the heart, the length of the shingle cut, hulled out and fell apart from the rest of the cut. This small diameter was covered with rosin and six well-defined cuts, evidently made by a sharp axe-like instrument, were plainly discernable. Between the surface of the smallest diameter and the surface of the tree, there were 283 year rings...The pine tree in its growth makes one of these rings every year...Two hundred and eighty-three year rings would take one back to 1621 (believed to be the year of the first Pilgrim Thanksgiving)—sixty years before LaSalle landed at Matagorda Bay, and while Cortez had barely finished the conquest of Mexico...

THE EARLY RAILROADS AND LOGGING TRAM ROADS

The first of the early railroads to reach Nacogdoches County was the Houston East and West Texas Railroad, at first a narrow gauge line, which built from Houston to Lewisville, five miles southwest of Nacogdoches in 1882. At that point, further track-building was delayed for almost a year, pending some right-of-way litigation. The Houston East and West Texas rails finally reached Nacogdoches in 1884, and the line was converted to standard gauge in 1894. By 1885, the following sawmill towns of Nacogdoches County were located on that railroad, namely: Lanana, Lola, Nacogdoches, Garna, Redfield, Appleby, Fitze, Garrison, and Attoyac.[3]

In 1901 the Texas and New Orleans Railroad began a new effort to complete its trackage from Rockland, Texas, across Angelina and Nacogdoches counties, and on to Dallas. The completion of the Texas and New Orleans line gave Nacogdoches County a new north-south route for moving lumber, where the Houston East and West Texas crossed the county in a general southwest to northeast direction en route to Shreveport. The following sawmill towns of Nacogdoches County were built along the Texas and New Orleans trackage after 1902, as follows: Tubbe, Clevenger, the Turner and Nabors sawmill (name of post office unknown), Nacogdoches, Sacul, Trawick, Cushing, and Caro.[4]

The Nacogdoches and Southeastern Railroad was chartered in 1905 as the logging road of the Hayward Lumber Company and its successor, Frost-Johnson Lumber Company. Before E. B. Hayward built his large sawmill there, Nacogdoches offered him the title to a railroad right-of-way to Oil Springs, which had been cleared for trackage twenty years earlier. Woden, an early Hayward logging camp, was located on the Nacogdoches and Southeastern, and in the center of Hayward's 50,000-acre tract of magnificent, old growth, long leaf pine trees. Eventually, the Nacogdoches and Southeastern Railroad extended 42 miles from Nacogdoches to Calgary, on the Santa Fe line in San Augustine County. Twenty-eight miles of the line belonged to Hayward's successor, the Frost-Johnson Company, which allocated trackage rights on its part of the railroad in exchange for log-tramming rights on the rails between Oil Springs and Nacogdoches.[5] This line connected with the Houston East and West Texas and the Texas and New Orleans trackage at Nacogdoches.

The Caro Northern Railway Company was originally chartered in 1894 as a logging tram road. It extended 16 1/2 miles from Caro (12 miles north of Nacogdoches) to Mount Enterprise in Rusk County, and the line was abandoned in 1934. The writer is not aware which sawmill company originally chartered the Caro Northern in 1894, but by 1904 the Whiteman-Decker Lumber Company hauled logs over its rails to its 100,000-foot mill at Caro.[6] The Caro Northern intersected the Texas and New Orleans trackage at Caro. (The year 1894 is probably an error.)

The tram roads of two out-of-county lumber firms also crossed parts of Nacogdoches County. The Angelina and Neches River Railroad, owned by the Angelina County Lumber Company at Keltys, had its eastern terminus at Chireno in Nacogdoches County. The logging tram of the Long-Bell sawmill at Lufkin, a part of which was owned by the Cottonbelt Railroad, ran across the southern tip of Nacogdoches County en route to its eastern terminus at Broaddus, San Augustine County. Several other tram roads of Nacogdoches County will be discussed with their respective sawmills.

NACOGDOCHES COUNTY'S ANTEBELLUM SAWMILLS

In the writer's history of sawmilling in Angelina County, he referred to East Texas' 'first' sawmill as being that of Wyatt Hanks on Ayish Bayou in San Augustine County, built in 1826. As it now appears, the 'first' may have been at Nacogdoches, or at least in Nacogdoches County. E. W. Winkler, in his 1902 article "The First Newspaper in Texas," referred to an advertisement in the first issue of the *Texas Republican* of Nacogdoches, dated August 14, 1819, which stated that a "Mr. Madden has engaged in building a saw and grist mill, and that this is a strange activity for a Spanish town." It is not known how long either Madden's or Hanks' mill were in operation, but by 1842, only the ruins of Hanks' mill dam and a few rotten components were still visible. Hence, Hanks' sawmill of 1826 can claim no more than second place among the East Texas' mills.[7]

In 1850, eight more Nacogdoches County water mills were recorded in the 1850 Census Schedule V, Products of Industry, for the census year ending June 1, 1850. These mills used only sash or upright saws, and an entire week's cut rarely exceeded 10,000 or 15,000 feet. In 1850, A. and C. Hotchkiss owned a saw and grist mill, powered by water, which cut the greatest volume of lumber in Nacogdoches County in that year. The mill, a $5,713 investment by the owners, cut 315,014 feet of lumber, worth $3,096, and ground 5,085 bushels of corn meal, valued at $3,813.

Another saw and grist mill in that county was operated by Ambrose Crain in 1850. His mill, which amounted to a $2,500 investment, was Nacogdoches County's second largest lumber producer in that year, cutting 300,000 feet, worth $2,000. His mill also ground 2,930 bushels of corn meal, valued at $2,140. A third saw and grist mill enumerated in Nacogdoches County in 1850 was that of W. B. Stivers. A $3,500 investment, his mill apparently operated for only a few months, during which he paid two employees $46 monthly in wages. Stivers' mill cut 70,000 feet of lumber, valued at $700, and ground 2,250 bushels of corn meal, worth $1,687.

Another mill of that year was Remonele and Vaight (latter spelling uncertain). With an investment of $3,000, the owners paid $40 monthly in wages to two mill hands. Their products included 200,000 feet of lumber, valued at $2,000, in addition to 2,857 bushels of corn meal, worth $2,142. Also in 1850, Bell and Martin's water-powered saw and grist mill amounted to a $2,000 investment by the owners. The sawmillers paid out $32 monthly in wages, but the volume of lumber cut indicates the mill probably cut lumber only during the winter months, a period when there was not much corn to grind. The millers cut 50,000 feet of lumber, worth $500, and ground 4,080 bushels of corn, valued at $3,060.

In 1850, Enoch Spivey cut only lumber at his water-powered sawmill. According to the industrial census, Spivey paid out $40 monthly in wages to two mill hands, but his mill probably did not cut lumber for more than four or five months during the year. A $2,000 investment, Spivey's mill produced 150,000 feet of lumber during the year, worth $1,500. Likewise, H. H. Carson's water-powered mill represented a $5,000 investment and cut only lumber. He paid $34 in monthly wages to two mill hands and produced during the year 200,000 feet of lumber, valued at $2,000. The eighth of the county's 1850 water-powered mills was that of L. N. Fall. The owner probably cut lumber for only a few months during the year, and he paid $20 monthly in wages to one mill hand. Representing a $2,000 investment, Fall's sawmill cut 80,000 feet of lumber during the year, worth $800.[8]

The 1860 Schedule V census of Nacogdoches County, which listed the county's manufacturers and products of industry, indicated that there were five sawmills in the county, two of which were steam-powered; two others were water-powered; and one was turned by a yoke of oxen. The writer can identify the approximate location of only one of the mills (which was also the largest) - that of Hardeman Brothers, located in the extreme southern tip near Etoile. The Hardeman mill was valued at $2,000 and used during the census year ending July 1, 1860. 2,500 logs, worth $1,250. A steam-powered mill, it employed nine men paid a total of $250 monthly. The mill cut 840,000 feet of lumber during the year, worth $8,400, all of which was sold locally.[9] One article noted that the 1904 terminus of the Nacogdoches and Southeastern Railroad near Woden was:[10]

> ...Within a few miles of the site upon which was once located one of the first sawmills ever operated in Nacogdoches County. The mill was located in a splendid pinery on Rocky Branch, and the boiler was brought from Natchitoches, Louisiana, on an ox wagon by B. M. Hall, he being paid $100 for delivering it safe to the owner of the mill, Major Black Hardeman...

No mill value was listed for the John Reid sawmill in Nacogdoches County. It was a very small, water-powered sawmill, which employed only two men. During the census year, it cut 180 logs, worth $100, and manufactured 40,000 feet of lumber, worth $400, all of which was sold locally. William Summers also owned a water-powered mill, which was worth $100 and also employed two men. During the census year, Summers' sawmill utilized 150 logs, worth $100, and cut 35,000 feet of lumber, worth $350, all of which was sold to local buyers.[11]

William S. Menard owned the only sawmill that was powered by a yoke of oxen. Menard reported that his mill was worth $100 and it employed only one person, presumably himself. He used 1,000 logs, worth $500, as raw materials and cut 100,000 feet of lumber, worth $1,000. Menard operated his mill the year round, whereas apparently Reid and Summers cut lumber only during the slack winter months. The latter mills were probably combination gin, grist, and sawmills.

The only other steam-powered sawmill in Nacogdoches County in 1860 was that of Wallace and Murph. Their mill, valued at $2,000, employed five men, paid a total of $150 monthly. During the census year, the partners utilized 2,000 logs as raw materials, worth $1,000, and cut 600,000 feet of lumber, worth $6,000, all of which was sold locally. The census gave no exact location of any of the Nacogdoches County sawmills of 1860.[12]

Only one of the 1860 sawmills survived to become one of the three mills recorded in the 1870 census. The Wallace and Murph mill of 1860 was owned by John W. Murph in 1870 and was located in District 5 of the county census districts. It was powered by an 18-horsepower steam engine, and the number of employees or wages paid were not listed. Representing a $3,200 investment, Murph's mill cut 300,000 feet of lumber in 1870, worth $4,500, and ground 2,200 bushels of corn meal, valued at $1,500.

In 1870, Samuel Knively's (or Knisely) sawmill was turned by a 26-horsepower steam engine and represented a $6,000 investment for the owner. In that year, the mill cut 500,000 feet of lumber, worth $8,000, and ground 2,200 bushels of corn meal, valued at $1,600. Knively's mill was located in District 1. Also in 1870, B. P. White's sawmill represented a $4,000 investment, was also in District 1, and was turned by a 25-horsepower steam engine. During the year, the mill cut 300,000 feet of lumber, valued at $4,500, and ground 1,650 bushels of corn meal, worth $1,500.[13]

NACOGDOCHES COUNTY MILLS FROM 1880 TO 1900

Of the sawmills enumerated in Nacogdoches County in 1880, two were still water-powered and four were run by steam engines. If any of them had been in the county in 1870, they had either been sold or the names had been changed. And apparently all of them had circular saws as opposed to the old sash saws of the Civil War era. In the 1880 manufacturer's census, E. Taylor's mill was located in Precinct 1 at Taylor Springs and was turned by a water turbine, that developed 15 horsepower. The owner had $2,000 invested in his mill, which had operated for only six months during the census year which ended on May 31, 1860. Taylor employed six mill hands, three of whom were over sixteen years old, and he paid them from 75 cents to $1.50 for a nine or ten hour work day. The mill produced for local consumption since no railroad had as yet arrived.

In the 1880 census, S. H. Stone's sawmill was the last of the water mills. Its "overshot wheel" developed fourteen horsepower and was located on a "branch of the Moral (creek)" in Precinct 1. Stone's mill represented a $1,000 investment for the owner, who cut lumber only for six months during the census year. Stone employed five mill hands, each paid from $1 to $2 for a nine or ten hour work day. The mill had a circular saw, but did none of its own logging.

Joseph K. Fisher's sawmill was located in "Enumeration District 50" of the 1880 census in Nacogdoches County. Fisher's steam engine developed twenty-five horsepower, and the mill did all of its own logging.

The owner employed six mill hands, all over sixteen years of age, paid from fifty cents to $2.00 for a ten hour work day. The Helpenstill sawmill was located in Precinct 5 of the 1880 census enumeration, and it used a 24-horsepower steam engine to turn a circular saw. An $800 capital investment for the owner, the mill ran for only six months during the census year which ended May 31, 1880, and it employed six mill hands, only two of whom were older than sixteen years.

T. C. Glover used a 24-horsepower steam engine to turn his circular sawmill, located in Enumeration District 3 of the 1880 Nacogdoches County census. Glover's mill was a capital investment of $2,500, and he did all of his own logging. He employed eight mill hands, only three of whom were over sixteen years old, and paid them from 75 cents to $3.50 (skilled) for a ten-hour work day.

James W. Power utilized a twenty-horsepower steam engine to turn his single circular sawmill, located in Precinct 1. Power had $1,800 invested in his equipment, did all of his own logging, and cut lumber for only six months of the 1880 census year. He employed five mill hands, only two of whom where above sixteen years, and paid them from $1 to $2 for a nine or ten hour work day.[14]

Until the tracks of the Houston East and West Texas Railroad reached Nacogdoches County in 1882, the few sawmills located there could cut lumber only to satisfy local demand. After the rails' arrival at Nacogdoches in 1884, several small sawmills were built along the line and immediately considerable quantities of lumber were being exported to Houston and elsewhere. Also, prior to 1894, the mills located on the H. E. and W. T. Railroad suffered an added burden in that the line's rails were narrow gauge, and no box cars of lumber arriving in Houston or Shreveport could be trans-shipped over a standard gauge railroad without first being reloaded into another car.

A list of that railroad's sawmills of April, 1893, revealed that the first and largest of the Nacogdoches County mills was that of the W. V. Carroway Lumber Company at Lanana, also known as the Lanana Mills, located on Lanana Creek six miles south of Nacogdoches on the Houston East and West Texas Railway. Its location was at one of the county's traditional water mill sites as far back as 1840. About 1870 the mill was reestablished as a combination saw, grist mill and cotton gin, cutting lumber for local use. It was converted to a steam, single circular mill to cut crossties for the railroad when construction arrived there in 1882. The Carroway sawmill cut 30,000 feet of lumber daily in 1893. Vestal

Carroway owned the mill until after 1893, after which it was sold to C. C. Galloway, who in turn sold out to Howard Lee in 1899. In August, 1899, the mill was auctioned for $1,350, and mill and the entire company town were dismantled and moved to Clawson, Angelina County.[15]

In 1893, the Lee and Manning Lumber Company owned a small sawmill at Lola, the next mill town between Lanana and Nacogdoches, which cut 15,000 feet daily. North of Nacogdoches, R. H. Lee owned a small mill at Garna, that cut 12,000 feet a day. The Fitze Lumber Company operated a small mill at Fitze, that cut 10,000 feet daily, and Simpson and Wragg owned a small sawmill at Attoyac, northeast of Garrison, on the county's extreme northeastern boundary, which also cut 10,000 feet daily. The Attoyac mill belonged to T. P. Wragg; his mill existed between 1893 and 1906, by which latter year, its cutting capacity had increased to 20,000 feet daily. Originally a water mill, the Simpson and Wragg sawmill converted to steam about 1895, and it cut red and white oak barrel staves for the European wine makers. On the railroad's list of rail stops of 1899, only the names of Lanana and Fitze still appeared, the mills at Lola and Garna having apparently already "cut out"and moved away. As of about 1894, W. V. Carroway had already sold his Lanana mill to Galloway and Wright. A 1900 list of Houston East and West Texas sawmills indicated no stops or sawmills at either Lanana, Lola, Garna, Fitze or Attoyac, although the Attoyac and Fitze sawmills apparently were still there.[16]

Another early sawmill was that of Blakey and Simpson, located seven miles southeast of Nacogdoches. Very little is known about that mill except that it existed in 1884 and cut 10,000 to 12,000 feet daily for local consumption. Nothing is known about the duration of its existence, whether water or steam-powered, or about its equipment.[17]

By 1904, Howard Lee of Garrison, Texas, was believed to have been the oldest sawmiller that was still active in Nacogdoches County, although that distinction might well have belonged to August Tubbe as well. Lee's mill at Fitze cut 20,000 feet daily in 1906, had no planer or dry kilns, and it cut crossties and timbers for the railroads and oil fields. Lee had already been sawmilling at a number of sites for about twenty years, and although Fitze was not listed as a Nacogdoches County sawmill site in 1900, it was certainly very active in that capacity in 1904, as two articles reveal:

> ...Mr. Howard Lee of Garrison, Nacogdoches County, ... owns a big mill near Garrison and has the reputation of having been in the lumber business in Nacogdoches County

> longer than anybody else there. He used to run a mill at Lola, on the Houston East and West Texas Railroad, and afterwards moved to Nacogdoches, and from there to Fitze. From Fitze, he recently moved his mill to Garrison...[18]
> ...Howard Lee, the well-known lumberman of Nacogdoches County, who last year (1903) moved his mill from Fitze on the Houston East and West Texas Railroad to a point four miles east in the country, has moved his mill back to Fitze, and is running on full-time as of yore...[19]

THE SAWMILLS AT CLEVENGER AND TUBBE

One of the earliest sawmill towns to be built after the Texas and New Orleans Railroad was completed from Rockland to Dallas around 1901 was Clevenger, eight miles south of Nacogdoches, where Joe P. Clevenger built a modern sawmill of medium capacity. There are several descriptions of Clevenger, one of which follows:[20]

> ...Clevenger Lumber Company—History of The Big Sawmill Eight Miles South of Nacogdoches—This new mill began operation in 1902...It has a capacity of 50,000 feet per day and has in the yard 700,000 feet. The logging camp is two miles distant, and the equipment of the three miles of tram (road) consists of fifteen tram cars and one locomotive. The mill has one circular saw, matcher, resaw, a cut-off saw, four Arkansas dry kilns, with a capacity of 25,000 feet per day. The engine is of 100 horsepower. There is a ten-year stumpage (supply of uncut trees). Forty houses accommodate the men and their families, of whom there are 50 on the payroll. The company has built a school house and church. A commissary adequate to the needs of the community is in evidence. Officers: J. P. Clevenger-owner; B. M. Hickman-manager; G. A. Dyer-mill foreman; M. P. Hale-postmaster and commissary manager; G. L. Watts-woods superintendent; B. J. Sisson-mill engineer; and C. A. Hawks-locomotive engineer...

The Handbook of Texas is in error in claiming that Clevenger was simply another name for "Tubbes" (sic: Tubbe), another small Nacogdoches sawmill town, four miles south of Clevenger, and very near the

Angelina River. In fact, each town had its own postmaster, and they co-existed together during the years between 1902 and 1912. Another description of Clevenger noted that:[21]

> ...South of Nacogdoches is the busy little mill of Joe P. Clevenger. The mill has a daily capacity of 50,000 feet of lumber, and is well-equipped with all the newest and most improved machinery. Mr. Clevenger's timber extends on both sides of the railroad; but he is cutting from the west side now, and has just finished laying two miles of big steel tram road, on which a regular railroad locomotive will be used in bringing logs to the mill...
>
> ...The mill employs about sixty men all the time, and there is enough timber in reach to keep the mill busy for five years, and sixty percent of this timber is of first-class long leaf variety. There is at present 1,320,800 feet of lumber on the yard at Clevenger, and the management is having no trouble finding a market for it, although prices are not as interesting as they were last year. Clevenger is surrounded by good farming land, which is well-watered and timbered. All the heavy pine timber is being cut, so if the land is ever placed on the market, it will cost but little money and labor to prepare it for cultivation...

It is not known exactly what year that the Clevenger Lumber Company cut out its timber and dismantled it mill, but it was probably between 1910 and 1915, depending on whether the owner were able to buy additional timberlands. In the 1910 city of Nacogdoches census (Prec. 1, res. 38), Joe Clevenger was still listed as a sawmill lumberman. One source observed that the Clevenger mill began in 1898 and cut lumber at a variety of locations—at a point two miles west of Nacogdoches in 1899, then eight miles east in 1901 and finally eight miles south of Nacogdoches in 1902. The same source added that Clevenger owned a circular mill, a planer complete with matcher, resaw and cut-off saw; and kilns, all energized by a 100-horsepower steam engine. Clevenger's maximum population was about 200. And one will note as well that the stumpage reserve of Joe Clevenger's timber indicated a maximum mill life of about ten years, as follows:[22]

> ...J. P. Clevenger has recently added new improvements to his mill near Nacogdoches and is ready to do his share of the New Year's business...

> ...Clevenger, the next mill on the (T. and N. O.) road, is one of the big mills of Nacogdoches County. It has a capacity of 75,000 feet daily and is operating four miles of first class tram road in the direction of the Houston East and West Texas Railroad to the west. The managers of the mill state they have enough timber in sight to run ten years...

The sawmill at Tubbe or Tubbe's Switch was located four miles south of Clevenger and twelve miles south of Nacogdoches near the Angelina River. Rev. August Tubbe began sawmilling in a water-powered mill on Dorr Creek in 1880. In that year, the Tubbe mill operated for ten months, employing seven hands (only two of them over sixteen) paid from $1 to $2 daily for a nine or ten hour work day. With $5,000 invested in his mill, Tubbe did all of his own logging. His first mill was a single circular saw, powered by "level turbine" water wheel, developing 24-horsepower at 150 revolutions per minute.

Although a smaller sawmill, every record of it indicated it did an active lumber trade, as some accounts have noted:[23]

> ...The next mill of any consequence is that of A. Tubbe in Nacogdoches County, near the line (southern boundary) of Angelina and Nacogdoches counties. This mill has a capacity of 40,000 feet daily and is doing a good business in addition to a good lumber shipping trade...
>
> ...Tubbe's mill is in Nacogdoches County, near the Texas and New Orleans crossing over the Angelina River. It has a capacity of 40,000 feet daily, employs thirty men, and has 500,000 feet on the yard. It is a small mill.

It seems probable that A. Tubbe might also have cut out and dismantled his mill around 1910 since no further mention was found of it. One early record of Sabine County compared the sawmill of Dave Meigs at Brookland to that of A. Tubbe in Nacogdoches County, noting that their mills were two of the last sawmills in East Texas that still ginned cotton, ground corn into corn meal, in addition to sawing lumber. Meigs mill even built wooden coffins sufficient to satisfy local demand. In 1904 Meigs' mill at Brookland was still a water-powered mill, but A. Tubbe is believed to have converted his mill to steam about 1901.[24]

One other sawmill, that of Turner and Nabors, was located midway between Clevenger and Nacogdoches. It was not operating as of November, 1904, "pending the adjustment of the business now in the hands of a receiver." Although the exact circumstances are unknown to this writer, the Turner and Nabors sawmill was dismantled in 1905 and removed to Beaumont, where it continued to cut lumber under the Turner and Nabors name for the next ten years.[25]

THE FOUNDERS OF THE HAYWARD LUMBER COMPANY

In 1901 the E. B. Hayward Lumber Company of Davenport, Iowa, purchased 23,000 acres of prime old-growth timber in the vicinities of Woden and Etoile, and within two years, that figure increased to more than 50,000 acres. About 1883, Nacogdoches had purchased land and cleared a right-of-way to Oil City for a railroad that was never built, and in 1903, a newspaper observed that: "...the right-of-way has been offered to the Hayward Lumber Company of...Iowa, if they would locate their milling plant here (Nacogdoches) and build and operate the rail (road)..."[26] E. B. Hayward quickly accepted the offer, and in 1905 he chartered the Nacogdoches and Southeastern Railroad as a forty-mile log tram road and common carrier to reach the Hayward Lumber Company's magnificent stands of long leaf timber.

E. B. Hayward was born in Essex County, New York, in 1842, and he served in the Union Army during the Civil War, acquiring the rank of major at age 23. Later, he moved to Davenport, Iowa, where he milled Wisconsin logs floated down the Mississippi River. In 1881, he bought a huge block of 100,000 acres of timberlands near Camden, Arkansas, and he became a partner of fellow Iowa investors in two sawmills in Arkansas. In 1904, E. B. Hayward organized the Hayward Lumber Company of Nacogdoches with himself as president; George W. Cable, Jr. of Davenport as vice president; George W. Cable, Sr. as secretary-treasurer; and W. W. Smith as general manager at Nacogdoches. Smith had formerly been the Kirby Lumber Company's mill superintendent in Kirbyville. Hayward had operated in three states under various company names, namely: as Adams and Hayward and E. B. Hayward, Son and Co. in Iowa; as Camden Lumber Company and Eagle Mills Lumber Company in Arkansas; and as Cable Lumber Company in Mississippi.[27]

In April, 1904, Hayward bought 8.26 acres of the J. M. Martinez Survey, one mile southwest of the public square in Nacogdoches, for his mill site, and construction began immediately for the new sawmill.[28] A few months later, a lumber journal carried the following article:[29]

> ...The throttle of a great engine was opened on Nov. 1 in the new and up-to-date sawmill plant at Nacogdoches...of

Hayward Lumber Company...The logging railroad of this company is laid with 45-pound steel rails...About ten miles of the road is completed; this will be extended as necessary, and will eventually penetrate a distance of 40 miles. The equipment consists of two locomotives and 33 logging cars. The log pond of the company has been made by excavating several acres, and the water supply is secured from a flowing stream...

...The sawmill structure is 40 by 186 feet...The machinery throughout was furnished by the Filer and Stowell Company...and consists of a single band (saw) and a Wickes 36-gang (saw), a heavy 72-inch edger...log kicker, log loader, Hill...(log turner), steam...carriage, 12-inch "shotgun feed," live rollers, slashers, transfer table, and trimmers...It is expected that the mill will have a daily capacity of up to 100,000 feet.

...The dry kilns...consist of three rooms constructed entirely of brick. The planer is 85 by 160 feet in size...and is equipped with one Hoyt sizer, three Hoyt matchers, one Glen Cove moulder, one Houston moulder, one band resaw, and one Hoyt edger...The yard has elevated tramways ("dollyways")...The power plant consists of 8 boilers set in three batteries...Water is obtained from two deep wells. The engine for the sawmill is a 24x48 (inch stroke) St. Louis-Corliss and that for the planer is an 18x32 (inch stroke) Buckeye...

...The electric light plant is being put in by the Ewing-Merkle Company of Shreveport and will be of sufficient capacity to light up the entire plant so that it can be run by night as well...A complete system of waterworks has been installed, including large fire pump and cast iron water pipe with regulation hydrants covering every part of the plant. On a 100 foot elevation nearby is placed a 35,000-gallon water tank for emergency use...

...In addition are being built the necessary lumber sheds for both rough and dressed stock, and a machine shop, a commissary building, an office building, and the usual complement of...(employee) houses will be speedily erected...

On the same date (Nov. 1), a Nacogdoches editor hailed Hayward Lumber Company as "Nacogdoches' largest and most extensive institution." The article noted that the new mill had a capacity of 110,000 feet daily, had a hotel and 50+ "nice cottages." The article observed as well that C. E. Wyman would be the new mill manager and C. C. Hayward would be the assistant mill manager.[30]

OPERATING AND LOGGING THE HAYWARD SAWMILL

Even before the Hayward sawmill sliced its first log, editors elsewhere were complimenting Nacogdoches' new addition to its economic well-being, one noting that:[31]

> ...The Hayward Lumber Co. of Nacogdoches have completed their mill and will start up about November 1st. The mill has a daily capacity of 100,000 and will give employment to a large crew of men. The company has built a standard gauge road into the rich pinery in the neighborhood of Oil City, sixteen miles east of Nacogdoches, and will extend the road as the necessity for it arises. A new town called Woden has sprung up out there and all the surrounding country is on a boom as a consequence. The tram road has been chartered and is known as the Nacogdoches and Southeastern. It is probable that it will ultimately be extended to a connection with the Cottonbelt Railroad near the new town of Broaddus in San Augustine County...

Eventually, that prediction proved almost correct, but not with the Cottonbelt road. The Nacogdoches and Southeastern rails finally formed a junction with the Santa Fe rails in 1925 at Calgary in San Augustine County.[32] Construction continued on their tram road in November, 1904, one editor noting that; "...The Hayward mill people of Nacogdoches County put on their log train for the first time last Monday. Everything worked all right. The company has purchased thirty log cars..." In January, 1905, an article added that: ...The Hayward Lumber Company ...has completed the construction work on the Nacogdoches and Southeastern Railroad to Woden, thirteen miles from Nacogdoches, and seven miles are now in operation..."[33]

Two articles of 1905 hinted that the Hayward mill continued to progress, reading as follows:[34]

> ...The Hayward Lumber Company...has ordered a locomotive of the Shay variety, which will be used on the spurs for logging the big mill. Large locomotives are used on the main tram line. The tram extension is being rapidly built in the direction of Oil Springs...
>
> ...The Hayward Lumber Company is putting in a new commissary. Their monthly payroll is $15,000. It "has brought several hundred new families into Nacogdoches..."

Nevertheless, the big Nacogdoches sawmill incurred an alarming rate of mill accidents during its first four months of its existence. And to some degree, its string of misfortunes would continue until E. B. Hayward sold out to Frost-Johnson in 1910. Some of its mishaps in March, 1905, included as follows:[35]

> ...Tom Weeks, while feeding a planer at the Hayward mill at Nacogdoches Saturday, accidently had his hand cut completely off. This makes about the twelfth man that has been hurt by accident at the mill, including one that fell from a scaffold and was killed outright...{Note: The mill had been operating for only four months. Read on for two weeks later.}
>
> ...From some cause or another, the band saw of the big Hayward mill at Nacogdoches was demolished last Thursday, and the edger ("bull" edger or gang saw) also was put out of commission for several days. When a big gang saw breaks, it is a dangerous customer and generally succeeds in cutting half a dozen men in two before it is stopped...

The string of ill luck continued during the years that followed, although not nearly as bad as the year 1909, and certainly contributed to E. B. Hayward's decision to sell out in 1910 and retire from active sawmilling. And certainly the year 1909 was the worst of all, as one sample of ill luck in January verified:[36]

> ...The blowing out of a cylinder head in an engine room of the Hayward Lumber Company of Nacogdoches, Texas, during the latter part of January resulted in the shutting down of the plant for several days' time. The machinery was apparently running smoothly when a cylinder head blew out, and the piston ran wild, doing considerable damage. Fortunately, there was no one close by at the time...

In March, 1909, the big Hayward sawmill burned down, but apparently the planer, dry kilns, and dry sheds were saved, since no damage to them was mentioned. A subsequent article stated that:[37]

> ...The Hayward Lumber Company...let the contract late in May for the machinery for their new sawmill, which will replace one destroyed by fire two months ago. The old mill was equipped with a band (saw) and a gang (saw), but the new one will be a larger and more modern mill, with double-(cutting) band equipment...
>
> ...The contract included two band mills of large size, with a full equipment of supplementary machinery...It is expected the new mill will be built and ready for operation some time in October. The debris from the old mill has been cleared away and...a small circular mill was installed by one company and is running night and day, cutting material for the new mill, and supplying the sales office. The capacity of the new mill will be 125,000 to 150,000 feet daily...

Even as the new sawmill was under construction, a severe wind storm struck Nacogdoches, threatening to destroy everything left at Hayward that fire had left standing, as one lumber journal article revealed:[38]

> ...A hard wind storm struck the town of Nacogdoches on the morning of June 3, (1909), causing $600 damage to the mill plant of Hayward Lumber Company. The long line of blow pipes, which run from the planing mill to the fuel room, was blown down by the great force of the wind, causing the planing mill to shut down for ten days while repairs were made. This concern is having a rather hard run of luck, losing their big sawmill only recently to fire...

HAYWARD MILL UNDER FROST-JOHNSON

In 1910 E. B. Hayward sold out his Nacogdoches sawmill, his extensive timberlands, and the Nacogdoches and Southeastern Railroad to the Frost-Johnson Lumber Company, a Louisiana and Arkansas timber conglomerate originally consisting of Enoch W. Frost, Edwin Ambrose Frost, R. L.Trigg, C. D. Whited, Clarence D. Johnson and George Alfred Kelley. Both of the Frosts and G. A. Kelley had joined with T. L. L. Temple in 1899 to found the Lufkin Land and Lumber Company, and when that firm sold out to Long-Bell Lumber Company in 1905, the partners were left with large sums of cash for immediate investment elsewhere. The owners had operated their single Arkansas and three Louisiana sawmills under the name of Frost-Trigg Lumber Company, with headquarters at Shreveport. Trigg left that firm in 1899, and his place in the company hierarchy was soon filled by Clarence Johnson. The Frost syndicate claimed control of seven companies and ownership of two billion feet of pine stumpage (uncut timber) in three states.[39]

After Frost-Johnson assumed ownership of the Hayward mill, managerial duties at Nacogdoches passed to C. E. Wyman and Harry Worth Whited, a son of one of the owners. In the 1910 city of Nacogdoches census, many of the key plant personnel of the Frost-Johnson mill were enumerated as follows: J. E. Stanfield, dry kiln foreman; J. M. Futch, yard foreman; Will Linder, sawmill foreman; C. E. Wyman, general manager; Sid Smith, office manager; W. C. Trupper, Charles Wyman, bookkeepers; Worth Whited, sawmill manager; Tom Dean, assistant manager; Roger Pierce, Chester Owens, Ray Owens, graders; J. A. Pruitt, planer mill foreman; T. H. Whited, sawyer; Jasper Wiggins, shipping clerk; Jesse Robinson, millwright; Dr. J. H. Reagan, mill physician; J. B. Heath, millwright; and Kate Biggers, boarding house owner.[40]

Frost-Johnson Lumber Company continued to buy up large blocks of timberlands, principally in San Augustine County, until its holdings for the Nacogdoches mill exceeded 100,000 acres. The Nacogdoches and Southeastern rails continued to expand until they reached 42 miles to the Santa Fe rail line at Calgary in San Augustine County. The company's steel gang was likewise expanding its tram's spur lines, removing rails from cutover areas and adding them in new stands of uncut timber. The Frost-Johnson company used both four-line, rehaul skidders and steam log loaders in the forests to log the Nacogdoches mill. After logging around Woden ended, the company built Camp Pershing at the mouth of the Attoyac River and added car houses for its loggers to live in. And when that area was cut over, the company moved its car houses to a new location, Camp Worth in San Augustine County, named for Worth Whited. Frost-Johnson employed 450 men in its mill and forest operations. Most of them lived in "mill town" on Hayward Mill Road, and they were paid in company coupons or mill checks for use at the commissary.

In 1925, the company name was changed from Frost-Johnson to Frost Lumber Industries. In 1928, Frost bought out the Waterman Lumber Company's sawmill at Waskom, Harrison County, and acquired another

47,000 acres of timberland (most of it already cut out). Later Frost joined with Peavy-Moore Lumber Company in owning and operating the Jasper County Lumber Company. During the 1920's, Worth Whited became the dominant officer of Frost Industries, guiding it through the depression and World War II. And although E. A. Frost lived to age 81 and died in 1950, the death of Whited in 1946 precipitated a company crisis that could not be resolved. As a result, Frost Industries sold out to Olin Industries in 1952, who in turn sold out to International Paper Company. The latter soon closed down and in 1954 dismantled the big Nacogdoches mill.[41]

THE SAWMILLS OF CARO

According to *Handbook of Texas*, the early sawmill town of Caro was located twelve miles north of Nacogdoches. It was at the junction of two railroads, the Texas and New Orleans road from Nacogdoches to Dallas, and the Caro Northern Railway, which ran sixteen miles north of Caro to Mount Enterprise, Rusk County. Beginning in 1904, the Caro Northern became the logging tram road of the Whiteman-Decker Lumber Company at Caro. *The Handbook of Texas* observed as well that the Caro Northern was chartered in 1894 (error), and was abandoned in 1934, but the writer has no information as to who its earlier or subsequent owners may have been. Another source states that the Caro Northern Railroad was chartered by the Whiteman-Decker company on September 15, 1907.[42]

The Whiteman-Decker Lumber Company was chartered in Austin on January 26, 1904, and was capitalized at $100,000. Its incorporators were Warren T. Whiteman, R. M. Decker, M. R. Schluter, M. W. Jones, and C. H. Morris, all of Winnsboro, Wood County.[43]

Although the Whiteman-Decker sawmill was always credited to Caro, its actual location was a half-mile from the rail junction, at a sawmill company town known as Wydeck.[44] Very quickly, a modern sawmill of 60,000 feet capacity was built at Wydeck. There were several news articles about the new Whiteman-Decker mill when it was first built, and two of them read as follows:[45]

> ...The big mill of the Whiteman-Decker Company of Caro, Nacogdoches County, on the Beaumont-Dallas division of the Texas and New Orleans (Railroad), is finished and ready for business. Their initial run (of lumber) will be made on the 16th (of Nov. 1904)...The equipment of this mill is modern in every particular, and

> it is one of th big lumber institutions of th East Texas pine belt...
>
> ...Wednesday was the day for auctioning off the town lots of Caro, the new sawmill town twelve miles north of Nacogdoches on the Texas and New Orleans. Caro is the seat of the big Whiteman-Decker sawmill that has recently been put in operation there and is now cutting about 60,000 feet of lumber (daily). Caro is therefore a sawmill town, but in building the mill and putting it into operation, a large section of country has been settled up that was hitherto somewhat isolated...

A few weeks later, C. C. Galloway, the veteran Nacogdoches County sawmiller, sold his mill, and a reporter added as well that:[46]

> ...The Whiteman-Decker people have the territory north of Nacogdoches on the Texas and New Orleans railway all to themselves now that C. C. Galloway has sold out...
>
> ...The Whiteman-Decker sawmill at Caro is running on full time, after spending a few weeks repairing and cleaning up during the Christmas holidays...

Six months later, some editors recorded in May, June and July that the Caro sawmill was still prospering as follows:[47]

> ...The Whiteman-Decker mill at Caro is running on full time and is making arrangements to build a tram road into the woods north of Linn Flat...The Whiteman-Decker mill at Caro is doing a good business these days...
>
> ...The Whiteman-Decker Lumber Company, with a mill at Caro in Nacogdoches County, has added another locomotive to its logging equipment... Our staff correspondent visited the mill of Whiteman-Decker...at Caro, Texas...It is a very large plant and thoroughly equipped with all up-to-date machinery for milling purposes. The company is capitalized for $650,000...(has) an extensive commissary plant, with stores, shops, etc. There are about 1,000 people in Caro, all dependent on these mills for a livelihood. The place has the appearance of a thriving country town, with laid-out streets and houses...

At some time in 1905 or 1906, the Whiteman-Decker sawmill was rebuilt and enlarged, greatly increasing its sawing capacity, probably with the addition of a band saw or a band and gang saws. When the Texas and

New Orleans Railroad published it list of sawmills in September, 1906, the list reported that the Whiteman-Decker mill's capacity had increased to 105,000 feet daily.[48]

In January, 1908, R. M. Decker and perhaps another investor sold out their interests in the sawmill to Saner-Ragley Lumber Company, and the restructured company became known as Saner-Whiteman Lumber Company. Its new investors were W. G. Ragley, John C. Saner, R. E. L. Saner, Warren T. Whiteman, and M. W. Jones. The Saner-Ragley Lumber Company was another early East Texas lumbering concern, having operated the Saner-Ragley sawmill at Carmona, Polk County, from about 1914 to 1945, and the Saner-Ragley sawmill at Ragley, Shelby County, in 1906. Warren T. Whiteman continued as mill manager at Caro, but he and R. M. Decker continued to operate the Whiteman-Decker sawmill at Farmersville, Louisiana, until at least 1922.[49]

The dreaded fire demon arrived at the Saner sawmill in Caro in May, 1909, but fortunately the fire was confined to a small area of the sawmill plant, as one editor observed:[50]

> ...The Saner-Whiteman Lumber Company at Caro, Texas, was visited by fire during the latter part of May (1909) that destroyed the boiler house of the dry kiln department. The building was destroyed, the boilers considerably damaged, but the engine escaped injury. Repairs were made at once, although the kilns were shut down for several weeks...

The last record known to the author of the big Whiteman sawmill at Caro came in 1910 when the Nacogdoches County census of that year was enumerated. The census of Caro listed the following key plant personnel of that mill, as follows: Warren T. Whiteman, owner and mill superintendent; William Kensey, steel gang foreman; R. V. Stewart, oiler; Harry V. Null, edgerman; G. W. Burgett, millwright; Dan T. Warren, grader; Earnest Brown, W. R. Walker, filers; Will A. Jackson, locomotive fireman; C. L. Gorman, Vernon Gorman, J. E. Anderson, locomotive engineers; Mary Chancey, boarding house owner; Lula Capland, telephone operator; John V. Crawford, Frank J. Harris, Walter Truscan, sawyers; V. L. Perry, sawmill foreman; James P. House, bookkeeper; James Williams, sawmill engineer; Clarence Baucom, yard foreman; Rube R. Coleman, planing mill foreman; W. Jacks, blacksmith; Dallas Carter, scaler; and Anna Newton, postmaster of Caro.[51]

The other large sawmill at Caro belonged to C. C. Galloway, one of the veteran sawmillers of Nacogdoches County as of 1904. Galloway bought out the Lanana Mills, a combination cotton gin, grist and sawmill on Lanana Creek, from W. Vestal Carroway about 1894, which Galloway continued to operate until about 1899. One article stated that W. S. Wilkins was superintendent of the Galloway and Wright mill on the Lanana until it was moved to Garrison. About 1899 Galloway joined with W. T. Wright in ownership of a planing mill and dry kilns across Bonita Bayou in Nacogdoches. In 1902 Galloway bought a huge tract of timberlands northwest of Caro, after which he built the Galloway Lumber Company sawmill at Caro. Situated at the junction of the Caro Northern and Texas and New Orleans railroads, his sawmill had excellent access to markets in Dallas, Nacogdoches, and elsewhere. Although the writer has not located many news items about the Galloway mill, a few of them follow:[52]

> ...The big mill of C. C. Galloway at Caro, with a daily capacity of 75,000 feet, is mentioned elsewhere in the Journal....
>
> ...The big mill of C. C. Galloway at Caro...near Nacogdoches is reported running on full time with abundant orders. Mr. Galloway is building a tram into a heavy pinery to the northwest of his mill...

Since Galloway had a planing mill at Caro as well, it does not seem plausible that he shipped rough lumber to his planing mill in Nacogdoches. Hence, the Bonita mill must have depended on small nearby mills that had no planing facilities. Nevertheless, only two months after the above news items were released, Galloway sold out his Caro sawmill and moved to Dallas, where he entered the retail lumber trade as Galloway and Garrison, in partnership with W. Y. Garrison. Two more news items which follow mention the sale of his Caro sawmill:[53]

> ...Mr. C. C. Galloway, the well-known mill and lumber man of East Texas, is selling his big mill at Caro...This comes as a surprise to lumber people, because Mr. Galloway has succeeded during the past two years in building and equipping one of the best and largest sawmills to be found between Nacogdoches and Dallas. He owns a large lot of pine timber in the northern part of Nacogdoches County and has built a good tram road to the timber...
>
> ...The good wishes of all East Texas lumbermen go with Mr. C. C. Galloway, who has sold out his big sawmill...and moved to Dallas, where he will open a wholesale and retail

> lumber business...

The writer has located no information about Galloway's successor at Caro.

THE W. G. HARRINGTON SAWMILL

The W. G. Harrington sawmill is generally credited to Nacogdoches when actually it was a considerable distance away, about nine or ten miles south of the county seat on the Houston East and West Texas Railroad. By the author's reckoning, the mill must have been at or near the Angelina River bottomlands, perhaps opposite the old Angelina County mill town of Michelli.

The earliest account of the Harrington mill indicated that it was built early in 1904, as one editor observed:[54]

> ...The following from a recent issue of Nacogdoches *Daily Sentinel* is about Harrington's mill, concerning which something has been said in the *Journal*...
>
> ...Harrington's mill is nine miles out of Nacogdoches on the Houston East and West Texas Railroad. The mill is now complete and is in first-class running order. The big whistle jarred this community last Monday morning at 4 o'clock for a full day's run, with J. J. Bowden, foreman; John Stephens at the carriage lever (sawyer); Jim Keaghey, fireman; J. W. Jordan, filer; and W. G. Harrington as general superintendent. The day's run was made successfully with 40,000 feet cut in ten hours. Plans are forth for the erection of a large capacity planer and the building of several miles of tram track...

The writer has no additional information about the Harrington mill's timberlands, tram road, or rolling stock. After eight months of successful operation, another news report of January, 1905 confirmed that:[55]

> ...The Harrington Lumber Company, which put in a mill ten miles south of Nacogdoches several months ago, has just completed a planing mill which makes this one of the best little lumber plants in that part of the country. It is a fifty thousand (foot) mill, backed up by an ample supply of timber. Mr. W. G. Harrington, the manager, formerly operated a planing mill at Nacogdoches, which he sold early last spring to W. T. Wilson. His mill is near Lanana,

> the site of the old Bermea Land and Lumber Company mill...
> ...The Harrington mill in Nacogdoches County has succeeded in having a (rail) switch put in, and is now ready to do a big shipping business...

In October, 1905, Harrington leased his sawmill to a competitor. The last newspaper article in the writer's possession presented a surprise ending to the Harrington mill story, that is, considering the coverage given to the mill's start-up, as follows:[56]

> ...Mr. W. G. Harrington, the well-known East Texas lumberman, who has been located in Nacogdoches several years, sold out his Nacogdoches interests some time ago and will move to Dallas, where he will probably enter the lumber business in a wholesale way...

Again no mention was published concerning the buyer of the Harrington sawmill, only that the transaction had been completed. The *Reference Book* of the Lumberman's Credit Association gave Harrington a poor financial rating, listing him as a "slow payer." Also the *American Lumberman* also confirmed the sale of the Harrington sawmill and the owner's move to Dallas. Nevertheless, the Harrington sawmill was still carried under its original name in a list of Houston East and West Texas Railroad sawmills, published in September, 1906. Its milling capacity was still shown as being 40,000 feet daily.[57]

OTHER CITY OF NACOGDOCHES LUMBER MILLS

The writer has already mentioned the Galloway and Wright Lumber Company, which existed across Bonita Creek from the old Nacogdoches business area from 1899 until perhaps 1905. C. C. Galloway and W. T. Wright operated a planing mill there, including moulders, matchers, shaper, cut-off saw, and dry kilns, using rough lumber purchased from the smaller mills nearby. The East Texas sawmill database credits that company with opening a retail lumber yard in that town in 1901, which was the first "major business in Nacogdoches to install and use (a) telephone for business purposes."[58]

Another early lumber concern was the Bonita Manufacturing Company, which belonged to Ireson Brothers. Since this firm also had no sawmill, the milling capacity of 10,000 feet, referred to in "Lumber Mills of Texas" of 1906, was apparently its planing mill capacity. According to the East

Texas sawmill database, the Bonita firm owned "planer, moulder, etc. for manufacturing sash and door frames, woodwork, interior finish lumber, cabinets and showcases, office furniture."[59]

In 1906 the Nacogdoches Lumber and Crate Company operated a sawmill with a 40,000-foot daily capacity. The firm belonged to the Nacogdoches Lumber Company, and it probably cut only rough lumber which was used to build crates for shipping agricultural products.[60]

Another Nacogdoches lumber firm was the W. T. Wilson Lumber Company, which was listed in 1906 with a daily capacity of 65,000 feet. The business was a planer mill only in March, 1904, when W. G. Harrington sold out to Wilson, but the new owner must have installed a sawmill soon after purchasing the business. Hence, it is unclear whether the daily capacity indicated was for sawing lumber or planing it. A credit reference book noted that Wilson also ran a wholesale lumber yard in Nacogdoches in connection with a planing mill and grain mill. It did not specify whether the "grain mill" was a grist mill for grinding corn into meal or a hammer mill for grinding animal fodder.[61] Also the East Texas database reported a Stern Lumber Company sawmill at Nacogdoches in 1906, but it reported no cutting capacity or other particulars about that mill. The Stern firm was not mentioned in the 1906 sawmill list, "Lumber Mills of Texas."[62]

Also in 1906, the Nacogdoches Oil and Development Company operated a sawmill at Hampton Switch, four miles southeast of Nacogdoches. The mill cut rough timbers only for the oil industry, and its circular saw cut 40,000 feet daily. Owned by investors of the Hayward Lumber Company and Nacogdoches Lumber Company, the mill was located on the Nacogdoches and Southeastern Railroad, the Hayward Company's (later Frost-Johnson) tram road, which ran to Oil Springs and Woden.[63]

One of the more recent sawmills of Nacogdoches was the J. E. Stone Lumber Company. Stone was originally a cotton ginner about 1925, and when the boll weevils decimated that business, he switched to sawmilling. In 1934, the Stone sawmill was rated at 40,000 feet daily for a ten-hour work day, and its equipment included a band saw, a gang saw, resaw, trimmer, edger, and other assorted machinery. In 1943, a Lufkin newspaper photo of the Stone sawmill noted that J. E. Stone Lumber Company also owned a sawmill at San Augustine. A 1947 graduate paper, "Lumbering In Nacogdoches County," did not list the Stone mill.[64]

Another news account of 1905 mentioned that the "Nacogdoches

Show Case and Manufacturing Company has been incorporated with an office and planer" mill. No other information about that firm has been found except that it utilized only hardwood in its products.[65]

Another later day milling firm at Nacogdoches was the J. S. Hunt Lumber Company. That sawmill existed from at least 1926 until 1946, but no other mill statistics have been found.[66] Still another sawmilling firm at Nacogdoches, the A. J. Cariker Lumber Company, arrived in town about 1929 or earlier. A Beaumont newspaper article of 1931 mentioned only the A. J. Cariker Planing Mill at Nacogdoches, but a Nacogdoches editor of 1934 noted that the Cariker Lumber Company operated twelve East Texas sawmills, including four near Nacogdoches, three at Cushing, three near Mount Enterprise (Rusk Co.), and one each at Caro and Appleby. The A. J. Cariker mill had a band saw, a planer mill, and dry kilns, and cut only pine timber.[67]

In the 1910 census of Cushing, Jesse W. Cariker was both a merchant and sawmiller at Cushing, and the other sawmill there belonged to Ambrose B. Martindale. If the writer can decipher the census enumerations accurately (based on the various locations of residences), two of Jesse Cariker's sons worked in his sawmill, and the other two worked for the Martindale mill. W. P. Cariker was filer and Dr. Fred Cariker was physician at the J. W. Cariker sawmill. Robert L. Cariker was mill engineer at the Martindale sawmill, and Albert J. Cariker, who was born in Georgia in 1876 and lived next door to A. B. Martindale, was manager of the Martindale sawmill.[68]

THE J. P. CARTER LUMBER COMPANY MILLS

The J. P. Carter Lumber Company was an outgrowth of the breakup of the old Emporia Lumber Company, of Emporia, Angelina County, and Doucette in Tyler County. Samuel Fain Carter, a wealthy Houstonian, was the president of that company from 1893 until 1906, and his younger brother, Joseph Press Carter, had served at first as Emporia's mill superintendent, and after his removal to Houston in 1900, as vice president of the Emporia Company. Also in 1900, the owners bought a huge tract of timberlands in eastern Tyler County and built the new Emporia (later Sunset) Lumber Company sawmill at Doucette. In 1906, when Emporia cut out its timber in Angelina County, S. F. Carter, the majority stockholder, decided to retire. He and J. P. Carter dismantled the Angelina County mill and split up their profits. In January, 1907, they sold out their Tyler County timberlands and the Doucette sawmill to Thompson Brothers Lumber Company of Trinity County for $500,000.[69]

The J. P. Carter Lumber Company was organized in December, 1905, with a capital stock of $50,000. In February, 1907, the stock was increased to $100,000, and again in October, 1907, to $150,000. The company investors were J. P. Carter, president and majority stockholder; Jonas Shearn Rice, vice president; and Ben. J. Barrow. Nothing about Barrow is known by the writer, but Colonel Rice was a patrician Houstonian, nephew of William Marsh Rice (benefactor of Rice University), as well as bank president and colonel of the Houston Light Guard in 1871. In 1881, he built the Rice sawmill at Hyatt, Tyler County, which he operated successfully for 25 years, dismantled in 1906, and moved to his uncut timberlands in Ward, Louisiana.[70]

J. P. Carter and Shearn Rice bought up large blocks of timberlands in Angelina, San Augustine, and Nacogdoches counties in 1906. The partners soon bought the old McConnico sawmill at Baber, Angelina County, which they rebuilt into a 60,000-foot mill, as well as a sawmill at Hillister (south of Woodville), probably the old Tyler County Lumber Company, at that moment in receivership.[71] About 1906 the firm also bought or built two other mills in Nacogdoches County and acquired the output of a third mill there. One sawmill was at Mayo, northeast of Appleby, about which Carter's biography carried the following notation:[72]

> ...Mr. Carter purchased a large tract of timber...forming the Attoyac River Lumber Company, a corporation of $300,000 capital stock, fully paid. This company has (1909) at Mayo, Texas, a fully-equipped planer, dry kilns, and other modern improvements, a standard gauge railroad tram, etc. It is one of the most modern plants in the South and manufactures 25,000,000 feet of high grade lumber annually (that is, 80,000 feet daily)...

The description of the Mayo sawmill strongly suggested that it was a new or fairly-new installation. The writer believes that at its beginning the new mill at Mayo may have been known as "Press," obviously taken from the middle name of the mill owner, J. Press Carter. After the second burning of the Emporia mill and the breakup of the Emporia Company in 1906, one article announced that the Emporia sawmill "will not be rebuilt, but that one will be erected at Press, a new station a few miles from Nacogdoches..." The Mahl sawmill was located on what is now Highway 259, about ten miles north of Nacogdoches, but the writer has no additional information about that mill. Carter is also known to have operated a planing mill at Sacul, which in 1906-1907 was purchasing the output of A. J. Caruthers planing mill. Carter's contract with Caruthers began about February, 1906 and is believed to have lasted until about February, 1907. A J. P. Carter letterhead of March, 1907, listed sawmills still at Mayo and Mahl, but the Sacul mill was no longer listed. Both the Mayo and Mahl sawmills are believed to have cut out and been dismantled no later than 1912.[73]

THE SAWMILLS AT SACUL

Other than the Caruthers mill mention above, there were at least six other sawmills in the vicinity of and credited to Sacul. Actually there was no sawmill in Sacul proper (all of them being at some distance out in the country) as one article revealed:[74]

> ...Mr. W. W. Foster of the Caruthers Lumber Company of Sacul...says...there is no sawmill at Sacul, but the A. J. Caruthers people have a yard, planer, and office there, and the cut of their mill is sold from the Sacul office. The daily cutting capacity of their (planing) mill is 40,000 feet, and this average will be increased in the near future as it is the intention of the management to put in a lot of new machinery...

Between 1897 and about 1902, A. J. "Jack" Caruthers had superintended the Clawson Lumber Company sawmill at Clawson, Angelina County, in partnership with three other investors of McLennan County, all of whom owned the Waco Lumber Company, a retail firm with several lumber yards in Central Texas. They also owned sawmills elsewhere in East Texas.[75]

The B and K Lumber Company was a steam-operated hardwood sawmill in the Angelina River bottoms, west of Sacul. Located on the Texas and New Orleans Railroad at Togo Spur, the 30,000-foot mill cut barrel staves and hardwood lumber for box cars before switching to pine logs, as one editor observed:[76]

> ...A big hardwood mill has been put in on the Angelina River on the line of Nacogdoches and Cherokee counties near Sacul...The hardwood mill near Sacul on the T. and N. O. (Railroad) is reported as doing a splendid business ...The B & K hardwood mill...has quit the hardwood

> business and is now cutting nothing but pine lumber. The company has a good mill and owns a great deal of fine timber on the Angelina River...

Also in June, 1905, there were two other sawmills near Sacul. The Morrison, McClennon, and Tingle facility was a 40,000 or 50,000-foot daily circular mill. However, it had no planing mill, and it sent its rough lumber to the Caruthers planer in Sacul to be smooth-finished. The Morrison sawmill was located about two miles north of Sacul. Also, S. F. Bacom, a Nacogdoches merchant, owned a sawmill three miles south of Sacul, its daily capacity unknown. One editor noted that Bacom's "large mill" was "cutting a great deal of lumber."[77]

In a record of the Texas and New Orleans Railroad's sawmills, published only a year later in September, 1906, neither the Morrison nor Bacom mills were listed, strongly hinting that probably both sawmills had either been sold or dismantled. However, three other small sawmills had been added to the number at Sacul. A. L. Bartlett and Sons' mill cut 8,000 feet daily there. J. L. Dobbs' sawmill also cut 8,000 feet daily, as did J. C. Marshall's small mill. Considering the 8,000 feet daily that each of them cut, the mills obviously had no planing mills attached and hence, sent their rough lumber into Sacul to be planed and marketed.[78]

THE SAWMILLS AT CUSHING AND TRAWICK

Like Sacul, Cushing and Trawick are located in northwest Nacogdoches County, near the Rusk County line, and between twelve and twenty miles northwest of Nacogdoches on Highway 204. The Texas and New Orleans Railroad was built through there about 1902, and very quickly Cushing became the location of the Jesse W. Cariker and Ambrose B. Martindale sawmills. At least three A. J. Cariker sawmills operated in that vicinity after World War I. J. W. Cariker, a Cushing merchant, owned one of the two earliest mills, and there were a number of news accounts about his mill, some of which follow:[79]

> ... J. W. Cariker, the sawmill and lumber man of Cushing, has used all the timber in reach of his mill by tram and is now building a new tram into a fine pinery south of Cushing in the Angelina River bottom...
>
> ...Good reports come from the sawmills around Cushing this week. Mr. Cariker's mill, two miles west of town is running on full time and handling a big business, with

> orders on the book for a month ahead. The new tram road is in operation and is giving the mill an opportunity of running all the time, without having to wait for logs...

A list of the Texas and New Orleans Railroad sawmills of 1906 showed the Cariker mill at Cushing as cutting 35,000 feet daily, but a Nacogdoches editor credited the Cariker milling capacity at 50,000 feet daily. The facility consisted of a circular saw and planers, but no dry kilns, and it cut lumber, crossties and bridge timbers.[80] Some of th key plant personnel at the Cariker mill included W. P. Cariker, filer; H. D. Perkins, planer foreman; Dr. Fred W. Cariker, mill physician; Tom J. Cariker, clerk; Jesse W. Carriker, merchant and mill owner; and Jesse C. Morrison, sawmill bookkeeper.[81]

Very little is known about Jesse Carriker's tram road except what appeared in the following quote:[82]

> ...J. W. Cariker who owns the big sawmill between Cushing and Sacul on the Texas and New Orleans Railroad...has completed the tram which runs from his mill to a fine body of timber in the Angelina River bottom...(He) bought a new tram locomotive and logging cars and is now whooping things up in a great style...

The Ambrose B. Martindale sawmill was built on the east side of the Douglass-Henderson road. The list of Texas and New Orleans sawmills of 1906 credited the Martindale mill with a daily cutting capacity of 30,000 feet, but a Beaumont editor of 1904 reported that: "...The Martindale mill at Cushing is busy working on full time and cutting 50,000 feet of lumber per day..."[83]

News articles of the winter of 1905 revealed that the heavy rains played havoc in the forests around Cushing, as one editor noted:[84]

> ...The recent spell of rainy weather has put the sawmills at Cushing...out of business for a few days at least. These mills are logged by teams in the woods, and the ground is so boggy that logging had to be suspended. A few days of sunshine will put everything in running shape again...
>
> ...The large mill of A. B. Martindale in the town of Cushing is also doing well in every way...

The 1910 census of Cushing, on the "east side of the Douglass-Henderson road," listed some of the key personnel at the Martindale mill, as follows: (pct. 7, res. 172) Ambrose B. Martindale, sawmill owner; (res. 170) Albert J. Cariker, sawmill manager; David R. Heath, sawmill foreman; Zeph F. Conway, sawyer; R. L. Cariker, Horace L. Roark, sawmill engineers; P. M. Wagner, postmaster at Cushing; and Dr. Wm. W. Ireland, mill physician.[85]

Another early sawmill at Cushing was that of W. F. Daniels, erected about 1902 to cut crossties and bridge timbers for the Texas and New Orleans Railroad. No mill statistics are available for the Daniels mill, but the writer believes its daily capacity was about 30,000 feet. A Beaumont editor observed that the Daniels sawmill burned down on November 6, 1909, and the fire was believed to have been incendiary in origin. The East Texas Sawmill Database noted the Daniels sawmill was not rebuilt. In the 1910 Cushing census (Pct. 7, res. 38), W. F. Daniels was still enumerated as a "millman, wholesale lumber."[86]

During the heyday of sawmilling there around 1910, the population of Cushing must have reached about 1,000 persons, with perhaps 250 men or more employed in the mills and woods. A business district quickly developed, and by 1910 George Carter and Lee James ran the town's newspaper; and Margaret and Ellen Wallace operated the town's hotel. Soon after the town was founded, one editor observed that:[87]

> ...The two big sawmills at Cushing are numbered among the town's most important business enterprises. They employ a large number of hands, who in turn spend their money with the home merchants, and thus keep the circulating medium at home...
>
> ...The Galloway and Garrison Lumber Company, with its sales office in Dallas, has bought a sawmill at Cushing... with a capacity of 20,000 feet daily and will put in a planer which will be ready for operation in sixty days...

Also the *American Lumberman* reported in 1906 that four new businesses were being erected there, as follows:

> ...Three sawmills and a planing mill are among the enterprises for Cushing, Nacogdoches County. The sawmills are being erected by Howell, Smith & Johnson; Ninty & Henslee and E. L. Roark, while the planer is being put in by Galloway & Garrison...

Nearby Trawick, also on the Texas and New Orleans Railroad and Highway 204, is about seven miles east of Cushing. It was likewise a center for several small sawmills. One of the earliest mills there was that of J. H. Summers and Sons, a firm which cut out its timber in 1909, dismantled its mill, and moved to a new timber stand in Walker County.

An editor published the following accout in 1909:[88]

> ...J. H. Summers and Sons, a well-known manufacturing and wholesaling company located at Trawick, Texas, is being dismantled at present...The mill is a 50,000 (foot) capacity affair, and has been cutting for many years at Trawick Their timber at that point has just been cut out...and the plant will be in operation at its new location in March. The plant includes sawmill, planing mill, dry kilns, and dry sheds...

The writer has a record of only two other sawmills at Trawick in 1906. At that time, James W. Van Sickle owned a small sawmill at Trawick that cut 15,000 feet daily. Van Sickle's mill was still in business in 1910 since the owner was recorded in that year's census as a sawmiller. The other small sawmill at Trawick in 1906 was that of W. E. Dunn, whose mill also cut only 15,000 feet daily. As the great forests were leveled around Cushing and Trawick, the rural population soon converted to an agricultural economy.[89]

SAWMILLING AT GARRISON AND APPLEBY

Appleby located nine miles northeast of Nacogdoches on the Houston East and West Texas Railroad, was probably the site of several sawmills after 1885, but the writer's records include the names of only three of them. One of the mills was that of Adam Caroway, whose family name, like that of Cariker, threads its way through all the sawmilling chronicles of early Nacogdoches County. Adam was probably related to the W. V. Caroway, who operated the sawmill known as the Lanana Mills in 1893, but the writer does not know the exact connection. In March, 1906, a news article revealed that:[90]

> ...Fire wiped out two sawmill plants last week...The Caroway mill was burned Tuesday night (prior to March 12), the loss being $2,500, with no insurance. The mill was sold the afternoon of the fire to J. F. Summers and J. W. Arthur, the new proprietors intending to start up the mill the next day. It is generally believed the fire was of incendiary origin...

The writer refuses to believe that $2,500 fire could amount to the entire loss of any mill that could cut 40,000 feet daily. Nevertheless, only six months later (Sept., 1906), Adam Caroway was credited with still owning a sawmill in Appleby, capable of cutting 40,000 feet daily, the largest mill in town. The writer has no information about a planer mill or dry kilns at the Caroway mill.[91]

Two other small sawmills have been credited to Appleby as well, although certainly other mills were located there over the years. Between 1897 and 1899, Nathan Payne owned a small steam circular sawmill there of 20,000 feet daily capacity. On November 3, 1899, the steam boiler exploded, destroying the mill, and the writer has no subsequent record of it.[92] The other mill located at Appleby in 1906 was the F. W. Duncan sawmill, which also had a capacity of 20,000 feet daily. The writer has no record of a planer or dry kilns at either mill.

An article in *American Lumberman* in 1906 mentioned another mill slated to be erected at either Appleby or nearby Sterne, as follows:[93]

> ...The Blount Lumber Company, with a paid-up capital of $60,000, has been organized at Nacogdoches by A. A. Blount, June C. Harris, Robert Lindsey, and Guy A. Blount. Machinery for both a sawmill and planer has been ordered, the intention being to equip a mill with a band saw which will give it a capacity of 75,000 feet a day. The mill will not give exclusive attention to pine, but will also manufacture hardwoods. The site for the mill will probably be at Appleby or Sterne on the Houston East and West Texas Railroad. The incorporators of the company own several thousand acres of fine timber land in the vicinity of Nacogdoches, which it is estimated will keep the mill running for fifteen years...
>
> ...The hardwood mill of the A. J. Oliver Lumber Company, which did not run a day since being built, changed hands at a private sale conducted by Hon. John Durst, trustee, of Tyler, Texas. The mill was bought by E. A. Blount of Nacogdoches, Texas, for $6,500. It is understood the plant will be moved to Nacogdoches County near Appleby...

Garrison is located in extreme northeastern Nacogdoches County, about twenty miles from Nacogdoches and near the Attoyac River bottoms. Named for Frank Garrison about 1885, there were also several other persons of the same family name who owned early stores and livery stables there. During the fall of 1904, Howard Lee, the veteran lumberman who had previously owned sawmills at Fitze, Garna, and Nacogdoches, moved his mill to Garrison. Evidently, Lee cut out his timber at Garrison very quickly, or sold out, since no mill for Lee was recorded at Garrison in the list of 1906.[94] Another source observed that W. S. Wilkins was superintendent of the Galloway and Wright mill on Lanana Creek until that mill was dismantled and moved to Garrison about 1900.[95]

Another mill there was that of J. H. Garrison and Son Lumber Company, which was in operation at Garrison from some time in the 1880's until at least 1906. One article of 1904 noted that:[96]

> ...J. H. Garrison and Son, the lumber people of Garrison on the Houston East and West Texas Railroad, have sold out their interests to W. V. Garrison, who is a partner of C. C. Galloway of Caro...The two will probably be consolidated...

In 1904 the Garrison sawmill cut about 40,000 feet daily, its equipment including a planer, circular saw, dry kilns, and dry sheds.[97] In March, 1906, Garrison Lumber Company sold out to Garrison-Norton Lumber Company, a corporation with $50,000 capital stock and three incorporators, J. A. Kelly, J. F. Adams, and W. V. Garrison.[98] The main mill of the new company was located three miles west of Rush (near Brookeland, Sabine County), and one source noted that the planing mill at Garrison would be moved to Rush about May, 1905, because the Garrison-Norton sawmill at Rush had "secured an option on the output of three or four mills besides their own and were looking for more..." Another article observed that the Garrison-Norton Lumber Company "has bought a 6,000-acre tract near its station at Pineland (Rush in Sabine Co.) and will build a modern mill to cut from from 40,000 to 50,000 feet a day..."[99]

Three other sawmills were located at Garrison in 1906. T. P. Wragg's mill cut 20,000 feet daily, and two sawmills there belonged to J. M. Cordell and Son. Since Garrison is near the hardwood bottomlands of the Attoyac River, it seems most plausible that the 30,000-foot mill of J. M. Cordell, Jr. was a pine mill, and that the 20,000-foot mill of J. M. Cordell, Sr. was a hardwood mill. Neither mill owned a planer.[100]

OTHER NACOGDOCHES COUNTY MILL SITES

Woden on the Nacogdoches and Southeastern Railroad, began as a logging camp of the Hayward Lumber Company, but very quickly two sawmills were also located there. In 1906 the Dial and Porter Lumber Company at Woden was cutting 20,000 feet daily. In 1905, a Beaumont editor published a news account about a new mill to be built at Woden as follows:[101]

> ...John Arthur and Adam Caroway, two experienced sawmill men, are arranging to put in a new mill at Woden, the present terminus of the Nacogdoches Southeastern Railroad, ten miles east of here (Nacogdoches). The mill will have a capacity of 50,000 feet a day, and is situated in a new territory. It has no competition for the timber except the Hayward Lumber Company...Some of the finest pine timberland in East Texas is situated near this mill...

Like Woden, Chireno was closely identified with the logging industry, and also with Angelina County, since Chireno was the eastern terminus as well as the logging camp of the Angelina and Neches River Railroad. The latter was the logging tram of the Angelina County Lumber Company of Keltys. The writer has been trying to determine if the Tilford-Hunt sawmill at Chireno in 1915 was the same as the Tilford-Hunt mill at Lacyville. Lacyville was described in the East Texas sawmill database as being six miles from Nacogdoches, whereas Chireno appears to be about twenty miles east of Nacogdoches, near the Attoyac River (Bayou). In 1915 the Tilford-Hunt mill at Chireno cut 35,000 feet daily of pine, white oak and red oak timber. The circular mill had a planer and dry kilns attached; and its daily cut amounted to 60% boards or lumber and 40% of large timbers. The Chireno sawmill could cut, plane, and kiln dry timbers up to 24 feet long, and it also cut crossties.[102]

The Tilford-Hunt sawmill at Lacyville shared the same ownership as the J. S. Hunt Lumber Company. Organized in 1904, the Tilford-Hunt Lumber Company's investors were A. W. Hunt, Mrs. A. W. Hunt, Lacy Hunt, and T. Tilford, and Lacyville, the mill town which sprang up around the sawmill (8 miles from Nacogdoches) was named for Lacy Hunt. In 1943, Lacy Hunt was president both of Tilford-Hunt mill at Lacyville and of the J. S. Hunt Lumber Company sawmill at Willis, Texas. He was also vice president of the Nacogdoches County Lumber Company at Nacogdoches. No mill statistics survive for the Tilford-Hunt steam sawmill at Lacyville.[103]

In 1899, the next stop north of Appleby on the Houston East and West Texas Railroad was Sterne, also located only a very few miles south of Fitze and Garrison. No sawmill was listed at Sterne in either 1893 or 1900, but a 1906 list of sawmills on the East and West Texas line indicated that W. A. Wright operated a 30,000 foot steam mill at Sterne. No other mill statistics are available. Wright was also associated with several other early mill sites in Nacogdoches County.[104]

The following news account was published in a lumber journal in 1909 as follows:[105]

> ...It is reported from Nacogdoches that George W. Cavin, the well-known lumberman of that place, has recently sold his two sawmills, one in Nacogdoches County, and one in Angelina County, and has bought another at Kirbyville which he will operate. He received $50,000 for his two mills and the adjoining timber, and he paid $10,000 for the Kirbyville mill...

In September, 1906, Cavin owned a small 20,000-foot sawmill at Nivac, on the Texas and New Orleans Railroad near Sacul, as well as a 15,000-foot sawmill at Dalmont, which the writer presumes was located in Angelina County. In the 1910 census of the City of Nacogdoches, George W. Cavin was enumerated at Precinct 1, residence 5, as a "wholesale lumber merchant" and as an "employer."[106]

In 1902, the Jeans Brothers Lumber Company (Will F. and Tom J. Jeans) moved their sawmill from Ora in Angelina County, to Clevenger, Nacogdoches County, where the brothers remained for only about one year before moving on to Rush in Sabine County.[107]

In 1901, Walter R. and Ellis Mast operated the Mast Mills Company, a steam-driven combination cotton gin, grist mill, and sawmill at Melrose, ten miles east of Nacogdoches. This mill probably had been a water-powered mill before 1900, but by 1901, the Mast mill and the Tubbe mill at Tubbe were probably the last two combination (gin, grist, and sawmill) mills left in Nacogdoches County.[108] In 1902, J. J. and T. D. Fulmer owned and operated Fulmer's Mill, a 20,000-foot steam circular sawmill, at Swift, a very small sawmill community of fifty people, located ten miles northeast of Nacogdoches in the Appleby vicinity. Fulmer's Mill also operated a planer and cut lumber and shingles for nearby residential consumption.[109]

Since "Spanish Bluff" (not to be confused with "Spanish Bluff" on the Trinity River) was located "about 12 miles southwest of Nacogdoches," it must have been on the banks of the Angelina River, and somewhere near the old ghost town of Angelina or Michelli (on the Houston East and West Texas). On August 6, 1914, the Sullivan Sawmill and Lumber Yard, located at "Spanish Bluff," burned to the ground, along with two million feet of lumber stacked on the yard. The editor of *Southern Industrial and Lumber Review* noted that the owners, T. E. Baker, Frank Sharp, and — Sullivan, planned to rebuild.[110]

The Ennis and Boynton Lumber Company operated a small steam circular sawmill and planer in the Attoyac River bottoms, north of Martinsville, in 1901. The mill cut pine and hardwoods, principally for local consumption. This company may have been the beginning of Boynton Brothers (Albert and Walter) Lumber Company, which cut timber at Boynton, Angelina County, 1904-1908; at Logtown, Newton County, 1908-1912; at Sturgis, San Augustine County (1918); before returning to Angelina County about 1920.[111]

A lumber trade journal reported in 1908 that Charles Popp operated an oak barrel stave mill "about ten miles south of Nacogdoches," probably at Lanana Creek on the East and West Texas line. In September, 1908 Popp sold his mill to C. W. Pope of Illinois. In 1905 a J. P. Popp of the Louis Verner Stave Company of Shreveport was seeking to establish a Texas branch of his company, which may have been the C. Popp stave mill. The Verner Company employed 700 men making white oak staves in five different states, all of which were exported to Marseilles, France.[112]

Another early sawmill was the R. A. Jones sawmill, located "seven miles east of Nacogdoches on the Logansport Road." Since no railroad was associated with the mill, the writer presumes it was not on the road to Garrison, but rather on Highway 7 toward Martinsville. A steam circular rig owned by R. A. Jones, J. W. Lacy, and T. A. Rocquemore, the sawmill cut lumber only for local consumption.[113]

The last sawmill the writer has found a newspaper record of was the T. M. Hooks Lumber Company, located somewhere in Nacogdoches County in 1931. In that year, the Beaumont *Enterprise* ran a full page of Nacogdoches articles, both city and county, and one article briefly mentioned that — "operating in the county are the mills of Frost Lumber Industries, the J. E. Stone Lumber Company, the A. J. Cariker Planing Mill, and the T. M. Hooks mill." No location or mill statistics for any mill was added.[114]

In 1907 the Cravens Lumber Company of Shelby County operated its sawmill at Timpson and its planing mill at Nacogdoches.[115] The 1910 Nacogdoches County census certainly indicated there were other sawmills in the county. In that year, J. T. Scogins and B. H. Scogins were operating a mill near Cushing (Pct. 7, res. 55, 301). Charles S. Spurlin also had a sawmill near Cushing (Pct. 7, res. 115). Thomas G. Vaught owned a sawmill near Garrison (Pct. 5, res. 40). William V. Scarborough ran a small shingle mill and employed three men, apparently near Chireno (Pct. 3, res. 274). The writer did not locate any sawmills at either Douglass or

Etoile, but assumes that some early sawmills operated in those towns as well.

A SUMMARY OF NACOGDOCHES COUNTY SAWMILLING

Somehow, whenever the writer approaches the sawmill summary of each East Texas county, he is a bit saddened, for he knows that that county, at some time in the past, has inherited a desert of cutover stump lands, or burned-over stump lands, in place of the magnificent, old-growth forest monarchs that once stood there. Although the writer did not see Nacogdoches County during the aftermath of the lumbering era, he does recall visiting Kountze, Woodville, and Jasper during the middle 1920's when the road to each town traversed through a waste land of stumps. And he recalls, too, asking questions in childlike innocence about the reason for the landscape of stumps and of being told, "Son, that's progress!"

Actually, the sawmiller only continued in a wholesale way what the Anglo-American pioneer had practiced in a modest manner since about 1825—girdling and killing the trees to clear land for cotton plantations. After all, to the farmer trees were a nuisance since they shaded crops from the sunlight and sapped the soil of its vital moisture. It was only after the arrival of the railroads that lumber had a route to market, and "King Lumber"quickly toppled "King Cotton" as the dictator of Nacogdoches County's economic growth. In fact, so many thousands of acres of forest had already been cleared for cotton production that perhaps Nacogdoches County's sea of unsightly stumps was not as visible as in neighboring counties.

It was only after the pristine East Texas wilderness had been shorn of its original beauty that a few sawmillers realized that the old "cut and run" tactics in the forest must end. Men such as W. T. Carter of Camden and T. L. L. Temple of Diboll realized that a sawmill town could last indefinitely if proper forest management were utilized. And the mill towns of Diboll and Pineland have long ago proven that there is no such item as "waste lumber," that even a handful of sawdust can be remanufactured into a marketable product.

Today it is totally guesswork to try to estimate how many sawmills have operated in Nacogdoches County in a century and a half, but the writer would guess about 150. In August, 1927, when there were 182 sawmills in 20 counties of East Texas, Nacogdoches County still had six large mills, with a daily cut of 210,000 feet, and thirteen small sawmills, with a daily capacity of 104,000 feet.[116]

Elva A. Cox, in her 1947 graduate paper, "Lumbering in Nacogdoches County" (S. F. A. University Forest Collections), wrote that:[117]

> ...In Nacogdoches County nine large sawmills and many small ones are operating...Besides these nine major plants and many small ones, there are two creosote plants, three furniture factories, one venetian blind factory, and one sash and door plant in the county. Some of the woodworking firms are: Caro Lumber Company, Garrison Lumber Company, Frost-Johnson Lumber Company, Nacogdoches Lumber Company, Summers Lumber Company, Stone-Crawford Company, Tilford-Hunt Lumber Company of Nacogdoches, Jowell Lumber Company, Still Planer Company, Sutton Lumber Company of Chireno...

At a glance, the reader can readily recognize some of the big sawmill firms—Frost Industries, Tilford-Hunt, Nacogdoches Lumber Company, and probably Caro Lumber Company and Garrison Lumber Company. Cox added that in 1947, two thousand persons in the county were still employed by the forest industries and they "thereby provide sustenance for about 10,000 individuals, out of the county's 35,392 population (according to census of 1940)." She added that a Nacogdoches Chamber of Commerce report of 1945 indicated that $6,500,000 worth of wood products had been sold during 1945, the last year of World War II.[118]

Although Frost Lumber Industries and most of the other large mills are now only dim memories of a distant past, one cannot underestimate the vital role that sawmilling and logging have played in Nacogdoches County. When virtually nothing but farming and forest industries were extant, the latter provided payrolls that added immensely to Nacogdoches County's growth and welfare. On a recent visit to Nacogdoches, the writer saw a small mill still sawing away on the outskirts of town, and it was a welcome sight to behold. Without a doubt, the second-growth forests will continue to furnish employment to scores of loggers, pulpwood and wood chip firms, and small sawmillers, and hopefully, with good forest management practices, these small forest industries will always remain an asset to Nacogdoches County.

Newton County

An early lumberman and his daughter pose beside a large pine harvested near Wiergate in Newton County.

NEWTON COUNTY

THE GEOGRAPHIC BACKGROUND

Beautiful Newton County, a civil and geographic entity in extreme Southeastern Texas, contains 941 square miles, and is bounded on the south by Orange County, on the west by Jasper County, on the north by Sabine County, and on the east by the Sabine River. Like its neighbor, Jasper County, Newton is a long and narrow county, about eighty miles in length, that is, north to south; and its width varies from about ten miles at its narrowest point on its southern border, to about thirty miles in the Toledo Bend area. Newton was cut out of Jasper County in 1846. It is one of the most heavily-wooded counties in East Texas. Originally, the county must have been solid forest, approaching 95 or more percent, but by 1860 many square miles of timber had been girdled, deadened, and been cleared for cotton plantations.

Elevation in the county varies from about 25 feet above sea level in the Deweyville vicinity to about 500 feet in the Blue Hills of the northern sector. The terrain varies from level lands in the south; to gently undulating slopes in the middle sector; followed by gradual and markedly undulating terrain in the Blue Hills region, a series of ridges and valleys, the latter often being the many creek beds. Some fifteen or more creeks crisscross the county, as follows: Griggs' Mill Creek, Quicksand Creek, Seybold, McGraw's Creek, Trout, Caney, Mill, Deer Creek, Big Cow Creek, Little Cow Creek, Big Cypress, Little Cypress, Nichols Creek, Whitman's Creek and Indian Creek. The many creeks, particularly those that are fed by springs or artesian flows and do not dry up in summer months, provided numerous sites for the many water-powered mills, which will be discussed subsequently.

Newton County lies wholly within the long leaf belt, which covers only about ten of the Southeastern-most counties. Despite the county's two billion feet (or more) reserve of long leaf stumpage (log scale measurement of uncut trees or logs) in 1875, many of which reached four feet in diameter and a hundred feet tall, Newton County also had an approximate half-billion feet reserve of short leaf and loblolly pines, perhaps a half-billion feet of marketable cypress along the Sabine River and creek lowlands; and an excellent reserve of marketable hardwoods - including white oak, red oak, pin oak, beech, ash, cottonwood, magnolia, tupelo gum, hickory, elm, and black walnut.[1]

THE NEWTON COUNTY RAILROADS

Certainly the most important of the early Newton County railroads was the Orange and Northwestern Railroad, organized in Orange in 1901, with Leopold Miller as president and chief investor; H. J. Lutcher, W. H. Stark, and others. As of December, 1901, the first thirty miles of O. and N. W. trackage intersected the Southern Pacific Railroad's rails at Orange; the Kansas City Southern trackage at Mauriceville; eventualy connecting with the Santa Fe Railroad at Buna. In 1905 the Orange and Northwestern began an ambitious project to connect Orange with Shreveport, but only the first 61 miles, to Call and Newton in Newton County, were finished at first. In 1917 the Orange and Northwestern line was extended to Wiergate, but the Shreveport project ended at that point. Bleakwood, Texas, became the junction point of the Orange and Northwestern trackage with the Jasper and Eastern Railroad.[2]

The Jasper and Eastern Railroad was chartered in 1904. Its seventeen miles of Texas tracks, built from Kirbyville in Jasper County to the Sabine River and beyond, crossed Newton County from west to east, creating two rail junctions, one at Kirbyville with the Santa Fe and one at Bleakwood with the Orange and Northwestern.[3]

The Kansas City Southern Railroad crossed extreme south Newton County in 1895, but the Sabine River railroad bridge was not completed until 1897. The Kansas City Southern created junctions involving Newton County when its rails intersected the Orange and Northwestern at Mauriceville and the Sabine and Neches Valley Railroad at Ruliff. The latter was chartered by the Peavy-Moore Lumber Company in 1921, with intent to build fifteen miles of rails from Deweyville to Beaumont, but only about two miles of rails were ever laid, to connect Deweyville with the Ruliff switch on the Kansas City Southern.[4]

Peavy-Moore Lumber Company also owned about forty miles of rails, a privately-owned tram road, which formerly belonged to the Sabine Tram Company of Laurel and Deweyville, prior to 1918.[5] And the Kirby Lumber Corporation also owned about forty miles of rails in Newton County, being the Cow Creek Tram Company, which formerly extended from Call to Salem on the Sabine River, and the rails of the old Newton County Tram Company, which extended from Call to the Trotti log camp.[6] The individual operations of the Sabine Tram Company, the Cow Creek Tram Company, and the Newton County Tram Company will be discussed separately.

NEWTON COUNTY'S EARLY WATER-POWERED MILLS

With the existence of fifteen or twenty Newton County creeks, many of which were spring-fed and not subject to drying up during summer droughts, it is small wonder that early Newton County had its share of water-powered mills as well. Many of them were combination grist mills and cotton gins, but frequently, as will be noted, the water mills were combined with a single upright or sash saw or a single circular saw, capable of producing from 3,000 to 5,000 feet of lumber daily.

According to the 1850 Newton County Schedule V, Products of Industry census, which is on microfilm at Lamar University Library, the following five water-powered sawmills cut timber in the county during 1850. With a capital investment of $2,500 Bennett Elkins cut 1,000 pine logs, worth $300, into 100,000 feet of lumber, worth $1,000. He employed 3 mill hands, paid a total of $36 monthly. John Tompkins' mill was worth $4,000. He utilized 1,000 pine logs, worth $300, and cut 100,000 feet of lumber worth, $1,000. He paid 4 mill hands a total of $48 monthly as wages. George Dougherty's water mill was worth $2,500. In 1850, he cut 1,000 saw logs, worth $300, into 100,000 feet of lumber, valued at $1,000. Dougherty paid 4 mill hands a total of $36 monthly. Cochran and Lewis' water mill was worth $2,000. The millers cut 1,000 pine logs, worth $300, into100,000 feet of lumber, valued at $1,000. The also paid 3 mill hands a total of $36 monthly as wages. Francis Wilson's water mill was worth $4,000. In 1850 he cut 1,000 pine logs, worth $300, into 100,000 feet of lumber, value at $1,000. During 1850, he paid 2 mill employees a total of

$24 monthly as wages. Since no railroads existed as of that year, all production of lumber was for local consumption.

Two of the early sawmills resulted from the migration of David R. Wingate and his brother-in-law, Alfred Farr, and their 151 slaves from Pearl River, Mississippi, to Big Cow Creek in Newton County in April, 1852. Both men were major plantation owners as well. In 1859 Wingate used 73 of his slaves (13 of them operated his sawmill at Sabine Pass) on his 600 acres of improved land to produce 350 bales of cotton; 1,500 bushels of sweet potatoes; and 4,000 bushels of corn, and Alfred Farr utilized his 67 slaves to produce only slightly smaller quantities of the same commodities.[7] Farr, who was a Methodist preacher, also built a water-powered combination cotton gin, grist mill, and sawmill on Big Cow Creek. Farr's first mill was an upright sash saw (similar to a cross-cut saw operated vertically), but one observer noted that "when I went there in 1894, they had a circular saw with patented teeth, which could be replaced when worn out." Another writer noted that "the (Farr) sawmill cut lumber until the Wiergate mill opened in 1918, and the grist mill ran until 1935." Over the years, a series of men owned and operated the old Farrsville sawmill.[8]

Between 1858 and 1862, D. R. Wingate owned the largest circular mill in Texas, at Sabine Pass, where he cut 30,000 feet daily until the Union Navy burned his sawmill, planing mill, and sash and door factory in October, 1862. After returning to Newton County, he built a small sash sawmill, powered by an 8-horsepower steam engine, on Big Cow Creek, but with minimal local demand for lumber and no outlet or means for shipping his lumber to market, Wingate ran his sawmill only intermittently before it caught fire and burned down.[9]

In 1935, Dean Tevis published a fine article in which Judge W. E. Gray of Newton County recalled the old water-powered mills of yesteryear, the facts about one that he recounted as follows:[10]

> ...the first mill north of Newton was Seale's Mill on Little Cow (creek). That mill was on the present site of Wiergate. ...Jodie Seale operated it for a time so sometimes you hear it even now referred to as Seale's Mill...Just before the (Civil) war, it was operated by...(my) father. He shouldered a musket, ... his wife Elizabeth running the mill during the war...The site now is known as Down's Mill though the mill itself has long been gone...

Josephine Peavy's Newton County thesis confirms that Seale's Mill was "one of the first in the county to saw lumber," possibly as early as 1845.[11] And indeed, Judge Gray's father, Simon Gray, was the operator of Seale's Mill in the year 1860 as is listed in Newton County's "Products of Industry" census of that year. The enumerator noted that the Simon Gray mill was turned by water power and cut 1,000,000 feet of lumber during the census year ending June 30, 1860. The lumber was worth $10 a thousand feet at the sawmill. Only one other water-powered, Newton County sawmill was enumerated in that census, that of W. C. Willis, which also cut 1,000,000 feet of lumber during the year. Since a census year was composed of about 310 working days, each mill needed to produce a daily average of about 3,250 feet to reach a total of a million feet.[12]

Two sources observed that John Blewitt's mill was in Newton County, although its products were enumerated in the Jasper County census of 1860. Perhaps the county line was vague and ill-defined in that era, particularly out in the forests. Josephine Peavy noted that "Blewitt's Mill was also located on Big Cow Creek." Dean Tevis' article, quoting Judge Gray, observed that "Blewitt's Mill was on Big Cow Creek, twin you might say to Little Cow (creek). It lay on the old Jasper-Newton road."[13] The Jasper County industrial census returns of 1860 reported that Blewitt's Mill was a steam-driven sawmill, which had cut only 75,000 feet of lumber during the census year. Perhaps the mill was used only intermittently to saw lumber, since Jasper and Newton counties produced nearly 7,000 bales of cotton in that year, and was used principally to grind corn and gin cotton. The lumber-sawing season would have been from December to February, since that was the "off-season" for farmers.[14]

Lee's Mill was a Newton County water-powered mill on Caney Creek, which also sawed lumber. Even the Newton County historians may not be aware that there were two "Lee's Mills" within the county. One Lee's Mill was that of Jesse Lee on Thickety Creek, about five miles northeast of present-day Kirbyville, and was the first lumber industry in 1883 in the Bleakwood community. But that was not the Lee's Mill to be found in *The Handbook of Texas*.[15]

The other Lee's Mill, in early days also known as Livonia, was located on Caney Creek, three miles northwest of Bon Wier and nine miles southeast of Newton. It was first built by the Windham brothers as a combination cotton gin, grist mill, and sawmill. One description of the Caney Creek mill, after it became the property of James R. and Moses Lee in 1879, observed that:[16]

> ...Lee's Mill was a water mill, which ginned cotton, baled it, and ground corn for the settlers from miles about. It cut lumber too, and it ran a 50-inch circular saw, which could go through any log like steam. This man's description of the way cotton was baled during the 80s (1880's) at Lee's Mill on Caney Creek is worth preserving.
>
> ...The power plant was a James A. Davis turbine water wheel, propelled by water which poured from a large dam, or pond, down through a tunnel to the wheel below. This wheel would drive a large circular saw through a log the same as steam power. This, with a 50-saw cotton gin, with a feeder and condenser, and the grist mill were all operated by one wheel. After the cotton was ginned, it was carried up a flight of steep steps in large baskets and put into the (cotton) press. After a part of the cotton was put in, then some one, usually my father or uncle, would get down into the deep box and tramp the cotton down...

It should not be implied that all the Newton County water mills also cut lumber, for such a combination mill was the exception to the rule. Throughout the nineteenth century, there remained only nominal demand for mills to saw lumber in Newton County, since without railroads, lumber could not be exported. And normally, the Sabine steamboats carried only cotton and freight, lumber being the lowest of revenue-producing items. However, the demand for the water mills to gin cotton and grind corn remained unsurpassed. Most settlers before 1900 built their homes and out-buildings out of six-inch pine logs, chinked with mud or clay and moss. Some of the other Newton County water mills mentioned by Judge Gray (but not necessarily mills that sawed lumber as well) included as follows: Watt Wilson's mill, Flem Albrighton's mill on Mill Creek, Joe Maddox' mill on Deer Creek, Jackson Youngblood's mill on Indian Creek, Adam Whitman's mill on Whitman's Creek, Down's Mill and many others. Like his father, Simon Gray, even Judge W. E. Gray had operated his own water-powered gin and grist mill for many years.[17]

THE EARLY INDEPENDENT LOGGING CONTRACTORS

Any pioneer crossing the Sabine River in 1850 most assuredly would have noticed the massive long leaf and cypress monarchs, which graced both sides of the river's banks. Since that pioneer most likely would have been a cotton farmer in search of prime cotton lands, his outlook toward trees may have been quite negative, since large trees sapped the soil of much of its needed moisture. To the cotton farmer, trees were things that needed to be girdled, deadened, and removed in order to clear the cotton fields of timber. However, a large, 25-foot log was also worth $1 each to the sawmills of Orange, and a tall, long leaf pine might produce three or four such logs.

Between 1850 and 1880, two types of loggers developed in Newton County and its neighboring regions. Deadened timber near a river or creek bank had value and could be sold to a logging contractor, who would assemble the logs in a creek "log boom" somewhere, "brand" them with his log brand much like cattle are branded for identification, and then float large rafts downriver to the Orange saw and shingle mills. In 1859 the Merriman Mills there bought 10,000 pine saw logs, each worth a dollar a log.[18]

Nothing, however, made the Newton County cotton farmer become a part-time logger during the winter months as quickly as losing a crop or for cotton to drop below ten cents a pound. The winter months were ideal for logging (particularly after a crop failure or low prices), because farm work was minimal at that time, and the Sabine River and creek water levels were highest, which made that the ideal time for floating logs downstream. If a crop had failed, two or three months of logging could enable the farmer to recuperate somewhat, buy cotton seed, and provide for his family. By the 1880's and 1890's, logging was becoming an institutionalized way of life, for it offered regular pay days and steady income, which farming often did not. By 1878, the daily appetite of Orange's "big log eaters" (sawmills) had grown to 350,000 feet of pine stumpage for lumber and enough cypress log stumpage to manufacture 265,000 shingles daily, and the Newton County logging contractors and their employees were hard-pressed to supply enough logs daily in such great quantities.[19]

Without a doubt, the "king" of Newton County private logging contractors was B. F. Van Meter, who stored his logs in Big Cow Creek. Van Meter was equally as well-known to the Beaumont sawmillers, for between 1875 and 1877, he and Colonel B. D. Crary operated the Yellow Bluff Tram Company, the tram trackage of which ran from Cairo in Jasper County to Yellow Bluff, located ten miles north of Evadale, where Crary and Van Meter's logs were skidded down the Yellow Bluff spillway into the Neches River. In fact, in February, 1878, the first locomotive in Jasper

County, long before any railroad ever arrived, "tooted" its way through the Jasper pineries on the Yellow Bluff tramway.[20]

Other well-known private log contractors of Newton County were G. W. Fisk, Rufus Wingate, and a host of others for whom only the surnames are known. The author will make a number of short quotes from the Orange, Texas *Tribune*, reprinted in the Galveston *Weekly News*, for the years 1877 through 1880, which will reflect so vividly how the Newton County economy suffered for lack of a railroad system, as well as how Newton County helped create the great lumber fortunes of the city of Orange, which for lack of a railroad, could have been retained locally. The quotes are as follows:

> ...Garner and Powers have a raft of 400 logs, which will be here (Orange) in a few days...Last reports state that Van Meter has 3,500 logs and (G. W.) Fisk 2,000 logs in the river in the mouth of Cow Creek...[21]
>
> ...Mr. (B. F.) Van Meter reached Orange Tuesday evening, bringing down something over 4,000 pine logs...[22]
>
> ...Mr. McSmith reached Orange Wednesday with two hundred tiers of cypress logs, some of them measuring 75 feet in length and seven feet in diameter...[23]
>
> ...Messrs. Palmer, Newton, Gilbert, and Marrs from Newton County arrived last Tuesday with 500 pine logs. Mr. Palmer says he will bring down another raft if the water holds good...[24]
>
> ...Mr. G. W. Fisk writes to Major Norsworthy that his camp put into the river last week 1,000 logs. Mr. Nash, Mr. Van Meter, and others are perhaps doing as well...[25]
>
> ...The largest run of timber ever brought down the river was turned loose Wednesday by Rufus Wingate. It is two miles long...[26]
>
> ...The log men are busy. Van Meter and his untiring ally, Capt. Bailey Vinson, are wrestling with logs and ties and are doing business on a large scale...Fisk is doing a good business and it is thought will furnish the Norris mill with pine logs...Rufus Wingate has carried down the biggest drive of pine logs ever set adrift in the Sabine River. Granger and Powers have moved their camp to Cow Creek and are putting in for Van Meter...Colonel Frank Powers is deadening cypress timber on the lands of Judge Wingate,

> and by the 10th of September (1878) will have between three and four thousand trees girdled. They are intended for (D. R.) Wingate's shingle mill...[27]
>
> ...Judge Wingate says any amount of ash timber can be obtained from the Sabine River...The northeastern markets have furnished all Texas with this very useful material, and its price has ranged from $45 to $50 per thousand feet. It can be supplied from this point for much less...[28]
>
> ...Mr. Rufus Wingate is expected down next week with about 1,800 tiers of pine logs...Mr. McKnight and Mr. Fountain came down this week with 900 pine logs...[29]
>
> ...Mr. B. F. Van Meter has 6,500 pine logs on the way down, which will soon reach Orange...Mr. G. W. Fisk has some six or seven thousand, which will arrive on the present rise. Many other timber men with smaller rafts are also on the way, some with as many as two thousand sticks (logs), and we think will deliver to our mills 20,000 pine logs...[30]
>
> ...Mr. Swinford purchased a cypress log 60 feet long, 37 inches in diameter at the small end and seven feet in diameter at the butt. This stick will make 50,000 shingles. It was brought down by Mr. McCorquadale and sold for $34. Mr. Van Meter brought down a pine stick 62 feet in length and 39 inches in diameter at the small end. This stick will turn out 2,000 feet of lumber. Eight men came down the river on Mr. Van Meter's big log...[31]
>
> ...Mr. Van Meter, who carries out his operations on Big Cow Creek, is chief of the timber men in the county at present. It is estimated that he has put in the (Sabine) stream since last spring about 70,000 logs. The timber business here is in its infancy. Van Meter has recently carried out to and down the Sabine River 23,000 logs...(The Newton County *Southeast Texan*)...[32]

There are many more quotes, but the number offered here should be sufficient to convey the message. Van Meter's 70,000 logs probably provided full-time employment for 250 loggers, but if that quantity had been milled by a Newton County sawmill, another 250 men would have found full-time employment as well. It was noted that in February, 1879, A. K. Norris' sawmill at Orange averaged cutting 40,000 feet of lumber from 100 logs each day in succession for thirty days.[33] Hence, if Norris' sawmill had been in Newton County, Van Meter's 70,000 logs would have amounted to well over a two-years' supply.

Van Meter's operation alone was quite large, but there were many others as well who carried down large rafts of logs to Orange. And even all of their operations dimmed in the face of the volume of logs marketed in Orange after Cow Creek Tram Company and Sabine Tram Company organized in 1889-1890, and during the 1890's, each tram company was dropping 1.5 million feet of logs into the Sabine River each week, or 78,000,000 feet of stumpage (totalling 156,000,000 feet) annually. A railroad finally reached Call Junction in 1895, and Cow Creek Tram's president, George Adams, finally "saw the light" and built his own Call sawmill. Although Kansas City Southern trackage finally reached Newton County and lumber could have been shipped westward throught Beaumont in 1895 (eastward after 1897), it was ten years later before Sabine Tram Company built its own sawmill at Deweyville, which would have retained at least a percentage of the sawed lumber profits in Newton County

Nevertheless, the independent logging contractors provided payrolls for perhaps 500 to 1,000 loggers annually, and the author would be remiss if he failed to recognize that fact and the economic security that was provided for those loggers. Yet it seemed such a pity that that one billion feet of Newton County pine, cypress, and hardwood logs could not have been retained in the county to succor mills at Newton, Bleakwood, Bon Wier, Call and Deweyville, rather than provide the feather beds upon which a half-dozen Orange lumber fortunes were consumated, only one of which ever built a sawmill in Newton County.

THE COW CREEK TRAM COMPANY

The Cow Creek Tram Company was organized in Orange about September, 1890, to harvest logs on a magnificent tract of 25,000 acres of timberlands, stretching west from Salem near the Sabine River to Call, Texas. The land was a legacy from Dennis Call, Sr., an Irish native and one of the earliest Orange merchants and cotton brokers, to his son, Dennis Call, Jr., who became the tram road's first president and one of its two chief investors. The other main investor and the tram's manager and 'guiding spirit' was George Adams of Newton, who also contributed a large block of pine lands, the exact size being unknown to the author. Dennis Call was to contribute his name to the sawmill town to be built at one end of the tram road,[34] and Salem, the old steamboat stop on the Sabine, would become the tram's headquarters.

By December, 1890, the first news account of the Call and Adams tram venture indicated that construction of the new tram road was well under way, as follows:[35]

> ...Some ten or twelve miles above Sabine Tram, near Blue Prairie, a lot of men are busily engaged in erecting houses and making necessary arrangements (laying tracks, etc.) for putting in the Cow Creek Tram (Co. at Salem), which when in running order, will add another great source for supplying Orange with timber...

During the intense competition which developed between the Cow Creek Tram Company, Sabine Tram Company, and Wiess Bluff Tram Company (the latter on the Neches River), it came to light around 1890 that log scalers on the Sabine and Neches rivers were using different methods for "scaling" or measuring the uncut footage or "stumpage" that a log or tree contained. Log scalers on the Sabine River used the "Ochiltree scale," whereas the log scalers on the Neches River used the "Herring scale" to measure logs, the latter method resulting in a 15% greater footage volume. Hence, the Orange sawmillers were buying up logs at a 15% premium, compared to what the Beaumont sawmillers had to pay.[36]

During the winter of 1891, the trackage of the Cow Creek Tram neared completion, allowing the proprietors to become a major supplier of saw logs at Orange by June, 1891. A month later, a Galveston editor observed that:[37]

> ...The (Cow Creek) enterprise was begotten in Orange early in the present year, and acts entirely as a feeder for this town, where it supplies under contract the Orange Lumber Company with logs, shipping them down the Sabine River from Salem, its logging center. It was not until subsequent to March 1 (1891) last that all of the material for use in its construction arrived in Orange...The road had then to be built, track and bridges laid, cars and engines put into running order...Since that time they have put into the Sabine River over 7,000,000 (feet) of pine lumber...
>
> ...Since its start, it has averaged daily delivery into the river of 126,000 feet, and on one occasion increased this to

> 189,519 feet...When we know that they had but one locomotive to work with, and only fourteen cars hauling, the work accomplished was unusually creditable.
> ...Within the last week or so, they have added six cars to their equipment and are beginning to use them now...The president and treasurer is D. Call, Jr.; Mr. George Adams, as vice president and business manager, attends altogether to the practical working of the company at Salem. Mr. John McKinnon (an Orange sawmiller) is the secretary...

A few months later, a news correspondent visited Salem and left another vivid description of the company, as follows:[38]

> ...this correspondent, in company with Mr. E. D. Downs, the popular and efficient sheriff of Newton County, paid a visit to this place, where is located the fine logging tram of the Cow Creek Tram Company...This company commenced operations December 1, 1890 (the date construction began). They now have in operation seven and a quarter miles of tram railroad, laid with steel rails, have one locomotive and 26 log cars. Last month the company put into the river 2,015,296 feet with one loading crew...Mr. T. J. Trotti {later to be the Kirby Company woods boss at Trotti log camp} is general foreman; Mr. Charles Smith, section foreman; Mr. J. C. Ferguson, saw boss; Mr. John Pollok has charge of the engine (locomotive); and the...good-natured Mr. Hal Gilbert is team boss. Mr. G. W. Smith is bookkeeper; Messrs T. W. Lee and A. J. Moss have charge of the store; Mr. William Zahlman, a native of Germany, has charge of the shop...

An article about Cow Creek Tram Company in March, 1892, observed that George Adams was president of the company, a position he would continue to hold until 1901, when all property of the Cow Creek Tram Company was sold to the Kirby Lumber Company.[39] In 1895 John Henry Kirby, however, was with the Texas Pine Land Association of Silsbee, and he was busy building a railroad from Kirbyville to Beaumont to supply logs to the Reliance sawmill at that point. Rail transportation became necessary since about one-third of the pine trees owned by the Texas Pine Land Association were "sinkers" - that is, trees so heavy due to high specific gravity or rosin content that they would sink in the river and needed to be moved by rail. As the trackage of Kirby's railroad, the

Gulf, Beaumont and Kansas City, reached Call Junction, only a few miles from the rails of the Cow Creek Tram Company, Dennis Call and George Adams decided to build a sawmill of their own - at a town Call would name for himself - and reap the additional profits that the Orange sawmills had enjoyed from sawing the Cow Creek logs into lumber.

THE COW CREEK SAWMILL AT CALL

Dennis Call planned efficiently for his new sawmill at Call in October, 1895, designing it to have a daily cutting capacity of between 80,000 and 100,000 feet daily. The mill was to be housed in a two-story frame structure, size 186 feet long and fifty feet wide, to be covered with galvanized sheet iron on the roof and all sides. A separate boiler house, size 33 feet by 50 feet, was designed to contain three tubular steel boilers, size 66-inch by sixteen feet, along with a 24 by 30-inch steam engine, developing 250-horsepower. Other equipment would include a fifty-foot log carriage with "shotgun" steam feed to carry the logs to 84-inch double circular saws. The mill would also contain a Hill drag saw to cut logs to desired lengths, a resaw, an Allis bull edger, a Glencove sizer, a large gang saw, trimmers, cut-off saws, overhead slashers, and a system of live rollers to carry sawed boards and slab waste to different parts of the mill. When completed, the mill was expected to be as modern as any of the mills in the towns of Orange or Beaumont.[40]

Work on the new sawmill building was "going ahead rapidly" as of January, 1896, and by April, 1896, all the machinery had arrived and was being installed.[41] In November, 1896, one news account recounted that: "Call-The new sawmill at this place is running regularly now, cutting 85,000 feet per day..."[42]

In June, 1897, a Galveston editor published another news account to the effect that Newton County's first large sawmill was sawing timber as follows:[43]

> ...Call, Texas-The Cow Creek Tram Company's mill here is running full-time, averaging 100,000 feet daily. This company is now cutting and will soon ship one bill of about 150,000 feet of clear heart lumber and timbers for a Galveston concern...They are also cutting on a big order of bridge timbers, stringers, and car sills for the Gulf, Colorado, and Santa Fe railway...They have completed a shipment of 300,000 feet of pine lumber to Sabine Pass for export...

Since the foregoing paragraph indicated the Call sawmill was progressing profitably, George Adams did a strange thing in 1898. He leased the Call sawmill (just as John H. Kirby leased his new Silsbee sawmill) in 1898 for three years to the Industrial Lumber Company of Beaumont, a concern with a large export trade with Mexico. The year 1897 witnessed much lumber overproduction and a general stagnation of demand in the lumber trade from sources other than the expanding railroads, and many sawmills shut down for many months or went bankrupt. The reason or reasons for leasing to Industrial became moot, however, for on June 6, 1899, the big Call sawmill burned to the ground.

A raging fire was discovered in the vicinity of the edger saw, which told mill foreman Jack Smith that it was pointless to try to save the sawmill. Besides, to make matters worse, the conflagration soon destroyed the waterworks and overhead tank, making it virtually impossible to combat the inferno. However, all sections of dollyways were quickly knocked loose and towed away in an effort to save the planing mill, dry sheds, dry kilns, and yard stock. Eventually, only the planing mill was saved, the dry kilns, dry sheds, and several million feet of yard stock going up in flames just as the sawmill did. The mill losses amounted to $100,000, but the two insurance policies totalled up to $75,000, leaving a net loss of $25,000. A few months later, the Call sawmill was rebuilt. The Cow Creek Tram owners at that time included Dennis Call, George Adams; M. T. Jones, a Houston retail dealer; and John McKinnon, who owned only one share of stock.[44]

"MILL G," KIRBY LUMBER COMPANY'S CALL SAWMILL

In time, the new Call sawmill was rebuilt as a single circular mill, and the lease of the Industrial Lumber Company expired in 1901. The Cow Creek Tram Company and its sawmill at Call were almost immediately purchased by Kirby Lumber Company to become a part of its vast, new system of sawmills. The following quote concerning the new, 100,000-foot Call sawmill is from Kirby's 1902 company book, as follows:[45]

> ...Call Station is located 2 1/2 miles from the Gulf, Beaumont and Kansas City Railroad...All switching is done by locomotives owned and operated by the Kirby Company...From Call the tram runs back into the timber lands from four to eight miles...The logging facilities at this point are ample to provide a daily feast for the saw...

...The mill is now provided with a single circular saw, with shotgun feed and the fastest return of the (log) carriage obtainable. This is supplemented by an 8-saw edger, which is often used as a resaw for splitting large timbers. The trimming from the edger saws are thrown on a bed of carriers (live rollers), which move slowly toward a gang of slashers (saws). After passing from the slashers, the waste material drops into the refuse carriers, which convey it to the burners (boilers). Leaving the edgers, the stock...passes farther to...an automatic trimmer which reduces it to the nearest even length...

...The machinery will be increased by the addition of a band saw...no addition will be made in the power plant. The engine used is a Corliss...equal to about 200 horsepower. It is fed by three boilers, 66 inch by 18 feet, and...the furnaces are fed by sawdust exclusively...

...The present dry kiln has a capacity of 30,000 feet a day. The plans for the improvement include the erection of two more of the same size...Close to the dry kiln boiler room is the electric light plant, which is used to light the mill when necessary to work by night...It usually is stopped at 10 (o'clock) at night, except when a night force is worked...

...The planer...is equipped with a No. 2 sizer and matcher, combined; four flooring machines, one inside and one outside molder, one picket header, two resaws, a two-saw edger or rip saw. The engine...is larger...having a 20-inch drive with a 30-inch stroke, and is supplied with steam by two 66-inch by 16-foot boilers...When the additional equipment...(is) installed, the plant will have a capacity of from 150,000 to 175,000 feet for each 10-hour run...The mill will then be a combination circular and band saw (mill)...

...Mr. Moore, the sawyer...is a thorough master of producing the most lumber with the least expenditure of nerve force. He,...to make himself understood, prefers a motion of his hand...The minute a log rests on the carriage, his eye glances at the cutting orders and the log is sized up...

By March of 1905, all of the improvements that Kirby had contemplated in 1902 had been completed. Kirby Lumber Company then transferred one of its veteran managers, B. H. Rice, who formerly had supervised the Village Mills plant, to Call to become the new sawmill superintendent of the Call facility. A Beaumont editor observed in 1905 that the Call sawmill:[46]

> ...came into the possession of the Kirby Lumber Company in 1901. It has a capacity of 160,000 feet (daily). All the requisites of a first class mill are in evidence here. The dollyways are in excellent condition, and the arrangement of the lumber in the yard speaks well for the yard foreman...The dry sheds have no superior...and are kept in perfect order...The yard around the boarding house is fenced and in point of fare, the table is not surpassed by any other boarding house in the lumber district...
>
> ...The commissary is spacious and carries a well-selected stock. A good school house, church, and amusement hall have been provided. The water tank is elevated 72 feet above the level of the mill, and is twenty feet high and twenty feet wide. It gives a heavy pressure on the hydrants conveniently located in different parts of the plant...
>
> ...The (mill) pond is artificial (man-made) and supplied with water from an artesian well. The dispensary, presided over by Drs. Swinney and Ogden, has all the appliances necessary...
>
> ...The supply of logs is the best and are hauled from the Front (logging camp) 20 miles distant...Ten cars of lumber are shipped daily...The lumber remains in the (dry) kilns from thirty to thirty-six hours.
>
> ...Fruit culture is carried on to some extent here. One 18-acre peach orchard has been yielding for several years. There is always an abundance of fruit and vegetables, nearly everyone having a garden...
>
> ...Mill personnel: B. H. Rice-superintendent; W. N. Holbrook-bookkeeper; J. W. Richards-shipping clerk; Allen Ashworth-assistant shipping clerk; W. C. Woodard-commissary manager; W. B. Surratt, S. N. Bateman, T. B. Pruitt-commissary clerks; Joe Grimaldi-mill foreman; Walter Coats-yard foreman; W. W. Richardson-woods

> superintendent; Sam Allen-planer foreman; Charles Allen, Jack Rathburn-sawyers; Frank Hutchison-filer; D. Sheffield-machinist; Charles Whitehead-electrician and sawmill engineer; Tom Holmes-planer engineer; L. Richardson, Mack Richardson, Jim Stelling-locomotive engineers; Dr. B. A. Swinney, Dr. R. Ogden-physicians...
>
> ...General Information: Capacity-160,000 feet per day; output, 1904-40,000,000 feet; feet of lumber in yard-10,000,000; stumpage reserve-25 years; miles of tram road-25; locomotives-1 N. G., 1 standard; saws-one band saw, one circular; kilns-2 standard; capacity of kilns-40,000 feet daily; waterworks-elevated tank; electric lights-100 incandescent; commissary stock-$48,000; men on pay roll-300; monthly wages-$16,000; sawmill engine-Corliss, 500 H. P.; planer engine-250 horsepower; planer capacity-120,000 feet daily; planer machinery-4 Berlin matchers, 2 molders, 1 resaw, 2 sizers, 1 edger, 2 cut-off saws, 1 picket machine...
>
> ...Company houses-200; one school house; one church; one community hall; population of Call;-1,500 people.

The Call sawmill produced 40,932,000 feet of lumber in 1907, a figure which ranked it among Kirby's four largest mills.[47] The Call plant produced considerably less than Bessmay, which was always Kirby's largest plant, but the Call output was still from 2 1/2 to three times greater than Kirby's eight smallest mills, which produced between twelve and seventeen million feet of lumber each annually. The Call tram, or Cow Creek Tram as it was still commonly called, extended for 25 miles, the complete distance to the Sabine River, from whence it turned north and ended in a fine timber stand east of Newton.[48]

Mill life at Call, especially before 1910, would have been very similar to that at Bessmay or later at Wiergate or Deweyville—somewhat primitive, frontier-like, except that Kirby permitted no saloons on mill property. The mill town was completely company-owned, with social life centering around the church house on Sundays, and the commissary and community hall in between. As time advanced, other features such as home electric lighting, a movie house, etc. were added. And the Kirby method of paying with mill checks, or "punch-outs," remained in force for most of the period before World War II. The Call pine sawmill burned for a second time in 1924, but it was soon rebuilt as a hardwood plant.[49] In

1948, Call still had an estimated population of 1,000 people.[49]

Perhaps the best record of social life in early-day Call came from the pen of Mr. John C. Gee, Jr., and is excerpted here with the permission of the publishers, as follows:[50]

> ...Nobody (at Call) had anything much because many of the breadwinners worked for two dollars and fifty cents a day. We worked ten hours a day, six days a week, and we all worshipped together on Sundays. One Sunday we heard a Baptist preach and the next Sunday we heard a Methodist. Callites were blessed by the fact that the lumber company had built enough houses on big garden plots for each family to have a home and a good-sized garden.
>
> ...All of us children had chores to do. We had fifty chickens running loose in the yard, and we had three cows...Monday was washday, and I mean all day Monday. The job of scrub-boarding the clothes was assigned to the girls...It was my job to keep the pot boiling and take the clothes out with an old broomstick...On Thursday we moved what little furniture we had and scrubbed and rinsed all the pine-wood floors in the house. Friday was baking day...We got up at 4:00 a. m., and we blew out the lamps and turned in about 9:30 p. m. We were so busy and so happy we never found out we were in the same boat with a majority of people in Call—NEEDY. (Ed.'s Note: one might say that everyone was poor, but nobody knew it!)
>
> ...Love was evident everywhere in our town. Everybody loved the baseball team...Doctor Beavers saw to it that the Chautauqua came to our town each year. Mr. Darden, our superintendent of schools, loved each one of us school kids as though we were his own...One of the pleasures we enjoyed on our daily one-mile trek to school was passing over the placid waters of the huge mill pond. By the way, the mill pond was overrun by giant bullfrogs and water moccasins...
>
> ...Talk about eating, we had some big ice cream socials down at the community room next to the Kirby Lumber Company office. We had domino parties and just all kinds of get-togethers there. Call's loving neighborliness really

> came to the forefront as we gathered in this place of relaxation. In 1925, the Company installed a tremendous radio in the community room, complete with B batteries and headphones...For the first time, the town of Call got the results of the World Series games and such events the same day the action took place...
>
> ...Our ease was interrupted when the Great Depression that hit the United States in 1929 filtered on down...Call was able to survive those terible times, I believe, better than any other place in the world. The sawmill shut down, of course, and all working for wages came to a halt. But the great sense of love that always held the community together naturally led to a great sharing of the little each person had. When a person killed a deer...it was shared with neighbors. To cap this off, Doctor Beavers dispensed rice from his silos in China, Texas, to each family every week...The final blow came in 1953 when the sawmill was closed for good. Life, as we had known it,...had come to an end.

What a beautiful description of sawmill life that probably could be equally applied to most mill towns in East Texas! As of 1950, apparently everyone in Call assumed the sawmill would be there forever. One of the last buildings built there was the red brick school house, built on FM-1004 near its crossing with the railroad. However, in 1951 when the Kirby Lumber Company decided to center all of its mill operations at a new centralized, all-electric plant at Silsbee, the fates of the Call, Bessmay, and Honey Island mills were sealed immediately. After the Silsbee plant was built, the Call sawmill closed down in 1953, and very quickly all plant facilities were dismantled and moved. The 300 houses there were sold, torn down, or else moved away, and Call disappeared almost as quickly as it appeared in 1896. Even the site of the old mill quickly returned to the forests it had helped to level, leaving little there as a reminder of bygone days, except a graveyard, the pine needles, the faint echoes of the big band saws, and of course, one lovely, little red brick school house.[51]

THE NEWTON COUNTY TRAM COMPANY AND THE TROTTI OR KLONDIKE LOGGING CAMP

Some time after 1892, T. J. Trotti, formerly the general woods foreman for the Cow Creek Tram Company, left his employment to become an independent logging contractor. His efforts were concentrated among a fine stand of virgin timber ten miles southeast of Newton and two miles from the Sabine River. The railway station at Bon Wier, on the Jasper and Eastern Railroad, was established specifically to serve the Trotti and Lee's Mill communities. At first, Trotti cut timber near enough to the Sabine River that the logs could be pulled by oxen and skidded into the stream. Gradually Trotti had to build a tramway and log town in order to get to the timber farther back from the river banks. A locomotive was finally carried up the Sabine River on the last voyage of Captain Tom Davis' old sternwheeler, the Dura; was unloaded at Belgrade, and was moved over a wooden rail tram road to a log camp then known as Klondike. In 1898, when Trotti teamed up with two other independent loggers, Sheriff E. D. Downs and Ed Ellington, to found the Newton County Tram Company, the name of the logging camp was changed from Klondike to Trotti.[52]

In 1901 the Trotti community and the Newton County Tram Company became a part of the Kirby Lumber Company system of log camps, and before the town was dismantled and abandoned when the timber was cut out in 1906, Trotti had a population of 700 persons.[53] A description of the log town of Trotti was recounted as follows:[54]

> ...The Kirby Lumber Company operated a picturesque logging camp in Newton County, at Trotti, about two miles from the Sabine River. No attempt was made to lay out the own, and therefore the streets wound in and out among the trees. The company and private interests shared about equally in ownership of the camp's shanties. The state maintained a free public school for the seventy-five children of the camp. About 100 men were on the (Kirby) Company's payroll, half of whom were single. Their wages ranged from $1.50 to $3.50 per day. These men cut and dumped into the river on an average of 3,000,000 linear feet of logs each month, which went to the Orange mills...

Specifically, the "Orange mills" was Kirby's "Mill D" at Orange, originally the old George Bancroft sawmill. After Kirby sold its "Mill D" at Orange to Miller-Link Lumber Company, the Trotti log camp shipped its logs by tram road to Call and Roganville until all the timber was exhausted.

The 1902 publication of the Kirby Lumber Company supplied only minimal information about Trotti, which follows:[55]

> ...Trotti is perhaps one of the most picturesque of the logging camps operated...At the time the selection of a location was made, it marked the western terminus of the logging operations then owned by the Newton County Tram Company. Since that time, however, the men of the saw have cut their way ten miles to the northwest, leaving in their wake a tangled mass of tree tops...The feature of the timber in and around the camp is...of a size eighteen to thirty inches (in diameter)...
>
> ...T. J. Trotti, the superintendent of logging operations...was the founder of Trotti, otherwise known as the Klondike, camp. Mr. Trotti has been engaged in cutting timber and supplying mills for 25 years. At the beginning of his operations, the logs were cut near the river and hauled to it by mules and ox teams...Mr. Trotti secured a locomotive and track and began operations on a larger scale...

As far as logging camps fared, Trotti seemed to as "permanent" as one ever became. Perhaps some loggers lived in crude "shanties" and tents, but a photo of the residence of a Mr. L. C. Steward at Trotti shows that of a very substantial and confortable home. No mention was found of a commissary, a hotel or boarding house, or a church, but one can wager that those buildings were just as present as the school house was. Nor was there any mention of a steam log loader in use at Trotti, but the Trotti research speaks of the use of mule and ox teams to pull, skid, and load logs. When the site was abandoned in 1906, the author presumes that the Trotti loggers were transferred to another of the many log camps owned by the Kirby company, especially in Jasper County.

OTHER NEWTON COUNTY MILLS, SAWMILL TOWNS, AND LOGGING CAMPS

Newton County today is the location of numerous ghost towns that were once thriving sawmill towns or logging camps. However, surviving knowledge about most of them is so negligible that the history of a number of them can be linked together to complete a single part or sub-chapter of this Newton County story. Such a ghost town is that of Howard's Switch.

Located in west central Newton County on the Orange and Northwestern Railroad, about five miles north of Bleakwood, Howard's Switch was the site of a sawmill established in 1906 by the Howard-Mixon Lumber

Company, owned by Thad Howard of Newton and Charles A. Mixon of Kirbyville. About 1909, the town had a population of 750, indicating that it was probably a medium-sized sawmill of 50,000 feet, with planer, dry kilns, mill pond, commissary, and boarding house.[56]

A reliable source, the U. S. Manuscript Census of 1910, verifies that the sawmill had been sold to J. W. Fleming, who resided in Howard Switch, and the mill functions were still being supervised by the ex-owners, Howard and Mixon. The mill employees were enumerated as residences 40 to 150, of Precinct 1, of Newton County, and some of the mill personnel were recorded as follows:[57] C. A. Mixon-sawmill superintendent; L. Kelley-mill foreman; E. E. Bean-sawyer; Dr. H. L. Willette-physician; M. W. Lloyd-filer; G. G. Wilson-millwright; Frank Kelly-sawyer; J. W. Fleming-sawmill owner; G. W. Bunch-sawyer; J. R. Kelly-manager, mill pond; T. Howard, mill foreman; and J. V. Sheffield-log scaler.[58]

Although Newton County was one of the last regions to feel the effects of the axe and the saw due to a lack of railroads, most of the timber resources there had already been swallowed up before 1900, leaving only a few small tracts of timber to be partitioned among a few remaining sawmillers. In 1889, George W. and J. B. Smyth founded the Sabine Tram Company and purchased a huge triangular body of timber in south Newton County, fronting for several miles on the Sabine River between Deweyville and Laurel and stretching westward almost to the Neches River, to include all of the Devil's Pocket region. Cow Creek Tram, the Newton County Tram Company, and the Miller-Link Lumber Company held huge blocks of land in central Newton County, most of which went to the Kirby Lumber Company in 1901 when both tram companies sold out. In north Newton County, Lutcher-Moore Lumber Company and Alexander Gilmer Lumber Company acquired tracts of timber totalling some 120,000 acres, Lutcher-Moore owning the lion's share of about 90,000 acres (that eventually Wier Long Leaf Lumber Company would harvest). Beginning in 1905, the Gilmer firm would log the western tract for its mills at Remlig and Jasper. By 1905 there was nothing left except a few small tracts of two or 3,000 acres, an amount sufficient to last a small sawmill for perhaps five years. Such a sawmill firm was the Boynton Brothers Lumber Company of Angelina County, who had purchased a small Newton County tract of timber, two miles east of Bleakwood, about 1905.

Before 1908, Albert and Walter Boynton were sawmilling near the

Nacogdoches County line, twenty miles east of Lufkin, when suddenly their log supply was exhausted. Their Angelina County sawmill was variously reported as being of 40,000 and 50,000 feet capacity before 1908, and their tram road ran into the Angelina River bottoms.[58] In 1908, Boynton Brothers Lumber Company dismantled their sawmill and erected it again east of Bleakwood at Logtown, which by 1910 had a population of 700 people. *The Handbook of Texas* noted that the Boynton Brothers sawmill at Logtown cut only 30,000 feet daily, but it was the same sawmill that they brought from Angelina County in 1908 and would take back to Angelina County after Logtown was dismantled in 1912.[59]

Boynton Lumber Company operated a short tram road into their timber tract of about 3,000 acres and employed about 100 men in their mill and logging operations. Some of their key plant personnel in 1910 included as follows: Jeff Mitchell, who operated the Logtown boarding house; Charles Gee-millwright; Walter Boynton-owner and sawmill foreman; Albert Boynton-owner and planer foreman; Moses Lee-commissary manager; Edgar Bone-yard foreman; James A. Giron-filer; Will Brown-log scaler; Lewis Bell-bookkeeper; and V. A. Allen-lumber grader. Since Logtown could boast of a commissary, a boarding house, and a population of 700, it probably also has a church, school and community hall.[60] After the Logtown sawmill was dismantled and returned to Angelina County in 1912, Logtown quickly became a ghost town and it soon returned to the forests that had fathered it and to which it belonged.

Adsul was another early west central Newton County sawmill town, probably in the Bleakwood or Howard's Switch area, because it was built in 1906 on the Orange and Northwestern Railroad. The Adams-Sullivan Lumber Company built a 40,000 foot sawmill there in 1906, and like Boynton Brothers at Howard's Switch, apparently owned only a small tract of about 3,000 or 4,000 acres, a log supply sufficient for about four years. According to *The Handbook of Texas*, Adsul never exceeded 500 persons in population because "the mill relied on a labor supply from the surrounding rural area."One record of 1908 reported as follows:[61]

> ...The Adsul Lumber Co. of Adsul, Texas, located on the line of the Orange & Northwestern Railroad, recently completed a deal for the purchase of a new sawmill plant, and a tract of standing timber, formerly owned by the Davis Lumber Company, about three miles from Adsul. The plant has a capacity of 20,000 feet daily, and is an entirely new and well-appointed plant. There is enough

> standing long leaf timber included in the deal to log the plant for five years...

Nevertheless, the census of 1910 certainly confirms that the Adsul plant could boast of a commissary, a short tram road, a planing mill, dry kilns, and probably a school and church. As is quickly discernible, the name "Adsul" was composed of the first syllable in each of Adams and Sullivan's surnames.

George Adams and Thomas Sullivan were the owners, and the immediate question which intrigues the author's mind is — was he the same George Adams who had owned the Cow Creek Tram Company? So far, the author doesn't know. Adams apparently operated a boarding house as well because seven of the plant's key personnel resided with him. The following key employees were located in Adsul's 1910 census, beginning with Precinct 5, residence 16, as follows: George Adams-owner and sawmill foreman; Leopard Adams-yard foreman; Thomas Sullivan-owner and planer foreman; Kathleen King-bookkeeper; Frank Croom-filer; Alfonso Scott-superintendent, log camp; Virgie Adams-commissary manager; Lloyd Mears-sawyer; Thomas Herring-sawyer; and Sam Herring-log scaler. With its timber finally exhausted in December, 1910, the Adams-Sullivan sawmill was dismantled and moved away; the mill site was abandoned and soon returned to forest, and in 1948, only about fifty rural residents still lived in that area.[62]

Hartburg became a post office, and later a sawmill town, on the Kansas City Southern Railroad in extreme southeastern Newton County in 1895. The railroad reached there in 1897, and in 1903 the Williams Lumber Company built a small sawmill there of about 30,000 feet daily capacity. Evidently, Williams had access to a much larger timber tract, for his mill at Hartburg survived until 1929. In 1910 the census listed the following key personnel at the Hartburg mill, as follows: George Williams-sawmill owner; Allen Surrough-millwright; John K. Wilson-sawmill foreman; L. Eaves-sawyer; and John Kinney-yard foreman. [63]

Named for John Hart, an Orange attorney, Hartburg managed to avoid becoming a ghost town and survived as a rural township. Another early mill there was the Barber Lumber Company, which in 1915 operated a yellow pine sawmill capable of cutting 30,000 feet daily. The Barber mill had no planer, and it manufactured lath, box shooks, ties, and shingles, in addition to yard stock. The Storm Lumber Company built a small, 20,000 feet mill there in 1940, and in 1948 the Hartburg oil field was discovered. A store and two churches survive there today, and Hartburg's children attend the Deweyville schools.[64]

Fawil was once a small sawmill town on the Jasper and Eastern Railroad (and on FM-363), five miles southwest of Bon Wier. Local legend maintains that in 1906, the sawmiller, Fonzo A. Wilson, tried to paint his name on a short board to place in front of his mill, and he was able to complete only his initials and the first three letters of his surname before he ran out of space on the board. When the railroad came through, it picked up the name 'FAWIL' painted on the board for the local railroad switch. F. A. Wilson's mill was quite small, about 10,000 feet daily, and it was eventually sold to W. E. Gray. Other sawmillers who succeeded Gray included John Ramsey, Joe Kinnear, and a Mr. Rayhill. After years of being only a ghost town, Fawil has received a 'lumbering rebirth' since Kirby Lumber Company built a plywood mill there.[65]

Woodmyer, or Woodmeyer was a typical Kirby logging camp, which was located on Armstrong Lake near the Sabine River. Its name was derived from J. F. Woods, the logging foreman there, and C. P. Myer, an early Kirby Company superintendent at the Silsbee mill, who rose to general manager of the Kirby system. The Woodmeyer camp had an "artesian water supply, a first-class commissary, a post office, and a school." The loggers at the site, established in 1911, sawed down and shipped 5,000,000 feet of stumpage monthly until the site was abandoned in 1918. Around 1915, Woodmeyer reached its maximum population of about 750 people.

Hazel Woods Wilburn described life in old Woodmeyer when she was a young girl. She recalled living with her family in two box cars, that were especially equiped and fitted out with windows and partitions. Her father and uncle took care of the camp corral, with its dozens of mules and oxen, located next door to the box cars. They harnassed and unharnassed the animals and fed and watered them in long troughs supplied for those purposes. Life for the Woods family at Woodmeyer consisted of caring for and milking three cows, feeding 25 chickens, gathering eggs, but the family had plenty of milks, eggs, butter, and chickens to eat. They even had a 100-pound block of ice delivered to them daily. The mule corral housed some sixty mules, in addition to the oxen. Mrs. Wilburn recalled baths in a large wash tub, and as at Call life was a daily routine of caring and providing for each other's needs, afflictions, and welfare.[66]

Old Salem on the Sabine River was one of Newton County's chief river ports in the old cotton riverboat days. It was also headquarters of the Cow Creek Tram Company, as is already related - where logs were skidded into the river daily to be floated down to Orange. After the tram company was sold to Kirby in 1901, the tram headquarters there was abandoned, although most of the personnel had already been transferred to the tram company's sawmill at Call in 1895.[67]

Harrell's Switch was an early rail siding on the Orange and Northwestern Railroad. Established in 1907 and named for a logging contractor, it was used for several years as a switch and shipping site for logs, poles, cross ties, and pulp wood.[68]

Bleakwood, Texas, boasted of a favorable location because it became the junction point of the Orange and Northwestern and the Jasper and Eastern railroads. Sawmill history there began about 1883 when Jesse Lee operated a water-powered mill and became postmaster there. From 1928 until 1942, the E. E. McDonald Lumber Company operated a small sawmill of 20,000 feet daily capacity in Bleakwood until his plant was dismantled and moved away. About 150 people still lived there in 1948, most of them subsisting from the cross tie, logging, stave, and pulp wood industries. Located on Thickety Creek, a small rural population will probably always survive there, and the Bleakwood schools are combined with Newton.

Bon Wier was an early station on the Jasper and Eastern Railroad, established to serve the Trotti and Lee's Mill communities. The town took its name from B. F. Bonner and R. W. Wier (see Wiergate), who were early Kirby officials. Three different logging camps were operated at Bon Wier, the first from 1918 until 1924; the second from 1927 until 1929; and the third from 1932 until 1935. In 1948 the Bon Wier Lumber Company owned and operated a small sawmill there, cutting 20,000 feet daily. With a 1948 population of 400, Bon Wier still ships some ties, logs, and pulpwood.[69]

Nevertheless, an article of 1908 indicated that Long Leaf Lumber Company operated a sawmilling plant at Bon Wier at one time, as follows:[70]

> ...The Long Leaf Lumber Company of Bon Wier, Texas, reports that it will tear down its present sawmill after the first of the year, and will build on the same site one of the finest circular mills that can be had. The work will be started early in January (1909), and the new mill will be built as quick as possible. The machinery for the new mill has been ordered, and the equipment is for a big circular mill that will cut about 80,000 feet daily, with entire

> additional machinery...
>
> ...This company is operating a circular mill of about 45,000 (feet of daily) capacity, and the increase in its business is the cause for its new project. This concern owns some fine tracts of long leaf yellow pine, and makes a specialty of sawn timbers, so that a circular mill is more valuable to them than a band (saw) mill would be. The output of this plant is sold through the sales office of the R. W. Wier Lumber Company of Houston...

Burkeville, probably the oldest community in and the former county seat of Newton County, has had populations varying from 200 to 1,000 over the years. For years it survived from the payroll of the nearby Wiergate Long Leaf Lumber Company, located two miles away. Burkeville has had several timber-related businesses located there over the years, including one sawmill in 1909, name not recorded.[71]

It is difficult to connect two sawmills in Newton County with any known place name or locality (other than Princeton). Also the mill, or mills, floated sawed lumber, usually only large timbers and cross ties, down the Sabine River to Orange , since no nearby railroad existed, as the following three quotes reveal:[72]

> ...The portable mill formerly owned and operated by Giles and Singletary on the Kaycee (K. C. S.) road, four miles south of Call Station, has been sold to Grubbs and Parker, and is now being dismantled, and a portion of the machinery will be used by them in the mill they are having built on Big Cypress Creek near the Sabine River...

A few months later, an Orange editor indicated that the new sawmill of Grubbs and Parker was being operated close to Princeton Bluff, as follows:

> ...Grubbs and Parker reached here yesterday with 300 sticks, fifty and sixty feet long on average of sawn lumber, from their mill at Indian Bluff. This is their third run this month, and they will furnish more later as they have just completed an important roadway that was a much needed improvement...
>
> ...Charles E. Parker of the Grubbs and Parker mill on Big Cypress came here (Orange) this morning and bought a wagon with spikes. He finds that a shoal just below their mill in the bayou makes navigation all the year round

> uncertain. To guard against low water, the company has decided to construct a plank road 1,000 feet to Princeton Bluff, from which point there is ample depth to run square timbers to Orange...[73]

The Grubbs and Parker mill sawed large timbers at Princeton Bluff primarily for the railroads throughout 1897 and 1898, and the author has no record of any other Newton County sawmill that floated its sawed products down the river to Orange. And an early sawmill at Cairo, Jasper County, did the same thing on the Neches River.

There were about five or six additional logging camps the author prefers to mention only briefly. Ruliff was the Peavy-Moore Lumber Company rail switch, where that company's tram road, the Sabine and Neches Valley Railway, reached a junction point with the Kansas City Southern rail line, a mile or so south of Deweyville. Thousands of log cars passed through Ruliff en route to the big Deweyville mill, and a like number of cars of finished lumber passed through there en route to market.[74] Camptown, another Peavy-Moore location, was a siding and loading switch on the Sabine and Neches Valley tram road, four miles northwest of Deweyville. It was abandoned after the big Deweyville sawmill burned in 1943.[75] Spear's Chapel in Newton County is said to have been the location of a small water-powered sawmill in pioneer days,[76] but the author has no additional information.

Palmetto, seven miles northwest of Newton, was a logging camp of the Palmetto Lumber Company. The camp was founded in 1916, and its maximum population reached about 250 before the camp was abandoned in 1926. The only Palmetto Lumber Company sawmill that the author has any knowledge of was built at Jasper in 1910, and apparently that company's tram road ended up at Palmetto in Newton County.[77] There were two Newton County logging camps named "Make-up."The first was located seven miles from Deweyville on the Peavy-Moore tram road, where log trains were "made up" before moving on to the mill. The second belonged to Kirby Lumber Company and was located on the Jasper and Eastern Railway, one mile east of Bon Wier. Two Kirby log camps were located at the latter, the first between 1911 and 1918, and the second between 1924 and 1928.[78]

Another log camp, named Quigley or Slade's Camp, was located in southeast Jasper County, but it shipped its train of 25 cars of logs daily to the Miller-Link sawmill at Newton before the latter closed down in 1909.[80] There are a few other ghost towns or infant communities of

Newton County, namely, Jamestown, Conn, Mayflower, Belgrade and Watson's Chapel, about which the author possesses no sawmilling or logging information at all.

THE BEGINNINGS OF WIER LONG LEAF LUMBER COMPANY

{One of the joys a historian may experience in the course of writing is to possess enough, or even "too much," material about his subject. Adversely, the saddest part is to have "too little," as is the case with the author's knowledge of sawmilling in the county seat of Newton. However, Wiergate and the Wier Long Leaf Lumber Company attracted reporters and writers like cheese attracts mice. By far, the most explicit (if sometimes tedious) body of writing about the Wiergate mill and community is the two chapters in Hamilton Pratt Easton's Ph.D. dissertation, "A History of the Texas Lumbering Industry." Dr. Easton spent parts of November and December of 1940 at Wiergate, and during his interviews, mill employees explained every detail about sawmilling to the interviewer. The author also has six long Beaumont *Enterprise* articles about Wiergate, two articles from *Gulf Coast Lumberman*, and several sets of memoirs, stories, etc., a body of writings extending from 1918 until 1955, twelve years after the mill had closed.}

There are certainly indications that at least two sawmills existed on the Cow Creek location in Newton County long before the mill town became Wiergate. According to a newspaper article of 1924, an old ex-slave name Pilgrim or "Uncle Pig" recalled that the first sawmill at or near Wiergate "was built by a man named Hilkins," (probably Elkins). Supposedly, it was a small water-powered, combination mill that ground corn meal, sawed lumber, and perhaps ginned cotton as well. The 83-year-old Negro lived on property belonging to E. F. Montgomery of Wiergate. The author also believes that Montgomery was operating a small sawmill on Little Cow Creek in 1910. The Schedule I census for Newton County in 1910 (Precinct 2, residence 244) revealed that E. F. Montgomery was by occupation a "lumberman, sawmill," although all his neighbors were enumerated as farmers.[81]

R. W. Wier, the man to whom the town of Wiergate owed its very fortunes and existence, was a long-time and established lumberman long before he came to Wiergate. Wier, along with C. P. Myer, Wiergate's first mill superintendent (who had contributed a part of his name to Woodmyer), had witnessed many years of employment together as Kirby Lumber

Company personnel, and in 1905, Wier, along with another Kirby Lumber Company officer, B. F. Bonner, contributed a part of each of their surnames to the naming of Bon Wier.[82]

In 1907, Wier went into business for himself when he built a 75,000-foot sawmill at Texla, Orange County, on the Orange and Northwestern Railroad. When Wier entered into an agreement early in 1917 with Lutcher-Moore Lumber Company of Orange to purchase and harvest that firm's huge timber tract in north Newton County, Wier sold his Texla mill to the Miller-Link Lumber Company. For Wier, it was a fortunate experience, for only weeks later the new Miller-Link sawmill at Texla burned down. After the Texla sawmill was rebuilt, Miller-Link sold out to Peavy-Moore Lumber Company of Shreveport, and during the 1920's Peavy-Moore operated sawmills at both Texla and Deweyville. Perhaps "purchase" is not the right word since Wier contracted to pay the Lutcher heirs a stipulated fee for stumpage and a small percentage of profits instead of outright purchase.[83]

When R. W. Wier moved into the Wiergate area in 1917, he organized the Wier Long Leaf Lumber Company in partnership with his brother, T. P. Wier of Houston, with R. W. Wier as president and general manager. Although the author does not possess a list of all the stockholders in the Wier Company, he is aware that two of the chief investors were B. F. Bonner, a long-time Kirby Company official, and Lutcher Stark, who was a chief stockholder of Lutcher-Moore Lumber Company in Orange. One article of 1918 observed that:[84]

> ...The Wier Long Leaf Lumber Company...purchased from the Lutcher and Moore Lumber Company of Orange 90,000 acres of the very finest long leaf southern pine timber...Wiergate is an entirely new sawmill town...When completed, it will be an entirely-modern mill town...R. W. Wier has been milling in a medium-sized way for a number of years (at Texla)...He has associated with him in this concern Mr. B. F. Bonner and several other well-known and successful East Texas lumbermen...

To build the larger Wiergate mill, Wier built at first a small "peckerwood" sawmill on Cow Creek to provide the lumber for the great variety of mill buildings, commissary, office, churches and schools, and a large quantity and variation of company housing, all of which would become the town of Wiergate in 1918. Earl Hines of Burkeville "began in the summer of 1917 hauling timber to build the 'Little Sawmill,' which was to saw lumber with which to build the big sawmill."[85]

By December, 1918, the new Wiergate sawmill was already sawing timber on its two band saws, although its huge double-circular saw lacked about one month to complete its construction. A large battery of "Moore Moist Air Dry Kilns" were completed and were already drying lumber. As of December, 1918, the planing mill also was not finished, but pending its completion in January, 1919, the first timbers of rough stock would be passing through the planers en route to the markets. *Gulf Coast Lumberman* reported that with the dismantling of the huge Central Coal and Coke Company (the 4-C mill) sawmill at Ratcliff, Texas, a few months earlier, the Wiergate mill would become the largest sawmill in Texas. Actually, it was the largest west of the Mississippi. It was also reported that Wier wanted to build somewhere closer to Burkeville, but the land owners around Burkeville did not want the mill hands living in their town. Although the railroad from Wiergate to Newton was more or less an extension of the Orange and Northwestern, Wier chartered a separate company, the Gulf and Great Northern Railroad, to build the last fifteen miles of track from Newton (which later passed to the Santa Fe).[86]

WIERGATE, THE BIG MILL AND TOWN

In her published memoirs about the sawmill and the town of Wiergate, Sylvia Larkin left an interesting description, as follows:[87]

> ...The town was named Wiergate after (R. W.) Wier. It consisted of the mill, planer, lumber (dry) kilns, sheds for storage (of finished lumber), a woods crew that included a lot of horses, grade and bridge crews; a skidder to drag and load logs onto trains, which hauled them to the mill and dumped them into a large pond...
>
> ...The town consisted of a commissary, drug store, doctors' office, a post office, barber shop, pressing shop, hotel, ice plant, depot, pay office, feed house, tool house, a large swimming pool with bath houses, an elementary school for whites, a school for the Negroes, two community houses (one for each color), and several churches, including a Baptist, Methodist and Pentecostal...

In 1940, H. P. Easton visited Wiergate for a week, and he left an informative description of the town, its businesses, mills and company housing, as follows:[88]

> ...The important buildings of the town, and almost the only ones except the dwellings, were the commissary and drug store, post office, lumber company office, schools, two churches, "opera house" and community house, a "hotel," an ice plant, and a small railway station. The churches, school, company store and office, and the community house were among the few painted buildings in town. Of course, the sawmill dominated everything. The three hundred houses were laid out in orderly rows to within about a quarter mile of the mill...
>
> ...The Negroes' shacks stood about a half-mile farther from the mill than did the homes of the white workers. Beyond the colored section, which was on a hill southeast across the railroad tracks, some individuals owned houses and other property. Though the town began with contingents of Mexicans and Negroes, because of the war labor shortage, the number of Mexicans has decreased in recent years. Their houses were inferior to those of the white workers...
>
> ...The houses in Wiergate were built during the rush of the first World War, by wartime labor, and were of rough clapboard construction. At first they did not possess ceilings, but the Wier Company provided most of them several years later. The homes were originally one-story shacks, without plaster and lacking inside finish, for they had only roof and wall boards. Batten strips covered the cracks between the pieces of rough lumber. As most of the cottages were unpainted, they looked distinctly weather-beaten. The company eventually painted a few of the houses and added sleeping porches. The company provided no stoves, but did put chimneys in some of the roofs. There was no plumbing in the Wiergate buildings to start with, except in the manager's home and the company office building, and the householders had to get water from the pipes outside. Those families who wished to do so installed plumbing fixtures at their own expense...

Hence, Larkin and Easton presented the essential details of a typical East Texas sawmilling town of the 1920's. As the time approached that the big sawmill would close down late in 1942, some significant changes had been innovated in Wiergate that will be recounted later.

THE GENERAL MILL OPERATIONS AT WIERGATE

C. P. Myer, the veteran Kirby employee with thirty or more years of sawmilling experience, remained at Wiergate for about six years until he died in a car wreck, and he was replaced by E. J. Booth. A Beaumont editor recounted in 1927 that:[89]

> ...Mr. Booth took the place of C. P. Myer, a member of the company and for a long time superintendent of the mill. Myer was always called 'chief' about the town and mill, and Mr. Booth fondly refers to him in that way today...
>
> ...Six hundred employees mean many children and young people in a town of this size...The company heads, largely led by Mr. Myer, planned recreation for the people of the town. They assisted in the building of churches, erected a splendid school with good teachers...and put up attractive recreational or company buildings for the children of every age. There are swimming pools and libraries and rooms for games. There are radios and phonographs; nearby is a movie show.
>
> ...The company store is a large and efficient institution. Wiergate itself is a pretty town. As the logging gang came through, they left some of the biggest, finest trees and these fleck (dot) the hillside of pretty Wiergate...

H. P. Easton, the Texas lumber historian, wrote that Wiergate's large mill pond, some ten acres in size, was the first item in town to attract his attention. Parts of the pond were cordoned off with "booms" of floating logs, chained together, which provided different services like a pasture fence. Sometimes the "booms" restrained the logs from being blown by a strong wind to the far side of the pond. One "boom" might even contain extra long and oversized logs, from which especially-large timbers and stringers could be sawed.

At the log camp, one log car might be loaded with "sinkers" (logs the specific gravity of which made them sink to the bottom), and the next car loaded with "floaters." As log trains approached the pond, there was an 80-foot stretch of rails (known as the "rollaway"), beside the spillway, that listed or tilted at about a fifteen degree angle to permit "floater" logs to run down the skidway. A restraining wall prevented the log car from overturning while on the "rollaway." It was necessary to "trip" the unloading chain so floater logs could spill down the skidway, but cars of sinker logs received special treatment elsewhere (too tedious to explain here). Cars of floater logs could be unloaded at the rate of one car a minute, and the constant crashing thunder or rumble of logs skidding into the mill pond could be heard for a mile through the forest.[90]

Sometimes Booth came to the pond personally to inspect and seek out the especially-large logs from which to saw special timbers, railroad stringers, etc., and a Beaumont editor described on one occasion those operations at the mill pond, as follows:[91]

> ...Charley, the "pond man," signals. A car is tripped, and a load of great fellows, nearly all "heart" pine, pitch into the water...E. J. Booth, the Wiergate superintendent, calls for the finest log in the pond. It moves slowly at first, reluctantly, towards the elevated skidway, which is almost the last step of the condemned log...Six men await it...They measure the stick carefully...count the rings to determine that this tree is 154 years old...They find the log is 48 feet in length and 29 inches in diameter...

Very quickly the log was locked in place on the 60-foot "shotgun-feed" log carriage before it struck the screeching band saw. A band saw is a marvelous sawmill tool, which turns on upper and lower rollers, called "dogs," eight feet in diameter. One "dog" is belt-energized from the main shaft, and the other rolls free, except that it is adjustable to make the saw blade as taut as possible. A band saw blade is an "endless chain" affair, a quarter-inch or less in thickness, twelve to fifteen inches wide, and about 35 or 40 feet long, with saw teeth on one or both edges. A "double-cutting" band saw has teeth on both edges, and it cuts on both the forward and return movements of the log carriage. Because a large circular blade is about three-quarters of an inch in thickness, the amount of sawdust that a band saw can save over a circular saw would be sufficient in two or three weeks time to build a large house.

The greatest fear in sawing a log consisted of invisible metal objects imbedded within the log — a nail, spike, or an ingrown piece of barbed wire. Such metal could easily shear off the band saw's teeth, or ruin the blade, and even in the Depression days of 1933, a new band saw blade cost $300. From the band saw, the log or lumber might pass through a dozen other saws, perhaps a drag saw to cut the log to proper length, gang saws, edgers and trimmers, resaw, cut-off saw, rip saw, or slasher saws, descriptions of which would be too tedious here. For an in-depth insight into the Wiergate saws, equipment, and general maintenance, the interested reader should consult the two Wiergate chapters of Easton's dissertation.[92]

In 1932, the Wier Long Leaf Lumber Company believed its log stumpage reserve, most of which was in Newton County (with small amounts in Jasper and Sabine counties), was sufficient still to last for fifteen years. However, ten years proved to be the accurate figure. One editor noted that Wiergate, with its 250,000 feet of daily capacity, was the largest of the four Newton County mills (only two of which were still operating), that could be considered "large," the others being Call, which was then closed down during the Depression; Peavy-Moore at Deweyville; and Miller-Link Lumber Company at Newton, which had shut down and been dismantled in 1909.[93]

As of 1937, E. J. Booth was still superintendent at Wiergate and Emmett C. Lee was the assistant superintendent. In 1937, R. W.Wier decided to modernize his dry kilns and some other mechanical aspects of the mill. At about the same time, a team of reporters and photographers from *Gulf Coast Lumberman* arrived to produce and leave what is perhaps the most glowing account of Wiergate that survives today, as follows:[94]

> ...When Wier built that great mill, equipped with three head-rigs - two double-cutting bands and a circular - (the bigest sawmill under one roof west of the Mississippi in the South), they likewise put in grand logging equipment and setup to supply the great mill with logs. So succssful has the woods crew operated that for 19 years, the mill has never caught up with the woods. It came mightly near to it one year, but never quite made it. Nineteen years for a big mill never to run out of logs for a single hour is some record...
>
> ...In the logging department, the Wiergate equipment matches that at the milling end. A two-line skidder hauls in the logs to the spur tracks, where a steam loader puts them on the cars and starts them to the mill. Three locomotives perform the woods services. The woods they log look more like a park than a forest. There is practically no underbrush in virgin long leaf timber. You can drive almost anywhere through their great forests in an automobile...
>
> ...Everything around and about the Wiergate plant is neat, clean, and in order. There is no refuse accumulations, no eyesores. The sawmill itself is a particularly open one, and

> the under part of the mill, instead of being a mass of low-hung timbers, pulleys, and machinery, is open, bright, and clean enough for living quarters. It is a grand looking mill, above and below, inside and out. . .
> ... About five hundred men live and move and have their being at the Wiergate plant. It is a particularly happy community. The company maintains a good hotel, a big general store, a big community house open to all employees and their families, a moving-picture theatre, etc...

In 1937, Wier contracted with the Moore Dry Kiln Company to modernize the company's several dry kilns to the "cross circulation fan type," and in so doing, increased their kiln production rate by 60% due to the reduction of time that lumber had to remain in the kilns. The *Lumberman* article recounted one incident to demonstrate the height of the trees in the Wier long leaf forests. In 1936, the company needed the tallest and stoutest tree they could find to use as a gin pole while erecting their new water tower. The woods crew located a tree 117 feet tall, 10 1/2-inches in diameter at the top, and with a taper so gradual that the butt end of the tree only measured 28 inches in diameter. After lifting tons of timbers as well as the new water tower into place without incident, the tree was then cut into logs and milled into lumber.[95]

T. L. Talbot, the mill foreman, estimated that the mill's hourly average cut was 20,000 feet, or 200,000 feet for a ten-hour work day. In 1929, the Wiergate mill cut 60,000,000 feet, its best month being 5,500,000 feet based on a 48-hour work week. About fifteen percent of the lumber shipped were rough timbers, but the remaining 85% consisted of kiln-dried, dressed lumber of all dimensions and grades.[96] Timbers emerging from the saw were "subjected to a chemical dip to keep them bright. . ." Often such timbers were sixty feet long, being special cut stringers for the railroads. The Wiergate planer mill's daily output of finished lumber was equal to the sawmill's daily cut, and included lathes, pickets, and all grades of flooring, siding, ceiling, shiplap, and square-surfaced lumber. Loaded dollies were pulled over the plank runways, called "dollyways," by mules and horses. Along the front of the planer and lumber shed were the lumber loading docks, with sufficient room for an entire train of box cars to load lumber simultaneously.[97]

The author would be remiss, however, if he failed to mention something about the Wiergate steam-generating plant. Eight large boilers, fired by sawdust, shavings, and other refuse, provided the steam on work days. On

Sundays, only sufficient steam was generated to run the engine that provided the water for drinking and fire protection, as well as the steam engine that turned the electric dynamo and provided power for shop machinery. The 750-horsepower main Corliss engine turned a 20-foot flywheel, the wide face of which contained semi-circular groves for the 1 1/2-inch ropes, which turned the mill line shafts to energize the band saws. The circular saw had its own steam engine, as did the planing mill and electric power plant.

Any sawmill as big as Wiergate would have an extensive filing room on the second floor, where two or more saw filers sharpened the band and circular blades, welded in new teeth, and generally repaired all saw equipment. Formerly, all saw filing was done by hand, but by 1940 an automatic machine did the filing. Nevertheless, saw filing was considered such tedious and nerve-wracking labor that the filers were given a special time off, mornings and afternoons, to recuperate. The machine shop at Wiergate contained every conceivable machine - lathes, milling machines, drills, shapers, wheel-boring and wheel-press machines, bolt cutters, etc., along with enough equipment and human know-how to build a locomotive. In fact, the plant machinists repaired the locomotives as well as all mill machinery. Other auxiliary work units included a complete carpenter shop and blacksmith shop.[98]

Because Wiergate employees were generally satisfied with their working conditions, there were no labor unions or contracts, no labor conflicts or strikes, at Wiergate. Between 1910 and 1912, long before the Wiergate mill was ever built, there were attempts at sawmill unionization in East Texas by the Industrial Workers of the World, but their many efforts were generally unsuccessful. Many sawmill fires in East Texas during those years were attributed to arson by union organizers, as well as acts of sabotage against the mill owners, such as driving spikes into trees to tear up the band saws.[99]

THE CLOSING DAYS OF THE WIERGATE MILL

By about 1939, Emmett C. Lee had replaced E. J. Booth as the last Wiergate mill superintendent. By June, 1940, after 21 years of sawing timber, it must have been painfully visible to all that at least 85% of the Wier forests had been cut out, and that the life of Wiergate as a mill town was fast ebbing. However, a Beaumont news article of that summer disclosed nothing but an upbeat attitude within the town. The population had grown to about 2,500, and an editor recounted that:[100]

> ...Here the Wier Long Leaf Lumber Company of Houston has built one of the largest mills west of the Mississippi River, which furnishes employment for approximately 700 employees and. . .it is with pardonable pride that E. C. Lee, manager of the lumber company, tells you that there has never been a day since the mill started when there has not been logs in the pond...
>
> ...The town itself is above average of what one expects of an East Texas sawmill town. There are the company-owned hotel and residences for the employees, a grammar school with a high school at Burkeville for the white students, and a high school in Wiergate for the Negro students. Community houses have playgrounds in connection (with them), swimming pools and lodge halls. The company has a general store, with a force of 13 employees and a sub-store in addition...
>
> ...The medical care of the town is in the hands of two physicians (Drs. McAlister and Baird) and a registered pharmacist (Hines), while the spiritual welfare of the white section of town is looked after by a Methodist and a Baptist Church, of which Rev. M. M. Davis and (Rev.) A. M. Funderburk are pastors, respectively, while in the Negro section there are five churches of various denominations. For amusement there is the Palace Theatre. That it is well-attended is evidenced by the change of program three times weekly...
>
> ...There is a Boy Scout troop at Wiergate, of which Q. Dudley is scoutmaster...This troop is the pride of Wiergate, and justly so, for these boys, like the East Texas pines, are the best, as evidenced by the enviable record they have made for themselves...Under the management of R. R. Warren, there is an amateur baseball team (the Wiergate All-Stars), one of the best in the Sabine-Neches area...

The day the last Wier log was scheduled to hit the band saw finally arrived about November 20, 1942. But the Wier office, store, and some other buildings remained open until the end of December, in order to allow the company to clear out its supply of sawed lumber, dismantle the mill, and wind up its local affairs. The company probably could have survived another year had caution and prudence reigned, but about January 1, 1941, some grades of lumber began to double and triple in price as the United States Army and Navy suddenly chose to erect about 100 military camps, which compounded an enormous demand for lumber so soon after the fiscal stagnation of the Great Depression. As a result, the Wier Company upped its standard work week to 55 hours and its lumber production by about forty percent. T. P. Wier advised a Beaumont editor that:[101]

> ...The company cut its last tree several weeks ago, and is now shipping out its final consignments of lumber. We don't know how long it will take to complete shipments... Last month...the last tree fell. A brief ceremoney marked the end of the cutting. The sawmill whistle blew for the last time...

And thus a great sawmill breathed its last breath and died, but for many, there were few outward sighs of regret. The old mill whistle ended up at Neches Butane Products Company in Port Neches. The new rubber plant at Port Neches was at that moment trying to hire 10,000 defense workers at wages up to $10.00 a day, and the shipyards at Beaumont and Orange were hiring any unskilled labor they could find. But some workers would soon have a different rendezvous with destiny, for the Newton County Draft Board was shipping many of Wiergate's young men out to fight the great battles of World War II. And some of them did not return.

One rubber plant worker at Port Neches L. B. Martin, felt quite at home whenever the old Wiergate mill whistle blew. Late in November, 1942, Martin, who was one of Wiergate's locomotive engineers, hauled the last load of logs from the forest to Wiergate's mill pond. The following month, when Wier Long Leaf Lumber Company sold the locomotive to The Lummus Company, the prime contractor for the new rubber complex, Martin followed his old engine to Port Neches and drove it for The Lummus Company. A year later, he began working for Neches Butane Products Company, switching tank cars with the same old locomotive he had driven at Wiergate. After he retired, Martin (now deceased) moved back to the piney woods that he loved so well, and as of 1993, his widow still resides at Hemphill.

THE SOCIAL ENVIRONMENT AT WIERGATE.

Whatever resemblance Wiergate may have born to other East Texas mill towns in 1920, the social environment there had changed a lot by

1940. That did not include racial integration, however, which arrived in East Texas at a much later period. Conditions, however, had probably changed greatly at Call, Bessmay, and Honey Island, all of them being nearby mill towns that would soon disappear as well. There have been several sets of memoirs written about Wiergate, and the theme that threads its way through each of them is that of a caring community, where each person is acutely aware and conscious of his neighbor's welfare as well as his own.

One set of memoirs are those of the town drug store manager, Earl S. Hines. Hines ran the company-owned drug store, which was a part of the commissary system. He knew the company physicians, Drs. F. E. McAlister, T. W. Martin, and W. L. Baird, probably better than anyone else. Hines filled their prescriptions, worked closely with them, and perhaps assisted some of them in suturing lacerations. He wrote affectionately of all three physicians, all of whom delivered babies in the homes and sometimes performed skilled surgery of a type one might expect to find in Beaumont or Orange.[102]

Hines also noted that "lots of home brew" had been made in Wiergate back in the Prohibition Era before beer became legal. The pharmacist likewise noted that only four commissary managers—H. C. Hopkins, E. E. Stringer, Falcon Mantooth, and G. A. Baber—had served the Wier Company during its 25 years of mill operations at Wiergate. Hines wrote that Stringer was the first manager to modernize the store and merchandise there with weekly specials. The former recalled that Stringer sometimes sold twenty pounds of sugar for $1.00; cane syrup for fifty cents a gallon; and bacon at ten cents a pound.[103] (When the *Gulf Coast Lumberman* reporter and photographer left Wiergate in 1937, they wrote of encountering an old-fashioned cane syrup-cooking operation on the edge of town, where cane was ground by a mule walking in a circle to turn the grinder, and the juice was cooked and skimmed in a huge iron vat.)

Another writer noted that the mill paid higher wages—$3 daily instead of $2—than other area sawmills; paid in cash rather than in company mill checks, tokens, or scrip; and had an annual payroll of $90,000. He added that while most of the mill employees were either Methodist or Baptist, Wier had brought in a number of German engineers and machinists from the northern states, who were Catholic—at a time when most mill town East Texans had never seen someone who was Catholic.[104] Two of the Germans were Paul Buhrner, the hotel manager and chef, and E. Emil Heisch, the master machinist.

H. P. Easton, who visited Wiergate in November and December of 1940, while writing his doctoral dissertation, left perhaps the best record of Wiergate's social environment of 1940, as follows:[105]

> ..In 1940, the houses in Wiergate have telephones, pianos, electric lights, drapes and curtains, and radios...The early electric light plant proved inadequate for both the mill and town, but the new equipment (of 1937) provided a surplus of electrical energy, some of which was wasted...There is no charge by the company for home lighting...The homes are now equipped with running water. Because there is no gas heat, some have large central wood stoves...(Wood from the slab heap was usually free fuel in a mill town.)
>
> ...Around the houses are grass-covered yards, gardens, trees, and shrubs. Pigs, cattle, dogs, cats, and chickens roam the streets at will...Wiergate streets are unpaved except for the county road. Possibly fifty percent of the townspeople own cows...The night milking is usually done by the wives because the husbands are too tired after a day's work...
>
> ...The amusement centers are the opera house and the community house...The opera house is a one-story structure with a balcony. It houses the community theater, which shows movies Mondays, Thursdays, and Saturday nights only...Before the opera house was built, the company used to give free open-air picture shows (movies)..
>
> ... Many entertainments are given in the community house. In the lodge rooms, the Eastern Star, the Odd Fellows, and other fraternal groups meet...Most school entertainments. . .are given in the school;...similarly a church or an individual wanting to put on a party or dance or other entertainment can use the community house...The men play dominoes and cards downstairs in the community house. After the white inhabitants of Wiergate were given their community house, the company built and furnished one for the colored people ...They hold their own entertainments in the school and churches and stage their dances in their own community house...

All in all, Wiergate did not lack for much that most of the larger towns had to offer. Amusements common to the larger towns of 1940 were usually found in Wiergate as well, although the facilities and variety of entertainment may not have been as elaborate. Hence, this was the Wiergate that came to abrupt end in December, 1942. Up until then, if mill employees lost their employment, there were usually other mill towns where one might find a job, but by 1943, most of the big sawmills and uncut stands of timber were gone. However, other sources of livelihood loomed in their places instead. Most Wiergate employees quickly found work in either Beaumont, Orange, or Port Neches, but once the war ended in 1945, they once more had to seek other means of employment when the shipyards and wartime industries closed down. For many, the sawmill way of life had come to an end.

WIERGATE — THE GHOST TOWN

In November, 1955, thirteen years exactly after Wiergate's last log went screeching through the band saw, a Beaumont *Enterprise* reporter visited the former location of Wiergate. "Still one of the most pathetic scenes in this pine-growing region," he wrote, "is the devastation left by the passing of a great sawmill." And somehow his immortal words rested on the lips of each of his readers.[106]

In 1955 Calvin Smith owned the only store still left in Wiergate, and if one scoured the countryside a couple of miles in every direction from Smith's store, he might be hardpressed to find thirty farm houses, the occupants of which still regarded themselves as residents of Wiergate. Each second Sunday of September was Wiergate "homecoming day," the day when all the former "Wiergators" came back to visit, reminisce, trade old tales and news of each other, or even scream out excitedly upon staring at some old photo: "Look! That's the old house I was born in!" Or perhaps: "Look, that's me at Susie's eighth birthday party!" The newspaper reporter wrote:[107]

> ...Today, grass has grown over the spots where more than 550 family-filled homes once stood; there were the white, the Negro, and the Mexican quarters. All but a few of the old roads and paths worn by the population...are dimmed ...since Wiergate died. Cows and horses graze around the areas where the three places of worship, the commissary, the ice plant, and the drug store stood...
>
> ...Bitterweeds grow where the theater once provided entertainment; the ghost town's two swimming pools are ugly gaps in the earth; all that remains of the big community

> house are the bare concrete block foundations that jut out of the ground where was once the center of Wiergate...A few stayed on in Wiergate...Some farm; some found employment at the D. K. Shaddock electric sawmill ...Wiergate, once used to the buzz and hum of many saws, is deafeningly quiet...

The *Enterprise* reporter made no mention of the log pond, but most probably the bull frogs still croak there. Nor did he mention the graveyard, where those who sleep the eternal slumber contribute to the "deafening" stillness. The author wagers, however, that the visitor of today, provided he or she has captured the nostalgia that the ghost town of Wiergate emits and has cupped his ear in the right direction, might still hear the faint murmurs of yesteryear's sobs and laughter, or catch the distant echoes of the big band saws' screech, as he slowly tiptoes through the pine needles where once the the town of Wiergate stood.

A HISTORY OF THE SABINE TRAM COMPANY

The Sabine Tram Company of Newton County owed its origin to two brothers, J. G. and George W. Smyth, Jr., formerly of Jasper County and later of Beaumont, who were sons of one of the signers of the Texas Declaration of Independence. Soon after the brothers' return from the Confederate Army, they settled at Beaumont, where in 1873, they founded the old Eagle sawmill. In 1883, they sold out to the Texas Tram and Lumber Company. With the proceeds from the sale, they purchased Jasper County timber lands and organized the Wiess Bluff Tram Company, which they sold to the Beaumont Lumber Company in 1888. Again with the proceeds, they purchased a huge, triangular block of timberlands, 28,000 acres bounded by Deweyville, Laurel (towns which they founded), and Sudduth's Bluff along the Sabine River, and stretching eastward beyond the Jasper County line. Early in 1889, the Smyth brothers founded the Sabine Tram Company and soon became the most successful of the Newton County tram companies and logging contractors, who were skidding logs into the Sabine River. The first of many published accounts about the Sabine Tram Company follows:[108]

> ...Laurel, Nov. 21 (1890)—The above-named place is situated about 23 miles north of the city of Orange and 2 1/2 miles from the Sabine River and is the seat of operations of the Sabine Tram Company. The place was so named by

Mr. J. B.(lewitt) Smyth, the polite and efficient bookkeeper for the company, in honor of Miss Laurel Gilmer, daughter of Mr. A.(lexander) Gilmer of Orange, one of the directors of the company, and the additional letter added as suggestive of the laurel growth there. This company commenced work on the 10th of January, 1889, and since that time have cut out quite an extensive tract of the country, and are now puting into the Sabine River an average of 130,000 feet of logs per day; and expect very soon to increase it to 150,000 feet (of stumpage) per day. Their logs are put out under contract with Messrs. A. Gilmer, George T. Bancroft and Sons Co., and the Orange Lumber Company...

...This company has 7 1/2 miles of railroad track with steel rails, two locomotives constantly in use, one in switching the loaded cars to the main line, and the other in carrying them to the river. And under the skillful management of Messrs. Pollock and Miller, they bound along (the tracks) at a lively rate...with a load of from 150 to 200 logs, and make an average of six trips per day...

...The company has about 120 men employed as teamsters, swampers, sawyers, track-layers, etc. Mr. T. D. Evans is general foreman and has a sub-foreman over each division: Mr. J. T. McMahon, foreman of the saw gang; with Mr. Willis McDonald as saw filer; and Mr. Ross Ferguson, a relative of Mr. G. R. Ferguson, vice president of the Nona Mills Company, as section foreman; while Mr. Jimmy Sanders looks after the teams...This company has never had a strike or trouble of any kind, and none is anticipated as the wages are up to standard, and they pay their men promptly once a month...

...This tram is commonly called 'Smyth's Tram,' as Mr. George W. Smyth is president; Capt. J. G. Smyth is vice president; Mr. J. B. Smyth is bookkeeper; Messrs. Allen and Charlie Smyth are head salesman and general manager of the store. It is due mainly to the efforts of Capt. J. G. Smyth that the citizens of Laurel have a splendid new school building...The school, with an enrollment of 65 pupils, with Professor Wilson, our ex-district and county clerk, as principal...

A month later, an Orange resident published the following account of his travels in Newton County, as follows:[109]

> ...The first twenty-odd miles above Orange, we came to the Sabine Tram Company, where there are some two or three hundred people dwelling in the great pine forest and busily engaged in felling and shipping the monarchs of the woods to our mills...The tram is becoming quite a little town. One of the first requirements of a prosperous neighborhood is the school house. This we found at the tram, which is taking on the airs of a city...

Very quickly an aggressive competition arose between the Sabine Tram Company and the Cow Creek Tram Company logging crews over which one could deliver the greatest number of logs into the Sabine River in one day, with a long list of claims and counter-claims as time advanced, as follows:[110]

> ...Laurel, Tex., Sept. 15—During the past week, the Sabine Tram Company of this place has put into the Sabine River with its tram railroad and one loading crew, 3,629 logs, which, by the 'Ochiltree scale,' measured 637,930 feet. This would be an average of 635 logs and 106,300 feet per day...The best day's work of the week was on Friday, Sept. 11th, when with one loading crew, they put in 849 logs, which scaled 167,903 feet in nine hours...Had the 'Herring scale' (the Beaumont or Neches River scale) been used last week, the 849 logs put in by the crew would have measured 193,000 (feet)...

As of May 1891, another Galveston reporter wrote another excellent article on the Sabine Tram Company, which was equally informative, as follows:[111]

> ...I am under many obligations for favors shown. . .when I came near being devoured alive by the innumerable fleas that have preempted the stable at Laurel...
>
> ...The post office was named Laurel after Miss Laura Gilmer of Orange... Last month this company put into the Sabine (river) 3,505,506 feet with two gangs of men, two locomotives, 28 (log) cars, and 11 miles of haul...They operate fifteen miles of tram...
>
> ...This company carries on its payroll about 150 men. T. D. Evans...is the general foreman;...W. J. Sanders, team

> boss; Willis McDonald, saw filer; J. T. McMahon, saw boss; G. R. Ferguson, track superintendent; William Thompson and J. J. Bevil, log scalers; John Miller and John McMahon, locomotive engineers; W. A. Smith, store manager. The blacksmith and wood worker (carpenter) are James Landry and Marion Smith, who are busy making cant hooks and grab hooks, which are bought from the North...
>
> ...This company has ten ox teams and four horse teams loading in the woods under Messrs. Ed Dickinson and Nat Hendrickson. The horses are large Texas-raised, Norman-Percheron stock, and are doing three times the work with a third less feed...With 28,000 acres of timber to cut from, it will require about 12 years time to exhaust this company's lands, that is, exclusive of the magnificent hardwoods, such as Tupelo gun (2 kinds), magnolia, laurel, sweet bay, hickory (3 kinds), Spanish oak, water oak, white oak, sweet gum, etc...

W. A. Smith, the store manager, eventually rose to woods foreman with Sabine Tram Company, eventually joined Peavy-Moore Lumber Company in 1919 when Sabine Tram Company sold out, and was still alive and active in 1940 when H. P. Easton. the doctoral student, interviewed him. His account of Newton County logging life is a treasure tale that the author can ill-afford to overlook. It is quoted verbatim as follows:[112]

> ...The men in the Sabine Tram Company camps in the years before the First World War were usually paid $1.75 to $2.00 a day. A few old hands who were partly disabled were kept on the company payroll at $1.50 a day because of past service to the company...
>
> ...There was a 'rough and ready' daily life in the company... camps and lumber towns in the early 1900's. The loggers were not educated, but were sturdy, light-hearted, and occasionally troublesome. In the company towns, the company superintendent was the final authority. Law and order was maintained officially by a deputy and justice of the peace, appointed by the company head. If a man were drunk and misbehaved, the officer usually fined him and merely told him to get out on the next train. When the large majority of the men felt the company head was fair-

minded, this type of supervision was sufficient...

...The justification for any apparent despotism on the part of the lumber company managers was that, as Smith expressed it: "With a bunch of men out in the woods somewhere, they'll run things for you if you don't, and they want to know what the rules are and who's boss. You can't be half-way firm about it! If the men were treated right, you could "work over" (fisticuffs?) a few trouble-makers, and have the sympathy of the rest..."

...Labor turnover in most of the lumber camps of the early years was not great. If a company were building a railway, it might require a carload of mules and 80 to 100 extra men. A few workers went from one camp to another, but found that conditions were only slightly better or slightly worse.

...Mill and camp-town managers had to watch out for labor trouble, especially when the I. W. W. (International Workers of the World) were active. Labor was often discontented and easily aroused in mill towns. The Sabine Tram Company, however, had little trouble at the Deweyville mill in the way of labor disputes or strikes, especially in the period before 1900. The I. W. W.. . .tried unsuccessfully to "organize" the mill, but, said Smith: "Where the men were reasonably satisfied, labor troubles did not happen. The main trouble was in those mills, whose owners were absent, and had superintendents looking after the men...

...Smith believed in letting the men themselves solve many of the problems...For instance a few loafers might be decreasing the output of the whole crew...Therefore it seemed better to let the men deal with such cases themselves. They knew who the loafers were and could be relied on to see that they did their share of the work...

...Like other mill and camp managers of his time, Smith had to be careful in the old days and carried a six-shooter regularly. When some of the men brought a quantity of whiskey to Deweyville where. . .whiskey was not allowed, Smith had to go up to "Bohemian Town" (?) and quell trouble. The foreign workers were the toughest. (One

> reporter wrote of two Finnish immigrants, who spoke no English) The company never employed Negroes until 1917-1918, when the World War caused an acute labor shortage in the mills. Then they hired both Negroes and Mexicans (at Deweyville)...[113]

W. A Smith's interview at Beaumont on November 28, 1940, was virtually priceless for the social history information it conveyed about life in an early Newton County logging camp or mill town, which is why the author chose to quote so much of it.

THE SABINE TRAM SAWMILL AT DEWEYVILLE

It appeared that the Sabine Tram Company prospered equally with most other tram companies and sawmills during the 1890's, although the company sometimes suffered as others did, and sometimes shut down, because of a stagnation of lumber demand between 1895 and 1897. In the winter of 1896, the tram company closed down for three months as one news account revealed:[114]

> ...Mr. George W. Smyth of the Sabine Tram Company... says that his tram, which is located at Laurel, Newton County, will not begin operations again...until March 15. The tram closed down December 15 because the company had too many logs in their boom at Orange...

In 1895 the owners of the Sabine Tram Company watched as John H. Kirby built a railroad from Kirbyville, through Call Junction, to Beaumont to deliver logs on. In 1896 the Cow Creek Tram Company extended its tramway to Call where they built a large sawmill during that year. At almost the same moment in 1895, the Kansas City Southern Railroad extended its rails across extreme southern Newton County between Dequincy, La., and Beaumont, but through service to the east and north had to await the completion of the Sabine River rail bridge in 1897. Since the tram's old headquarters at Laurel stood in the midst of a large cutover tract and was neither on the river (2 1/2 miles distant) nor near the railroad, the Sabine Tram owners decided to abandon Laurel and build a new town nearer to the railroad, which they named Dewey (the 'ville' came later) to honor Admiral George Dewey, who had just captured Manila Bay in April, 1898, during the Spanish-American War. In January, 1899, the following quote observed that:[115]

> ...The Sabine Tram Company's sawmill at Dewey is rapidly

> approaching completion. It is thought the machinery will be in position and bolts on by the end of next month...

By 1904 J. Blewitt Smyth, son of the founder and former bookkeeper at Laurel, was general superintendent of the Sabine Tram Company. In 1899, following the building of the 150,000-foot Deweyville sawmill, a new company investor and director, C. E. Walden of Beaumont, joined Sabine Tram Company as sawmill superintendent. By 1919, Blewitt Smyth was president of Sabine Tram Company, and Walden was vice president.[116] Other Sabine Tram sawmill personnel of 1904 included B. R. Moses, general sales agent, and W. C. Gray, planer mill foreman.

Various newspaper accounts reported a variety of large lumber orders received by the Deweyville sawmill in 1904. By then the mill was enjoying a large foreign market. Excerpts of news articles in the author's possession list "a big order of railroad stuff," a Texas order "for immediate delivery of 2,000,000 feet of lumber," and "a big order for kiln-dried saps and German primes (?) for the export trade."[117] About 1902, the Sabine Tram Company bought out the Juanita Lumber Company, across the river from Laurel, and several thousand acres of its timber lands located in Calcasieu Parish, Louisiana. The Sabine Tram operated the Juanita sawmill as well for several years, and in 1904, the Deweyville sawmill was tramming Louisiana logs for 34 miles over the Kansas City Southern tracks to Deweyville.[118]

The following quote of 1904 refers to Sabine Tram Company's tramway either near Deweyville or near Juanita in Louisiana, probably the latter. (W. A. Smith was woods foreman during that period, and the town name of 'Smith' probably came from him.) The author will make no attempt to explain the towns in the quote, entitled "A Sawmill Town," which follows:[119]

> ...The town of Smith is situated on the main line of the Sabine Tram Company's tram, six miles west of Smith's Junction, on the Kansas City Southern railway. From Smith's Junction to Deweyville is 34 miles. Deweyville is not exactly on the K. C. S., but about 2 miles to the west on the banks of the Sabine River...From the main line of the tram near Smith, a spur has been built to Juanita, where the Sabine Tram Company has another big mill in operation...
>
> ...All the people who work on the trams and in the woods contiguous to the towns live at Smith, which has grown into a pretentious village of 1,200 inhabitants, with splendid

> church and school facilities. Nearly all of the employees have been with the company for quite a while and might be considered permanent. . .The logs that are sawn into lumber at Deweyville are hauled over the main line of tram from the woods west of Smith to Smith's Junction, and thence carried thirty-four miles over the Kansas City Southern to the mill at Deweyville...

By 1919, the Sabine Tram Company, in addition to the 150,000-foot mill at Deweyville, owned 107,000 acres of timber lands in Newton and Jasper counties in Texas and Calcasieu Parish in Louisiana, as well as locomotives, dozens of log cars, scores of horses and oxen, and 37 miles of steel rail tramways.[120]

Up until 1917, the Deweyville sawmill had cut only pine timber or cypress. Sabine Tram Company continued to tram deeper into its Louisiana forests in search of pine and cypress logs, whereas over 200,000,000 feet of virgin hardwood forests, including white oak, pin oak, ash, magnolia, beech, gum and hickory, stood on the company's magnificent timber lands adjacent to the sawmill. In April, 1917, Sabine Tram began building a 50,000-foot hardwood mill at Deweyville. The contract completion date was June 1, 1917, and the company hired A. O. Davis, formerly superintendent of the Beaumont Lumber Company's hardwood mill, to be the new Deweyville hardwood mill foreman as of June 1st. The hardwood mill there had been in operation there for only two years, however, when the Sabine Tram owners decided to sell out to the Peavy-Moore Lumber Company of Shreveport.

In July, 1917, however, the Sabine Tram Company would have to endure a visit by the dreaded fire demon, which wiped out its entire plant as follows:[121]

> ...To replace the large sawmill at Deweyville recently destroyed by fire, the Sabine Tram Company will at once erect two sawmills at that place, one for cutting yellow pine and the other a hardwood plant to cost aggregately $150,000. Both plants will be in operation within four months...The company contracted with Lufkin Foundry and Machine Company for the machinery for both mills...
>
> ...To secure material for erecting the two plants, the company has begun the erection of a small sawmill with a capacity of 25,000 feet a day, the machinery having formerly been in the company's plant at Juanita,

> Louisiana, but was dismantled some time ago...A large concrete building will contain the boilers, engines, machine shops, and blacksmith shops, and the two mills will be located on either side of the building. The yellow pine mill will be equipped with both a circular and gang saw and will have a capacity of 125,000 feet a day. The hardwood plant will have a single-cutting (band) saw and gang saw and its capacity will be 100,000 feet a day...

THE PEAVY-MOORE SAWMILL AT DEWEYVILLE

The backbone of the Peavy-Moore Lumber Company of Shreveport, Louisiana, was Anderson Jasper Peavy, an Alabama native born August 8, 1866, but reared in Lufkin where he was a contemporary and associate of J. H. Kurth and Simon Henderson of the Angelina County Lumber Company. After advancing to stockholder, secretary and treasurer of the Lufkin Foundry and Machine Company, Peavy entered the sawmilling business for himself at Shreveport in 1910. Other officers of Peavy-Moore Lumber Company were R. J. Wilson, vice president; R. T. Moore, vice president; S. G. Samples, vice president; and John S. Welch, secretary-treasurer.[122]

Jasper Peavy, whenever he bought or built a new sawmill, usually founded a new company with one of his company officers, the entire group serving as an interlocking directorate of several Peavy companies. Together, they owned the 200,000-foot Peavy-Byrnes Lumber Company sawmills at Kinder and Wynfield, Louisiana; the 250,000-foot Peavy-Wilson Lumber Company sawmills at Peason, Louisiana, and Holipaw, Florida; a Peavy-Welch Lumber Company sawmill; and two Peavy-Moore sawmills at Deweyville and Texla (the latter in Orange County), Texas. Peavy also organized and chartered three railroads, one including the Sabine and Neches Valley Railroad between Deweyville and Ruliff in 1921.[123]

The purchase of Sabine Tram Company by Peavy-Moore Lumber Company in September, 1919, proved to be the largest timber and sawmill transaction in East Texas since Kirby Lumber Company was founded in 1902. The sales contract included the merchantable timber on 107,000 acres of land in Texas and Louisiana, containing a half-billion feet of pine stumpage and perhaps as much hardwood as well; two sawmills at Deweyville; 37 miles of tram road and all rolling stock; as well as logging facilities, equipment such as tools and livestock, and buildings. Peavy-Moore was capitalized at $800,000, but the undisclosed sales price was much more than that.

Sabine Tram retained titles to all cutover land (107,000 acres); the George W. Smyth retail lumber yards; a manufacturing conern; and the Uvalde Rock Asphalt Company. The company composed in 1919 of four officers and stockholders, Blewitt Smyth, C. E. Walden, Frank Alvey, and R. F. Cheesman, all of Beaumont, the Sabine Tram Company had grown from a $40,000 investment in 1888 to a company of 1919, capitalized at $800,000, with a surplus and undivided profits of $2,225,000, plus the undisclosed sales price. Sabine Tram Company was a genuine success story.[124]

In 1921, Jasper Peavy chartered a Texas railroad with intent to lay trackage to Beaumont, but the Sabine and Neche Valley rails never advanced beyond Ruliff.[125] Peavy-Moore Lumber Company also operated the 75,000-foot Texla sawmill (in Orange County on the Kansas City Southern Railroad) until 1929 when the timber there was exhausted.[126]

On September 22, 1922, as Peavy personnel operated the old Deweyville sawmill, consisting of two double-circular and one gang saw, the mill caught fire, resulting in a total loss. Since Jasper Peavy carried adequate insurance to replace it, he quickly turned to his old company, the Lufkin Foundry and Machine Company (which had become a sawmill manufacturer), for relief and replacement of the mill. The new Deweyville mill structures consisted of a two-story sawmill building, 180 feet by 64 feet in size, plus a lathe mill, power plant building, electric dynamo and ice plants, as well as planing mill and dry kilns. Every item was designed by Lufkin Foundry and its engineers.

The new Deweyville mill consisted of two double-cutting band saws and a large gang saw, with the usual auxiliary equipment, two "shotgun-feed" log carriages, drag saw, edgers and trimmers, resaw, cut-off saws and slashers, and a complete lathe mill and box factory. The power plant consisted of "H. S. and G. twin engines, 20-inch by 30-inch stroke, supplied with steam from four 350 H. P. water tube boilers." With all the mill auxiliary machinery, live rollers and conveyers, drive shafts and belting, gears and carriages, automatic log turners, steam kickers and inclined log haul-up, the Deweyville plant was guaranteed to be the most modern in East Texas. The new mill was completed in April, 1923, but alas, it burned down for the second time three years later. A new Lufkin mill of exact duplication was shipped to Deweyville in a record 58 days.[127]

Some of the key Deweyville mill personnel of 1923 included Joe Dixon, millwright; Fred Burdette, plant superintendent; I. F. Pennington, planer foreman; F. M. Carrier, filer; D. L. Cobb, sawmill foreman; W. A. Smith, superintendent of manufactures; S. A. James, dry kiln foreman; W. S. Siler, superintendent, box factory; and O. P. Barnes, millwright.[128]

The new, ultra-modern Deweyville sawmill of 1926 cut throughout the Great Depression, to the best of the author's knowledge (who visited there many times), although the writer is reluctant to say that the mill did not shut down at some occasions between 1930 and 1934. The author drove several truck loads of lumber away from the mill, beginning in 1936. During that period, Peavy-Moore logged or purchased logs from far and wide to feed its mill's 250,000-foot daily "sawdust appetite." The author recalls when Peavy-Moore logged the Gilbert Woods at Fannett, Texas, about 1938, and remembers the great number of flat cars side-tracked there, being loaded with logs for the Deweyville mill.

By 1940, Cecil Smith (who subsequently would own the Smith-Feagin sawmills at Kountze and Woodville) had been promoted to superintendent of the Deweyville sawmill. Mrs. Josephine C. Peavy, a Deweyville school teacher who wrote her M. A. thesis, "A History of Newton County, Texas," in 1942, interviewed Cecil Smith on several occasions. Mrs. Peavy's husband was A. J. Peavy, Jasper Peavy's nephew, and he was the Deweyville commissary manager until his death in 1941. He was replaced by his father, J. E. Peavy (Jasper Peavy's brother), until the mill was sold in 1942. Clark Allen (who later would own the Allen-Peavy sawmill at Kountze in partnership with Bill Peavy of Shreveport) served as assistant to Cecil Smith. Mrs. Virginia Peavy of Lufkin recalls other key Deweyville personnel of 1940 as including Guy Baber, who ran the commissary prior to A. J. Peavy; Johnny McGee, Frank Allen, Stewart Murphy, and G. S. Murphy, but Mrs. Peavy does not recall what positions they occupied.[129]

Josephine Peavy's thesis is about the only surviving source of material about Deweyville and the Peavy-Moore Lumber Company before the mill burned for the third time in 1943. She wrote that Jasper Peavy purchased some Newton County timberlands long before he bought the mill at Deweyville. In 1942 a spokesman for Peavy-Moore claimed that the company had more timber available then than five year earlier because it was tramming logs from longer distances, presumably from distant points in Louisiana as well. The company planing mill was likewise noteworthy because of its artistic moulding, the reworking of wood into picture frame stock, "table stock, toy stock, crate and box stock, bed slats, pickets, lathes, tool handle stock, broom sticks, and coathanger stock."[129a]

In many respects, Deweyville was a typical company-owned mill town. No statistics have been found on the number of mill employees, but the writer presumes, based on Deweyville's population of 1,500 in 1940 and the daily capacity of the mill, that at least 250 men worked in the mill plant and perhaps an additional 200 were loggers in the forests. Since the mill trammed for such long distances, some logging undoubtedly was contracted to private log contractors. In 1940, Peavy-Moore sawmill was cutting at a reduced rate of 30,000,000 feet annually, much less than its known capacity.[130]

The town of Deweyville was divided into two parts, with one side of town for white employees and the other side for Negro employees. All property in town belonged to the company, including both mills and all mill facilities, the Sabine and Neches Valley Railroad, several hundred employee houses; the commissary, with its meat market and drug store; the community house, the hotel, the electric plant, and the water system.[131]

Deweyville also had a school system with 259 students and nine teachers. There was also a company-owned church building, used by both the Methodist and Baptist faiths, and the lumber company provided some financial support. There were also a Sunday School and other church-related activities, under the auspices of the Methodist pastor, Rev. W. S. Winkle, and the Baptist minister, Rev. H. C. Smith.[132]

The owner, Jasper Peavy of Shreveport, served as president of the Southern Pine Association of New Orleans in 1926, and in 1933, he helped draft the Southern Pine Association's lumber code of trade and labor practices, outlined in President Franklin Roosevelt's National Recovery Act. Peavy died in 1942,[133] at a time that Peavy-Moore Lumber Company was ready to dispose of its Deweyville facilities. World War II had already begun, with its resultant labor shortages and a rush of sawmill personnel to better-paying defense industries. In 1942 the company sold out to the E. L. Kurth Lumber Company of Lufkin (E. L. Kurth being the son of Jasper Peavy's old friend, J. H. Kurth), but the ill-fated mill burned down for a third time a year later. In 1948, the Williams Lumber Company, formerly of Hartburg, was operating a 30,000-foot mill at Deweyville, but no other mill statistics are known. Although the town's population dropped from 1,500 in 1940 to 400 in 1950, Deweyville escaped becoming a ghost town, as Call and Wiergate did, and will probably always survive to serve its rural and commuter population.[134]

SAWMILLING AT NEWTON, TEXAS

Josephine Peavy's thesis mentions the following sawmills: "....William Davidson's mill in Newton; the Miller-Link (sawmill), Newton's largest steam mill...."[135] Only a pittance of information survives about either mill; in fact, the greatest paucity in the writing of this paper concerns the lack of surviving knowledge about early sawmilling within the city limits of Newton. Nothing is known about when Davidson's mill began operations there (it was certainly after the railroad arrived in 1905), or when the mill was dismantled, but the names of some mill personnel appeared in the 1910 census, beginning at precinct 1, residence 309, as follows: William Davidson, owner of sawmill; James Broderick, foreman of steel gang (track builders); Dr. B. A. Swinney, company physician; A. A. Fuller, bookkeeper; and J. E. Davis, millwright.[136] No information survives about the size of the sawmill.

The writer can not identify one sawmill in Newton County, which may have been in the community of Newton. In its October, 1909, issue, *Southern Industrial and Lumber Review* reported on page 41 that J. D. Gober, a sawmill financier, had asked for a receiver for the Southern Lumber and Manufacturing Company of Newton County to protect a $5,000 loan owed to him. The article said that Southern Lumber owned a mill and timber in Newton County worth $15,000, but no exact location was mentioned.

The Miller-Link Lumber Company sawmill at Newton was the largest mill, and yet has the least bit of known information surviving about it, despite the fact that the author owns two separate "Histories of the Miller-Link Lumber Company," in both Texla and Orange, which note only that the Newton mill was dismantled in 1909. It was certainly founded subsequent to 1905, which is when the Orange and Northwestern rails arrived in Newton. Leopold Miller, the president of the Orange and Northwestern, and J. W. Link, both of whom were Orange entrepreneurs, owned the mill. However, the writer has recently located two short articles about the start-up of the Miller-Link mill, as follows:[137]

> . . . A deal was consummated on September 1st (1908) whereby the Miller-Link Lumber Co of Orange became owners of the Newton Sawmill Company plant at Newton, Texas on the line of the Orange and Northwestern road. The plant is a very modern 50,000 capacity mill, with planer attached....

> . . . The Newton Sawmill Company is a concern that was formed just at the time the panic struck a year ago (1906-1907), and they purchased the Newton plant and proceeded to modernize it. It has never been operated under their ownership. The stockholders were G. M. Sells, H. B. Jackson, Sam C. Tremble, J. A. Holland, D. A. Ford, and W. E. Lester, all Orange citizens except the last.
>
> . . . The mill is now in operation under the new management, and is running to full capacity....Mr. D. A. Ford...will remain as manager of the plant under the new ownership. He was general manager and a partner in the company before it was sold to the Miller-Link Lumber Company. The output of this company is being sold through the Orange office of Miller-Link....

In January, 1909, Miller-Link employees were busy modernizing the new Newton sawmill, adding a Filer and Stowell shotgun-feed log carriage, tripper, and log turner; new live rollers, etc. In the same article, Miller-Link reported that they had just purchased "some enormous tracts of timber" to support their milling operations in Orange and Newton. Hence, it is totally inconceivable that Miller-Link could have shut down and apparently dismantled the Newton sawmill during the same year, but such was stated in a history of that company.

According to *The Handbook of Texas*, the Williams Lumber Company cut lumber at Hartburg from 1903 until 1929, and at Deweyville, beginning in 1948.[138] Apparently, the Williams sawmill was erected at Newton about 1929, following its dismantling at Hartburg, and the following quote recounted that:[139]

> ...Like most East Texas towns, Newton has a sawmill (in 1930), which although small does a good business. The mill is owned and operated under the firm name of Williams Lumber Company. Lumber in the yard at present is valued at several thousand dollars. The mill has not been running for some time due to the depression in the lumber market...
>
> ...The box factory here has a capacity of 700 dozen baskets a day or 8400 a week. Day and night shifts of approximately 100 (persons) are employed by the Denver concern. The payrolls of this new and thriving business run close to $350 a day...These baskets are used by the growers of the

Rio Grande valley, Idaho,...

In 1949, the city of Newton still had its box factory, a stave mill, and two sawmills (names and mill statistics unknown), with a combined mill capacity of 75,000 feet of lumber daily. The May-June, 1964, issue of *Texas Forests and Texans*, page 27, carried a large advertisement of Newton Lumber Company, Limited, as "wholesale manufacturers of southern yellow pine" lumber. While no mill statistics were given, the mill officers consisted of Claude H. Williams, manager; W. L. Williams, sales manager; and Nathaniel (Mickey) Williams, timber production manager.

Between about 1905 and 1930, Newton County also had a thriving "naval stores" industry, the tapping of pine trees for sap or rosin and its distilling into turpentine. Several turpentine camps which employed hundreds of men, often of foreign descent, existed, and much information about them survive.[140]

A SUMMARY OF NEWTON COUNTY SAWMILLING

For a county that was so beautiful and so richly blessed with probably three billion feet of virgin forests in 1880, it is so sad that Newton County derived and retained so little of its greatest and richest treasure house. Oh, it cannot be denied that over the decades, Newton County loggers and many sawmill workers enjoyed the payrolls and some benefits of steady income, especially when crops failed and cotton prices plummeted. But the profits of their labors, and sawmilling and logging was the hardest of labors, were skimmed off and floated down the Sabine River to Orange and Beaumont and Houston. It was a matter of fact that lumbering, for the most part, tended to enrich the investors in the larger towns of neighboring counties, rather than create banks in Newton County where the wealth could be retained locally. Perhaps there was no other way it could have happened — the author simply doesn't know — since sources of wealth were definitely needed to underwrite any lumbering project. Much of the causes date back to the Civil War, cotton as a Newton County way of life, and the economic prostration left in East Texas as a result of losing the Civil War.

Newton County was totally dependent on outside wealth to build the county's first railroad, the Orange and Northwestern, in 1905, and yet it was the profits of Newton County logs floated down to Orange that created that wealth. In that age, a thriving timber industry was wholly dependent on the railroads to move its lumber products. Without wealth, there was no hope to attract the wealth that financed sawmills, veneer plants, box factories, turpentine plants, paper mills, or any other forest industry.

Railroads are not as vital today for moving timber products as they once were, and large trucks generally can provide those services adequately to sustain the forest industries of today. Fortunately timber, unlike oil, is a renewable source of wealth that hopefully Newton County residents can rely upon for decades to come to generate the wealth needed to sustain the county's economy. So much of the county's wealth of an earlier age was utterly wasted due to bad forest practices or ill-conceived methods, machinery such as steam skidders, used in the forests of yesteryear.

The out-of-county sawmillers of that age practiced "cut and run" forest techniques, which "grabbed' the profits and ran, leaving behind in the county the ugly, cutover stump lands upon which no thought of reforestration was ever practiced by most of them. Fortunately the timber processors of today realize that in order to skim timber profits tomorrow, reforestration must be practiced today. And every bit of tree trimmings, etc., has a use somewhere in today's industry rather than be abandoned in the forest. It is up to the Newton County residents to reverse that process totally — that while earning a profit is what attracts wealth and is a legitimate part of our American way of life, it must be done within the framework of Newton County's welfare first, because the unborn generations of that county will also need those forests so that they too can subsist and harvest the bounties that those beautiful forests provide.

Orange County

The early headquarters of the Lutcher-Moore Lumber Company stood beside the Sabine River at Orange in Orange County.

ORANGE COUNTY

THE ANTEBELLUM SAWMILLS

Orange County is still a vision of sylvan serenity today, with its forests of long and short leaf yellow pine trees in the northern sector, and elsewhere, its panorama of scenery broken only by the deep and cypress-lined, moss-draped bayous and occasional marshy or hardwood lowlands. To visualize how the county must have appeared before the arrival of the Anglo-American settlers in 1830, with their axes and crosscut saws, one need only to replace mentally the trees of today with the stately pine and cypress monarchs of yesteryear, some of them often four feet in diameter and a hundred feet tall. Although Orange County at no time had more than about one billion feet of marketable timber within its boundaries, its future treasures were the forty to fifty billion feet of pine and cypress trees which lined both sides of the Sabine River for two hundred miles to the north.

By 1900, however, the long leaf pine reserves of Orange County had been reduced to a half-billion feet.[1] And elsewhere along the banks of the Sabine, three major tram roads, the Sabine Tram Company of Deweyville and Laurel, the Cow Creek Tram Company at Belgrade and Salem, and Lutcher and Moore's Gulf, Sabine and Red River Railroad in Calcasieu Parish, Louisiana, upped their delivery of logs in time to 100,000,000 feet of "stumpage" (scale or log measure of uncut timber) annually to feed the ravenous appetites of Orange's "big log eaters."

Orange like its counterpart of Beaumont in Jefferson County, was most fortunate in that it had water access to the Gulf of Mexico and did not have to depend entirely on rail transportation. The Texas and New Orleans Railroad arrived in Orange (that is, for the second time) on November 20, 1876.[2] Bi-directional train service into Louisiana did not arrive until five years later, but that did not matter to the sawmillers because Orange's lumber markets during the 1870s-1880s were only at Texas points west of Orange, that is, in South, Central and West Texas.

Many of Orange's earliest lumber products were hand-hewn, but there was a water-powered sawmill built in the county in 1837. In 1891, Joseph Harmon, whose father, John Harmon, settled in Orange in 1830, reported that:[3]

> ...the first mill built in what is now Orange County was put up by Robert E. Booth in 1837, and was run by water, cutting 2,000 feet a day, a remarkably large cut for that day. After that...Robert Jackson erected a steam mill at the Narrows, eight miles above here on the Louisiana side...

In a list of Republic of Texas postmasters published in October, 1840, R. E. Booth was named as being the postmaster at "Mount Holland," several miles northwest of Orange, which as of that year, was one of only two post offices located in present-day Orange County.[4]

Another article, entitled "The Mills of Orange," stated that Booth built his mill there in 1836, prior to his service in the Texas Army (of which the writer owns a copy of his military bounty certificate). It likewise states that Payne and Bendy built the first steam sash or upright sawmill at the Narrows of the Sabine River in 1841, using an engine and boilers from a sunken steamboat, which they sold to Robert Jackson in 1844, who soon tore it down and moved it to Orange. "It was the first mill to make sawdust in Orange," noted the same article, which the writer has serious doubts about since he has noted other errors in "The Mills of Orange."[5] Jackson's mill was not listed at Orange when the the Schedule IV, Products of Industry, census was taken in 1850, and a Nacogdoches *Chronicle* article, written by a correspondent who visited in Orange in 1853, noted that: "...Captain Jackson has one (steam mill) now building (here)." Hence the writer believes that Robert Jackson's mill was moved to Orange in 1853, an opinion also voiced by an article in the New Orleans *Democrat* of 1889.[6] The steamboat engine and boiler probably came from the steamboat *Rufus Putnam*, which sank at Eaves Plantation, south of Belgrade, on January 6, 1840. This was the only steamboat wreck on the lower Sabine River as of 1841, and its site was plainly marked on the Texas-U. S. Boundary Commission Map of 1840.[7]

There were a large number of cypress shingles being exported through Sabine Pass as early as 1839, and from all indications, all of them were being manufactured at hand-hewn shingle mills in Orange. In 1839 the schooners *Santa Anna* and *Waterwitch* carried loads of shingles from Orange to Galveston, and in 1842 the steamer *Patrick Henry* cleared Sabine Pass with a cargo of hand-sawn lumber.[8]

In 1850 the only lumber industry at Orange, recorded in the Products of Industry census of that year, was Marvin Delano's shingle mill. Delano processed 300 cypress logs, worth $300, by hand into 576,000 shingles, valued at $1,152. He employed three men, each paid $20 monthly, and one woman, paid $12.[9] In 1850, the only steam lumber industry in Jefferson County (of which Orange County was then a part) was the Spartan Mill Company of Sabine Pass.

The first steam sawmill at Orange, erected in 1851, was built by two prominent local citizens, William Smith, who served terms in the state legislature, and John Merriman, who was Orange County's wealthiest resident in 1860, with assets totaling $60,000, including 49 slaves.[10] In 1853, a visiting Nacogdoches newspaperman observed that:[11]

> ...Messrs. Smith and Merriman have a large steam mill in successful operation; it is a masterpiece of workmanship. Capt. Jackson has one now building...Mr. Smith has one at the East Pass...Yet there is room for twenty more, and they will find ready sale for all the lumber they can make...

In 1855 Dr. William Hewson built the Empire Mills on the Sabine River, south of Orange, that reputedly cut 8,000 feet daily and worked ten men.[12] The Empire Mills, said to have been "the best in the state," were deliberately set afire on May 31, 1856, by a Moderator posse of Orange's Vigilantes, who sought to flush out two murderers, Sam Ashworth and Jack Bunch, who were believed to be hiding in the sawmill, and who had just killed Deputy Sheriff Samuel Deputy on the Sabine River. The mill conflagration was a $15,000 loss, which included 100,000 feet of stacked lumber in which it was believed the murderers were hiding.[13]

Another source observed that W. C. Brazee and James Woods built a sawmill, consisting of one circular and one upright saw, at Orange in 1857.[14] A Galveston editor of that year also confirmed that Brazee, Woods and Company was one of four steam sawmills located at or near Orange that had a daily output of lumber averaging from 3,000 to 4,000 feet daily. The other three lumber mills were John Merriman, Robert Jackson, and R. A. Neyland and Company's steam shingle mill.[15]

Between 1857 and 1859, Henry R. Green, a Galveston News correspondent, who was also a Beaumont school teacher, made three trips to Orange, and his accounts paint word pictures of matchless prose still unequalled in their descriptions of the sawdust city on the Sabine River. In 1857, while Orange still bore the name of Madison, he wrote that:[16]

> ...Madison sits like a fairy upon a bewitching crescent formed by the Sabine, the houses neatly painted, stretching like the links of a chain clear around the bend of the river, so that standing at the uppermost part of the town, the eye is presented with a continuous line of fascinating cottages..."My God," I could but exclaim..."What a magnificent view! Just look at those pretty houses, so whitely-painted, ensconced like a duck in a nest of roses..."

Two years later, during his second visit to Orange in 1859, Green exclaimed upon visiting a local lumber mill that:[17]

> ...the spoke-shaving machine in (John) Merriman's Mills takes my beaver (hat) with the greatest facility and the least expense. Let lumber fall within its reach, and it comes out a wagon forthwith! Here is the greatest shingle and sawmill exporter in the state, constituting the chief pursuit of its inhabitants. The town is filling up with newcomers, which already reach to nearly one hundred and fifty families...

The principal source of information for Orange's earliest lumber industries, not otherwise dependent on some one's memories, are the Schedules IV, Products of Industry, of the decennial census. Three Orange County firms were cutting timber with steam-driven equipment during the census year which ended on July 1, 1860, but strangely missing for the entire year was the sawmill of Brazee, Woods and Company, as follows:[18]

> ...John Merriman's Steam Sawmill, Orange, Texas. Capitalization: $23,000; raw materials and value: 10,000 saw logs, worth $10,000; employees: 8 men; wages paid for the year: $1,920; product and value: 1,104,000 feet of cypress and pine, worth $13,800.
>
> ...Robert Jackson's Steam Sawmill, Orange, Texas. Capitalization: $7,000; raw materials and value: 500 logs worth $1,500; employees: 4 men; wages paid: $100 per month; products and value: 300,000 feet of cypress and

> pine, worth $3,000.
>
> ...R. A. Neyland's Steam Shingle Mill, Orange, Texas. Capitalization: $4,000; raw materials and worth: 1,000 cypress logs worth $1,000; equipment: steam engine and shingle shaver; employees: 6 men; monthly wages: $180; product and value: 3,744,000 cypress shingles worth $11,252.

In 1857, there were 6,120,000 shingles exported at Sabine Pass, or about twice as many as were manufactured by Neyland's shingle mill in 1860. Since there is no indication that any shingles were being manufactured at Beaumont as of that year, the writer must assume that millions of shingles were still being manufactured by hand at Orange during 1859.[19]

Such was the city of Orange in 1861 when the great tragedy of the American Civil War made of Orange County an economic waste land, even though not a single shot was fired in anger there. Three Orange Confederate companies were quickly enrolled, who marched away to fight. A large number of them never returned, most falling victim to disease, although some of them were killed in battle. And as if the town and county had not suffered enough, Mother Nature added insult to injury on September 13, 1865, when a huge hurricane slammed ashore, killing and maiming several people at Orange and reducing the town to rubble and debris except for three or four residences.[20] In the county's census enumeration for 1870, the population had dropped by several hundred, and it took Orange County an entire decade to recover from the devastation of the war and the winds.

In 1866 A. T. Chenault and John McGehee built a sawmill at Orange, but the latter soon sold out his interest to Chenault. The owner soon converted the business to a cypress shingle mill. One article noted that:[21]

> ...Charlie Chenault's...father was the first to introduce the manufacture of sawed shingles (at Orange). He started a Muzzy upright shingle machine near where the office of the Bancroft Lumber Company now stands in 1867 and averaged at first 10,000 shingles per day. The machine was eventually crowded up to a capacity of 15,000, and when the word went around town a year or more afterward that it had cut 20,000, it made a sensation...The old Muzzy was afterward superceded by a rotary (machine) that soon wore out and was put in the scrap pile...

THE ALEXANDER GILMER LUMBER MILLS

One of the Horatio Alger stories of Orange is that of Alexander Gilmer, an immigrant Irishman, who arrived in Texas without a dollar to his name, and who died a half-century later with a million-dollar fortune. The 1860 Orange census listed Gilmer as a 32-year-old ship carpenter with $3,000 in real and personal assets.[22] One source made the following observations when Gilmer died in July, 1906, as follows:[23]

> ...While the wealth of the deceased (Gilmer) will today amount to more than a million dollars, he came to America like the most of the sons of Ireland without a dollar, and he is one of the few who became immensely rich in the timber business...

Despite his wealth, Gilmer lost five sawmills to fires at Orange , the first one burning in 1869 and the last one burning in 1899.[24] In addition he lost "five loaded lumber schooners in storms along the Texas coast," as well as one cotton schooner captured while running the Union blockade. In addition to the five sawmills that burned, he also lost in the fires "millions of feet of lumber, in all aggregating approximately the value of a million dollars."[25]

An early article of 1871 noted that: "...Mr. A. Gilmer has launched his new schooner from the ship yard of Linford and Co...She is as handsome boat as there is in the trade and is an honor to her builders..."[26] During his lifetime, Gilmer built around fifteen lumber schooners.

His first timber venture came in 1867 when he bought a two-thirds interest in James Woods sawmill, which stood on the Sabine River, where later the Orange Lumber Company sawmill would be built. The little sawmill could cut 7,000 feet of "boards and scantlings" during a work day of 14 or 15 hours, but it could cut crossties at the rate of 10,000 feet a day. The owners paid the sawyer $2.50 a day, but in that Reconstruction age, in addition to the sawing, the sawyer "filed the saw and kept the engine in repair, laced belts whenever they broke..(and) looked after the sorting and piling of the lumber." The little sawmill burned down at midnight on April 1, 1869.[27]

Woods was very dejected after the fire because they had no insurance, and everything that he owned had been invested in the mill. He told Gilmer, "Sandy, I'm wiped out!"

"Forget it, Jim," Gilmer responded. "We'll build a bigger and better mill than it was on Conway's Bayou, where we'll have plenty storage room for a year's supply of logs. And you'll still have a one-third interest by letting a share of your profits go each month to pay for it. That way, you'll get square on your feet again and owe no man a cent!"

The next day, Gilmer wrote C. B. Lee and Co. of Galveston and told them to send Gilbert McDonald, that company's expert boilermaker and machinist, who would repair the damaged boiler and engine and re-erect them once more on Conway's Bayou. W. H. Bell of Galveston, who represented a Cincinnati sawmill manufacturer, arrived in Orange to sell Gilmer a single circular mill of latest design. Bell even guaranteed his mill would cut 15,000 feet a day, or it would cost Gilmer nothing. The latter turned to Woods and inquired, "What do you think of the deal, Jim?"

"Sandy, it's your money!" the partner responded, "So it's for you and Bell to decide what to do! My guess is you'll get a sawmill fer nuthin, 'cause there never was a mill made that can cut 15,000 feet from the logs we got on this river!" (In 1865, the Page circular sawmill was designed only for the second-growth timber of the Eastern Seaboard. Its 54-inch blade would choke down in any logs larger than sixteen inches.)

The next return trip of Gilmer's schooner from Galveston brought back Bell, his crew of men, and his new sawmill to be erected on Conway's Bayou. Weeks and months later, the day arrived to saw the first timber, and by 3 o'clock in the afternoon, only 11,000 feet had been sawn when "some of the journals got hot and the mill had to be stopped to correct bearings and adjust idlers." Woods laughed again and predicted that they had just acquired a new sawmill for nothing. The next day, Bell began again and by sundown, he and his men had cut 15,000 feet of lumber with ease. So Gilmer, realizing that Bell had won the bet, wrote him a check drawn on Ball, Hutchings, Sealy and Company of Galveston.

Eventually the little Conway Bayou sawmill's cut, with edgers added, swelled to 20,000 feet a day and to 25,000 feet on crossties. In January, 1877, Woods and Gilmer sold out to Charles H. Moore, a lumber dealer of Galveston, and Eberle Swinford of Orange. Woods moved to Newton County, where he soon died, but Gilmer promptly built a new shingle mill for himself on the Sabine River.[28]

Early in 1877, the Galveston *Daily News*, in an article captioned "The Sabine Trade," casually announced that:[29]

> ...We learn that C. H. Moore of Galveston and Eberle Swinford have purchased the mill and lumber interests of Mr. A. Gilmer, one of the largest mill owners of Orange. Messrs. (H. J.) Lutcher and (G. B.) Moore, and Mr. T. A

> Rathwell, all of Williamsport, Pennsylvania, representing some of the largest sawmill interests in Central Pennsylvania, have been traveling for the past week among our pineries on the Neches, Angelina, and Sabine rivers, and are well-pleased with the quantity and quality of pine. They think of coming to Texas...

Even a century in retrospect, the writer feels inclined to advise that editor that "Messrs. Lutcher and Moore" finally quit thinking about the proposition and arrived in Orange to stay!

In addition to his role as lumberman, Alexander Gilmer was one of the six largest merchants in Orange.[30] He immediately built a modern shingle mill, and by August, 1877, was shipping an average of 800,000 shingles weekly.[31] During the census year ending July 1, 1880, the Schedule V census, Products of Industry, reveals the following statistics:[32]

> Alexander Gilmer Shingle Mill, Orange, Texas. Capitalization: $35,000; employees: 35 men throughout year; daily hours worked: 9 in winter, 11 in summer; daily wages: skilled, $3.00, unskilled $1.50; annual wages paid: $10,500; months in operation: 12; equipment: two 4-gang saws, 1 circular saw; 3 boilers, one 150-horsepower steam engine; raw materials and value: cypress logs worth $15,000, mill supplies worth $500; product: 14,000,000 shingles; product value: $32,900; origin of logs: swamps of Sabine River—mill did no logging, but shipped shingles on its own schooners.

Around January, 1882, Gilmer leased his shingle mill to Sheriff G. W. Michael, who had just recovered from flesh wounds from buck shot received during the Orange riot and assassination attempt of the previous August.[33] Early in December, 1882, the A. Gilmer shingle mill burned to the ground with a loss of $35,000, but the mill carried a $15,000 fire insurance policy. Sheriff Michael managed to save all lumber at the mill, along with 5,000,000 shingles which were stacked near the mill.[34]

After the burning of the Gilmer shingle mill, the owner replaced it with a lumber mill and operated it himself thereafter. The following description of Alexander Gilmer appeared in a Galveston newspaper in 1887, as follows:[35]

> ...When Mr. A. Gilmer, the owner of one of our largest lumber mills, came to our city, it was a mere village of some fifty or 100 inhabitants. He was little more than a boy

> and was almost penniless, but not waiting for an easy position to be offered him, he went to work at some of the hardest work known to lumbermen - that is, running timber on the river and making hand-made shingles....Mr. Gilmer began from the first to accumulate property about him, and before he was many years in Orange, he was one of our leading merchants. Some ten years ago, he built a most magnificent mill on the site of the present one. About five years ago, that mill was burned to the ground, but Mr. Gilmer at once rebuilt his mill, which today is one of the grandest lumber manufactories in the South. Mr. Gilmer's mill furnishes employment to some 65 or 75 men and cuts from 75,000 to 80,000 feet per day...

An article of 1888 noted that Gilmer's mill was making about 40,000 shingles daily in addition to lumber.[36] A year later, a New Orleans correspondent visited Orange and observed that:[37]

> ...A. Gilmer, soon after the first attempt of D. R. Wingate, built his now famous mill. Like all of his predecessors, Mr. Gilmer's first mill was small and unpretentious, and probably its cost would not have exceeded $15,000. Like the Wingate mill, this one was destined to burn. Undaunted by his great loss, Mr. Gilmer immediately rebuilt and the present Gilmer mill is rated as one of the best in Texas, and is probably worth $100,000. It cuts about 90,000 feet daily...

Beginning in 1889, Alexander Gilmer began acquiring large acreages of timber, at first in north Orange County. In 1890 he became a large stockholder, along with J. G. and G. W. Smyth, when the Sabine Tram Company was organized in South Newton County, and the first tram company town and headquarters, Laurel (Deweyville was the second), was named for Gilmer's daughter, Laura.[38]

In 1891, a Galveston editor took note once more of a scene that could only have occurred on the matchless Sabine River, observing that:[39]

> ...Once more Orange was treated to a spectacle last Wednesday which made its oldest inhabitants rub their hands with glee....Eight schooners, among them several three-masters with all their canvas spread, majestically sailed up the beautiful Sabine...Their point of destination, with one exception, was Gilmer's Wharf...

It seems plausible that most of the vessels must have been a part of Gilmer's own schooner fleet.

Alexander Gilmer was to face the horrible fire demon three more times during the decade of the 1890's. On February 12, 1891, the Gilmer dry kilns caught fire and were destroyed with all contents, a $9,000 loss partially covered with a $4,000 insurance policy.[40] The dry kilns were quickly replaced.

However, the loss of the dry kilns in 1891 was minute compared to the conflagration of two years later. On March 15, 1893, the Gilmer sawmill suffered a $150,000 loss by fire, the worst that Orange had ever seen up until that year. The entire mill, the dry kilns, and 5,000,000 feet of stacked lumber were consumed, and although the planing mill was heavily-damaged, it was saved through the heroic efforts of two Negro employees, who endured the intense heat while spraying streams of water on the planers and other machinery of the planing mill. The loss was partially covered by a $22,000 policy.[41]

Despite the immense conflagration, a net loss of $128,000, Alexander Gilmer immediately replaced it with a new double-cutting band sawmill of 100,000 feet capacity. The last Gilmer fire at Orange occurred on October 3, 1899, only one day before the big band sawmill of Dennis Call, Jr., another Orange businessman, burned at Call. If anything, the Gilmer inferno was even worse than the fire of six years earlier. Very quickly, sawmill, planing mill, dry kilns and millions of feet of stacked lumber were consumed by the flames, and it was only by the "tireless work and heroic efforts of employees and other citizens" that the D. R. Wingate Lumber Company was saved. Again the net loss was staggering, about $110,000, for Gilmer had only a $40,000 policy to cover a part of the loss of $150,000.[42] A Beaumont newspaper predicted that "Sandy" Gilmer would rebuild as usual, but the newspaper was mistaken. Gilmer had had his fill of fires at Orange.

Alexander Gilmer's decision not to rebuild at Orange, however, may have been motivated more by his supply of timber there than any other reason. He soon became the president and chief stockholder of the Gilmer Lumber Company at Lemonville, about ten miles north of Orange, where he owned a 100,000 foot mill and built a standard gauge tram road that connected his mill with the new Kansas City Southern Railroad.[43] Because of his 50,000-acre tract of timberlands that he owned in North Newton County, "Sandy" Gilmer's last sawmill venture in 1905 was a 100,000-foot band mill at the town of Remlig (which is Gilmer spelled backwards), in the extreme northeast corner of Jasper County, adjacent to the Sabine and Newton County lines. His son-in-law, H. S. Filson, was the superintendent of the new Remlig mill and would later supervise all Gilmer mill activities.[44]

Thus ended the fascinating lumber career of Alexander Gilmer although Orange would have other similar careers which follow. Gilmer died while on a voyage to New York City in July, 1906, and his body was returned to Evergreen Cemetery in Orange, where he is buried beside his wife. As one writer summed it up, he lost a million dollars to fires, schooner wrecks, and other disasters, but he still managed to acquire and leave a million-dollar estate when he died.

THE D. R. WINGATE LUMBER COMPANY

David Robert Wingate was another pioneer Orange settler, who lost five sawmills to fires, although only three of the fires occurred in Orange County. His first sawmill that burned in 1846 was located on the Pearl River, Hancock County, Mississippi.[45] His second loss to fire occurred in October, 1862, when the Union Navy came ashore at Sabine Pass, and burned the Wingate and Company sawmill, planing mill and millwork plant.[46]

After having returned to and having lived on his Newton County plantation for several years, Wingate moved to Orange County in 1874, at age 54, where he owned 33 acres of land on the Sabine River, adjacent to John Fielding's shipyard.[47] When he arrived in Orange, there were only three small mills in operation there, W. B. Black's sawmill (which in 1876 was dismantled and sold to the Beaumont Lumber Company); R. B. Russell and Son's steam shingle mill; and Eberle Swinford's Phoenix Mill; and all the products of those three mills had to be shipped by water because the Texas and New Orleans Railroad from Houston did not reach Orange until 1876.[48]

In 1873, a year before Wingate moved his family to Orange, he bought a half-interest in Swinford's mill. The Phoenix Mill, which they operated under the firm name of Wingate and Swinford, was located on the Louisiana side of the Sabine River, and it manufactured shingles, lumber, and lathes. By 1878, the Phoenix Mill's new rotary shingle machine could cut 80,000 daily. Wingate, however, soon tired of a business of which he was not the sole owner, so in 1877, he sold his interest in the Phoenix Mill to Eberle Swinford's other partner. Thereafter, C. H. Moore and Eberle

Swinford operated two mills at Orange, the Phoenix or "Upper" Mill and the Conway's Bayou or "Lower" Mill, which the partners had purchased from Gilmer. The two mills had a combined daily output of 160,000 shingles.[49]

Construction on the new D. R. Wingate sawmill, located on the Sabine River adjacent to Josiah Jordan's Excelsior shingle mill, began in October, 1877, and was completed by July, 1878. Its first bill of lumber was 250,000 feet of crossties, cut at the rate of 35,000 feet daily.[50] The new mill had cost $50,000 to build and its daily cutting capacity was equal to seven car loads.[51]

During the census year ending July 1, 1880, the mill statistics of the Wingate Lumber Company were recorded in the county's Schedule V, Products of Industry, as follows:[52]

> ...D. R. Wingate Lumber Company, Orange, Texas. Capitalization: $35,000; employees: maximum, 50; average, 45; daily work hours: 11 throughout year; daily wages: skilled, $3.00; unskilled, $1.50; annual wages paid: $20,000; months in operation: 12; equipment: one 4-gang saw, one circular saw; 3 boilers, one 125-horsepower steam engine; raw materials and value: cypress and pine logs worth $60,000, mill supplies worth $5,500; product: 10,000,000 BF of lumber, 2,000,000 shingles; origin of logs: Sabine River and tributaries—mill did all of its own logging.

Like Alexander Gilmer, economic misfortune seemed to stalk the old sawmiller's footsteps wherever he went, and on November 29, 1880, the new Wingate sawmill burned to the ground. A newspaper article carried a report of the conflagration as follows:[53]

> ...We regret to learn of the burning of Judge D. R. Wingate's mill at Orange on Saturday last. We learn that the carelessness of the watchman in leaving a boy in charge of the mill was the cause of the disaster. This was one of the finest mills in the South, and Judge Wingate's loss is estimated at $50,000. The Judge is one of the best men in this section, and we heartily sympathize with him in his severe loss...

With little more than a sigh of regret, Wingate and his mill hands began clearing away the debris and started rebuilding. By May, 1881, a much larger Wingate sawmill and attached shingle mill were back in operation.

The new mill cost over $60,000, and during the 1880's, it cut between 70,000 and 90,000 feet of lumber and 75,000 to 125,000 shingles daily.[54]

During the 1880's, Wingate recouped his losses rapidly, for markets were stable, profits were high, and lumber demand exceeded supply. Most lumber milled in Orange was shipped to Galveston and points in Central and West Texas. After 1890, the lumber economy was less palatable due to a severe depression and lumber overproduction. However, the Texas sawmills along the coast survived by developing foreign markets, and the expanding railroads continued to purchase large quantities of cross ties, bridge and trestle timbers, and lumber for depot buildings.

A newspaper article of 1888 left the following description of the D. R. Wingate Lumber Company as follows:[55]

> ...Another mill we would like to mention if that of D. R. Wingate, who. . .has acres covered with lumber and is turning out thousands of feet daily. The capacity is about the same as Gilmer's...Mr. McKeever is in charge of the machinery, Mr. Rufus Wingate is superintendent, and Mr. Sam Swinford is the head man in the main office...

In June, 1890, Judge D. R. Wingate faced for a second time the total loss of his large Orange sawmill by fire. The conflagration began in the mill's boiler room on June 1, 1890, and within fifteen minutes the flames had spread to the second floor of the sawmill. The planing mill and most of the stacked lumber escaped the inferno, thus reducing the total loss to about $50,000. A $25,000 fire insurance policy reduced the net loss even farther to about $25.000.[56]

At first, Wingate planned not to rebuild, since he was already 71 years old and his personal interests were turning to rice farming. However, he finally relented and allowed the firm of D. R. Wingate and Company to be reorganized as a joint stock company, under the supervision of his son-in-law, B. H. Norsworthy, and another Orange lumberman, John McKinnon. By 1895, controlling interest in Wingate's mill, as well as the adjacent Orange Lumber Company sawmill, passed to a Houston holding company, composed of Colonel M. T. Jones as president and A. F. Sharpe as secretary-treasurer.[57]

D. R. Wingate died in Orange on February 15, 1899, only a couple of days short of his eightieth birthday.[58] As a result, he was not alive in April, 1902, to witness the third and last loss by fire of a D. R. Wingate sawmill in Orange.[59]

THE A. K. NORRIS SAWMILL

Around 1875 the northern timber reserves, especially in Wisconsin and Michigan, were totally depleted, and sawmillers in those regions, such as Lutcher and Moore, viewed East Texas, and especially Orange, as preferable for a future sawmill site. A. K. Norris of Chicago was another of those who came from the North to East Texas in search of sawmilling opportunities, and in April, 1877, he "brought his machinery (to Orange) from Wisconsin to put up a new sawmill." A year later, an editor noted that the mill of A. K. Norris and Co. was "running at full tilt."[60]

The Norris mill also had its own sash and door factory as was verified by the Schedule V, Products of Industry, of the Orange County census for the year ending July 1, 1880, as follows:[61]

> ... A. K. Norris Sawmill, Orange, Texas. Capitalization: $20,000; employees: maximum 50, average, 45; daily work hours: 11 1/2 throughout year; daily wages paid: skilled, $3.50; unskilled, $1.50; annual wages paid: $20,000; months in operation: 12; equipment: two 4-gang saws, 1 circular saw, 3 boilers, one 100-horsepower steam engine; raw materials and value: logs worth $60,000 and mill supplies worth $5,500; product: 10,000,000 feet of lumber; product value: $95,000; origin of logs: Sabine River-mill did no logging. Norris sawmill re-manufactured 40,000 feet of its lumber in its sash and door factory, which employed six men.

In the spring of 1879, the Norris mill was experiencing a shortage of both logs and labor within the mill as were most of the Orange mills. In September, 1878, Thomas and Van De Mark leased "the new mill recently erected at Orange by A. K. Norris and formerly known as the Norris Mill. This mill is built with the most improved mill machinery, planing mill attached, and has a capacity of 30,000 feet per day."[62] In October, 1879, the A. K. Norris mill was cutting timber under lease to W. F. Stewart and Company, lumber dealer of Galveston (who was C. H. Moore's partner).[63]

Two years later, a New Orleans paper recorded the loss to fire of the Norris sawmill on November 19, 1881, as follows:[64]

> ...The large mill of A. K. Norris burned last night about 3 o'clock. The loss is between $25,000 and $30,000. A large amount of lumber owned by W. F. Stewart and Company

> of Galveston also was burned. The fire caused quite a loss to the town as a large number of men were thrown out of work...

The A. K. Norris mill was never rebuilt.

THE LUTCHER AND MOORE LUMBER COMPANY

In April, 1877, only three months after the owners' earlier visit to East Texas, previously mentioned, H. J. Lutcher and G. Bedell Moore of Williamsport, Pennsylvania, were back in Orange, erecting their large Star and Crescent sawmill on the Sabine River. An editor observed during that month that: "Lutcher and Moore. . .are putting up the largest sawmill in the state, capable of turning out 100,000 feet of lumber every 24 hours."[65] Four months later, the same newspaper recorded that the sawmillers "are putting up the finest steam sawmill on the river, if not in the Southwest, which is now about ready for business and will have a capacity for turning out about 100,000 feet of lumber a day."[66]

In October, 1877, when the Star and Crescent mill's picket machine was placed in production, a correspondent made the following comments:[67]

> ...Tuesday we saw Mr. Lutcher's picket machine at work and are forced to the conclusion that it is the best thing extant. It is very simple in construction, but strong and durable, and turns out a splendid picket, all exactly the same length, and cut at the rate of about 6,000 per day. The pickets will be placed upon the market at $12 per thousand...

The 22-gang saw at the Star and Crescent mill was one of the sawmilling marvels of the 1870's, which was long before single or double-cutting bandsaws came into general usage. As soon as a 36-inch log, surfaced twelve inches apart on two sides, was fed into the gang saw, it dropped 22 size one-inch by twelve-inch boards in one minute's time.

Like the Norris mill, Lutcher and Moore also manufactured windows and doors. The 1880 Products of Industry census reveals the following mill statistics:[68]

> ...Lutcher and Moore's Star and Crescent Mill, Orange, Texas. Capitalization: $60,000; employees: 60 men throughout year; daily work hours: 11 1/2 summer and winter; daily wages: skilled, $3.50, unskilled, $1.50; annual wages paid: $35,000; months in operation: 10, idle two months; equipment: one 22-gang saw, 1 circular saw, 2

> boilers, one 250-horsepower steam engine; raw materials and value: saw logs worth $75,000, mill supplies worth $5,000; products: 15,000,000 BF lumber, 7,500,000 wood lathes; value all products: $150,000; origin of logs: Sabine River-mill did all of its own logging. Star and Crescent mill reworked 40,000 feet of its lumber at its sash and door factory, which employed 15 men.

The steam engine had a unique history of its own, which will be related later. Lutcher located the steam engine at the Levingston Shipyard, where in 1865 the upright marine engine had been removed from a Confederate gunboat before it was scuttled in the Sabine River.

Immediately upon their arrival at Orange, Lutcher and Moore made phenomenal purchases of timberlands in East Texas as well as in Calcasieu and Beauregard Parishes in Louisiana. While the writer has no exact information on their holdings there, they must have approached 500 square miles (260,000 acres) or more. As of 1905, the two Lutcher-Moore sawmills at Orange were sawing 300,000 feet of stumpage daily, equal to the cut of thirty acres of virgin timber. This amounted to fifteen square miles a year, and in the 1905 history of Lutcher-Moore Lumber Company, they quoted a stumpage reserve of forty years, that is, 600 square miles.[69] The company also acquired huge acreages of virgin cypress forests along the Mississippi River.

In March, 1885, Lutcher and Moore founded the Louisiana tram road that would supply their lumber needs for the next 45 years, as follows:[70]

> ...The building of a log railroad from Niblett's Bluff northeast, a distance of several miles, is another large enterprise on foot by the wide awake lumber dealers, Lutcher and Moore. Niblett's Bluff is sixteen miles north of here, and the road will extend northeast. From this bluff, logs will be towed down, abandoning the necessity for crowded booms from one rise of the river to another.....The above firm are extensive land owners along the proposed line, and the road will be extended as the timber supply is exhausted. Material for constructing the Gulf, Sabine and Red River Railroad is now in transit...

A few days later, the first locomotive, rails, and tram cars left Orange aboard a steamboat en route to Niblett's Bluff.[71]

After 1893, Lutcher-Moore soon tired of operating the tram and logging operations, and they contracted with private timber companies thereafter to log the mill. The first contractor, the (J. C.) Arbogast and (J. E.) Craddock Timber Company, made their logging camp headquarters at Stark, Louisiana, and under them, the Gulf, Sabine, and Red River Railroad quickly grew into the largest tram road in the Southwest.[72]

By 1905, the Lutcher-Moore mills were being logged by the Sanders-Trotti Tram Company, with its headquarters and logging camps at Stark and Fields, Louisiana. The number of loggers employed by the tram company varied from 250 to 500, and the large number of oxen and mules used in logging operations forced the contractor to engage a steamboat full time, hauling corn, forage, and stores to Niblett's Bluff. As of 1905, the equipment of the Gulf, Sabine and Red River Railroad included 100 miles of main tracks and laterals, ten locomotives, and 151 log cars, making it certainly one of the biggest logging trams in the nation.[73]

Another description of the giant sawmill company appeared in 1887, as follows:[74]

> ...At the extreme southern edge of the town rises the grand pile of machinery, consisting of double circulars, two sets of gang saws, lathe mills, picket saws, planers, electric light machinery, and repair shops belonging to Messrs. Lutcher and Moore, with a mill of their own cutting 75,000 feet per day. This mill takes the cut of several mills on the East Texas Railway...and sells in Mexico, Texas, Kansas, and the other states. Mr. Lutcher...is ably assisted by Mr. Emmet Buehler, his efficient and obliging foreman. Mr. Frank J. Drick...is at the head of accounts...

A year later, another Galveston correspondent visited the Lutcher-Moore sawmill and reported that:[75]

> ...they actually shipped 54,482,000 feet during the year 1887 and are completing a contract for 40,000,000, the largest order ever placed in the South...One of their (dry) kilns will be the largest in the world. They carry more manufactured lumber than any firm this side of Chicago...

The same article noted that in 1888 Lutcher and Moore were utilizing three locomotives, 175 loggers, and 80 tram cars along the their 20 miles of tram road in order to dump 250,000 feet of stumpage in the Sabine River daily. Another 200 worked in the Orange mills, and the payroll for the mill and tram footed up to $13,500 monthly.[76] An article of 1890 observed that the lumber firm shipped 78,000,000 feet in 1889, again indicating how much lumber they were buying from the sawmills on the East Texas

Railroad.[77] By 1893 the Lutcher-Moore sawmill capacity had increased to 125,000 feet daily or about 3,000,000 feet monthly.[78]

In 1890 the Lutcher-Moore Lumber Company was officially organized and incorporated. In 1901, G. Bedell Moore, having long resided in and acquired valuable investments in San Antonio, sold out his interests to others in the firm. By 1905, the sawmill firm's officers were H. J. Lutcher, president; Dr. E. W. Brown, vice president; W.H. Stark, secretary and treasurer, and F. H. Farwell and John Dibert, general sales agents. Stark was the active manager and superintendent of all mill operations. Stark and Brown were married to Lutcher's two daughters, respectively, Miriam and Carrie.

In order to avert buying up the output of the other mills, Lutcher decided to buy another large mill at Orange, each the equal of the other, and one to be known as the "Upper Mill" and the other as the "Lower Mill." In January, 1902, the company bought the L. Miller Lumber and Shingle Company, remodeled it, and that plant served as the Lutcher-Moore "Lower Mill" until it burned on August 1, 1915. The Miller mill purchased by Lutcher contained a double circular and band sawmill, with a large gang saw. It was on that site that the new "Lower Mill" was built with a much larger capacity.

A 1916 history of the firm reveals that Lutcher and Moore Lumber Company, although it began sawmilling at Orange in 1877, was officially organized and incorporated in Texas in 1891, as stated above. Each mill cut 150,000 feet daily, and each possessed one band saw and one double-circular. Each mill had 10.8 million feet of lumber on the yard, and each mill shipped 10.2 million feet in September, 1905.

Other company personnel in the main office in 1905 included D. A. Ford, bookkeeper-timekeeper; A. L. Dees, assistant bookkeeper; Harry Ortmeyer, invoice clerk; and T. E. Johnson, stenographer.

Personnel of the "Upper Mill" included S. T. Edwards, mill foreman; Ed Cox, planer foreman; A. E. Beattie, dry kilns foreman; G. D. Jones, yard foreman; Allen Miller and Montie Percy, sawyers; A. I. Stevenson, river foreman; and J. F. Carrol, scaler. The planer mill consisted of a 275-horsepower marine engine, four matchers, two moulders, one sizer, one double-matcher and floorer, two edgers, and one picket machine.

Other engines at the "Upper Mill" included one 250-horsepower engine which ran the bandsaw; one 300-horsepower engine which ran the circular and rest of the mill; and one 50-horsepower which ran the dynamo. The "Upper Mill" also had 42-foot boilers of 800-horsepower capacity, dry kilns of 60,000 feet capacity, and employed 190 men.[78a]

The old marine engine had at first been installed in the old Trinity River cotton steamer *Josiah H. Bell* in 1855. In 1860, the *Bell* was hauling rails and crossties for the Texas and New Orleans Railway on the Neches and Sabine Rivers, and in 1862, the *Bell* became a Confederate cottonclad gunboat, which mounted a single 64-pound rifled cannon manned by Lt. Dick Dowling and his company. In 1865, the engine of the *Bell* was removed at Levingston's Shipyard, and the *Josiah Bell* was scuttled four miles south of Orange to prevent its capture by the Federals.[79] As of 1905, the proud old engine had been in continuous use without shearing so much as a pin.[80]

Personnel of the "Lower Mill" included J. B. Channing, mill foreman; Lewis Smihil, planer foreman; L. M. Pond, shipping clerk; G. W. Curtis, dry kilns foreman; E. R. Maxwell, scaler; and T. W. Freight and W. Burgess, sawyers. The "Lower Mill" had one 400-horsepower engine and boilers. Both mills had elevated water tanks and sprinkler systems throughout. The "Lower Mill" employed 170 men. In 1905 the mills were working ten hour days, and payday was every Saturday. Lutcher-Moore Lumber Company paid out $22,000 in wages every month.[81]

After G. B. Moore's removal from the company and the death of H. J. Lutcher in 1912, active management of Lutcher and Moore Lumber Company passed entirely to W. H. Stark, the company's secretary-treasurer. By 1902, lumber sales at the vast Lutcher-Moore industries was in the hands of an able young manager, Frederick H. Farwell, general sales agent. As Stark grew older and chose to remain less active in the company, Farwell was promoted to secretary-treasurer. In 1920 he became vice presidrent and general manager, and in 1946, sixteen years after all Lutcher and Moore timber-processing at Orange had ceased, Farwell became president.[82]

Several years earlier, Lutcher and Moore Lumber Company had purchased 60,000 acres of cypress timberlands at Lutcher, Louisiana, in the Mississippi River bottoms north of New Orleans, and at Donner, Louisiana, west of New Orleans. The firm operated two cypress sawmills in those towns, under the corporate lnames of L and M Cypress Lumber Company at Lutcher, as well as (John) Dibert, Stark and Brown at Donner, Louisiana. The ultimate value of their Louisiana timber holdings would eventually exceed $2,000,000. Other companies Lutcher and Moore would either own or control included Red Cypress Door and Sash Company, the Orange Mercantile Company, the Orange Ice, Light, and

Water Company, the L and M Turpentine Company; also an interest in the Yellow Pine Paper Mill Company; two railroads, an export line, and a fleet of lumber schooners. [83]

After 1920, logging on the Sabine River dropped off sharply, but it was the Great Depression's stifling of lumber demand in 1930 which prompted Lutcher and Moore to silence the band saws and close their two large mills after 53 years of continuous lumber processing in Texas. Earlier the company had executed an agreement with the Wier Long Leaf Lumber Company of Wiergate to log Lutcher's huge timber reserve of nearly 120,000 acres in north Newton County, but Lutcher and Moore were not actively engaged in manufacturing lumber there. For many years after 1930, the Lutcher and Moore Lumber Company existed entirely as a real estate, mineral rights for leasing, or timber sales firm, but eventually all of its assets were liquidated. The Lutcher, Stark and Brown families have bequeathed an abundance of philanthropic gifts to the City of Orange.[84]

THE RUSSELL AND SON SHINGLE MILL

In 1866, Robert B. Russell of Orange purchased the old Robert Jackson sawmill, which was powered by the old marine steam engine apparently removed from the ill-fated steamboat, *Rufus Putnam*, when the latter sank at Eaves' Plantation 1n 1841. According to one source, Russell soon converted the sawmill to cut 25,000 feet of lumber and 125,000 shingles daily.[85]

A biographical sketch of Robert B. Russell is imperative at this point. He was born in Utica, New York, in 1817, but at an early age, he moved to San Augustine County, from which point he enlisted in the Texas Army and fought at the Battle of San Jacinto. The only San Jacinto veteran to reside at Orange, he is buried in Evergreen Cemetery, where in 1973 the author applied for and received a state marker (paid for by his two grandsons) from the State of Texas that now rests on Russell's grave; the author paid for its installation.[86]

Russell moved to Orange in 1854, where he ran a hotel and also served as the Confederate postmster until 1865. In 1877, after R. B. Russell had been manufacturing cypress shingles for eleven years, a Galveston editor reported that: "....Under its present management, Russell and Sons has sawed over 50,00,000 shingles."[87]

Despite the statement in the *Southern Industrial and Lumber Review* (Fn.85) about the Russell mill's cutting capacity, the mill discontinued manufacturing lumber immediately and its daily cut of shingles loafed along at about 75,000 throughout the 1870's. Several statements in a Galveston newspaper observed that:

> ...Russell and Sons have 4,500,000 shingles on hand... Russell and Sons mill is averaging 70,000 shingles daily.[88]... The shingle mill of R. B. Russell and Son started up again Monday and is now under full headway, turning out 75,000 shingles a day.[89]...Russell and Son's mill is now averaging 400,000 shingles a week. They made a sale of a half-million to one party last week.[90]...The foundation and framework of Russell and Son's mill is finished, and the weather-boarding is now being put on...The mill of R. B. Russell and Son, which stopped running nearly two months ago in order to have a new foundation put under it and to rebuild the whole framework, started up yesterday.[91]...

The following statistics for the R. B. Russell mill were recorded in the Schedule V, Products of Industry, census for the year ending July 1, 1880, as follows:[92]

> ...R. B. Russell and Son Shingle Mill, Orange, Texas. Capitalization-$20,000; employees-maximum, 50; minimum, 45; daily work hours-9 1/2 winter, 11 in summer; daily wages-skilled, $5.00, unskilled, $1.50; annual wages paid-$8,000; months in operation-5, idle for repairs, 7 months; equipment-one 4-gang saw, one circular saw, two boilers, two 50-horsepower steam engines; raw material and value-cypress logs worth $12,000, mill supplies worth $2000; product-10,000,000 shingles; product value-$25,000; origin of logs-Sabine River and tributaries. Mill did no logging, but shipped some shingles on its own schooners.

The life of the Texas hero, Robert B. Russell, soon came to a sad end late in November, 1880, when a Beaumont editor observed that:[93]

> ...Death of A San Jacinto Veteran—Our readers will be distressed to learn of the sad death of Mr. R. B. Russell, one of the best and most prominent citizens of Orange. One day last week, while looking at one of the (box) cars, which had been loaded with shingles at his mill, some of his men, not noticing him, ran another car down the track, and Mr. Russell was caught between the bumpers and received

> injuries from which he died last Tuesday morning...

Although no other mention of Russell's schooner fleet has been noticed elsewhere, except in the 1880 census, it's obvious that the company owned its own schooners for transporting shingles to Galveston and elsewhere. Generally, the 1880's were years when lumber and shingles exceeded supply, and the shingle firm, redesignated Russell Brothers, prospered greatly. It is obvious, too, that sometime after 1880, Russell Brothers' mill was overhauled and new and improved shingle machines installed, for in March, 1889, their daily shingle cutting capacity was reported as being 125,000.[94] In June, 1890, the Russell shingle mill burned to the ground and was never rebuilt, probably because the Sabine River cypress timber supply was almost exhausted, having already caused other shingle makers to cease operations.

In July, 1891, an unnamed Orange historian wrote that Russell Brothers shingle mill was "still in operation as a cypress shingle mill."[95] The editor of Galveston *Weekly News*, however, quickly took the historian to task, observing that:[96]

> ...this historian who writes from Orange cannot be safely credited as he states that the shingle mill of R. B. Russell "is still in operation as a cypress shingle mill." As a matter of fact, the Russell shingle mill was burned down several months ago and has never been rebuilt. The mill site was sold last week to Messrs. W. H. Stark and Dr. (E. W.) Brown, sons-in-law of Mr. H. J. Lutcher, for $5,000...

THE JOSIAH JORDAN SHINGLE MILL

Another of the early shingle mills of Orange was that of Josiah Jordan, who built his Excelsior Mill on the Sabine River in 1876, adjacent to the D. R. Wingate Lumber Company sawmill. Jordan used three single block machines to cut an average of 90,000 shingles daily.[97] Jordan named his shingles the Excelsior brand; hence his mill became known as the Excelsior Mill. One news account of 1878 reported that: "...the Excelsior Mill, with its new boilers and machinery, will start up perhaps the last of next week, adding as much as 60,000 daily to the above figures."[98] In 1879, the mill statistics of the J. Jordan Shingle Mill, recorded in the Schedule V, Products of Industry census, of 1880 were as follows:[99]

> ...J. Jordan Shingle Mill, Orange, Texas. Capitalization-

> $15,000; employees-maximum, 35; average, 30; daily work hours-10 summer and winter; daily wages paid-skilled $3.00, unskilled $1.50; annual wages paid-$8,000; months in operation-11, idle one month; equipment-one 5-gang saw, one circular saw, two boilers, one 50-horsepower steam engine; raw materials and value-logs worth $12,000, mill supplies worth $200; product-10,000,000 shingles; product value-$25,000; origin of logs-Sabine River and tributaries-mill did no logging of its own.

According to *Southern Industrial and Lumber Review* of 1902, Jordan closed down his shingle mill in the summer of 1889 because the price of sawed shingles had dropped at the very moment when cypress logs became both scarce and expensive. His shingle mill was probably closed even earlier because in a list of the Orange County shingle mills published in March, 1889, the Josiah Jordan shingle mill was no longer listed.[100]

THE THOMAS BANCROFT AND SONS MILLS

A lumber journal of 1902 wrote that one of the earliest of the Orange County sawmills became a reality when:[101]

> ...Thomas Bancroft and Sons built a mill here (Orange) in 1877 and made lumber and shingles. The mill had a capacity of 50,000 feet of lumber and 90,000 shingles (daily). They were the first to introduce dry kilns to this place...

Fortunately, George Bancroft wrote a 21-page typescript about 1920, outlining his family's many activities while sawmilling at Orange, and his memoirs, along with the writer's research, verify that while the Bancroft mills made both lumber and shingles during their first years at Orange, the company did not manufacture both lumber and shingles at the same time. Bancroft wrote that:[102]

> ...In 1873, my father and two brothers formed the partnership of T. Bancroft and Sons, and with their earnings, bought from a Mr. Woodell (sic: Burton and Woodill), a sawmill located on the Louisiana side of the Sabine River...and installed in this mill the first flooring and moulding machines brought to Orange...
>
> ...At that time the Sims Mill was operating on the Louisiana side of the river at the entrance to Phoenix Lake,[103] and the

> Woods mill (Woods and Gilmer) was operating on Conway's Bayou (south of Orange); and on the Texas side of the Sabine River, the (R. B.) Russell mill (was) at Orange, and the (W. B.) Black Mill (was) at the mouth of Adams Bayou. Transportation being entirely by water, boats would take cargoes from the mills on the Louisiana side...for delivery to ports on the Gulf Coast...
> ...Cypress was the only timber being worked in the mills, and this mill turned out cypress flooring, moulding...at the rate of 15 or 20,000 feet per day...(On) Sunday, March 26, 1877, the sawmill of T. Bancroft and Sons was destroyed by fire...

George Bancroft added that since outlaws were hiding from officers in a nearby marsh, he believed the outlaws set fire to the mill to create a diversion. Obviously, the Bancrofts manufactured only lumber in their first mill that burned, and their second mill was only a shingle mill. Bancroft noted that Henry Thompson, a wealthy Orange merchant and wholesale shingle dealer, offered to loan the Bancrofts money to rebuild with; he would also furnish all cypress logs and buy the mill's total output, provided the owners would build adjacent to the railroad on the Texas side of the river. Bancroft added as well that:[104]

> ...This shingle mill was built in 1877, and equipped with a circular saw rig to cant (saw) timber to size for shingle bolts...The bolts went up a belt elevator to the second floor...where the shingle machines were located. The shingles dropped from the machines onto sorting tables...and passed into bins to be packed into "bunches" (bundles)...This shingle mill had a capacity of 75,000 shingles per day...

On the initial runs of the new shingle mill, "Bancroft and Sons" were "making between 35,000 and 40,000 shingles per day..."[105] Apparently, the earlier statement about the Bancroft dry kiln was correct as well. In March, 1879, another editor observed that:[106]

> ...Bancroft and Sons' mill is getting under full headway. It turned out 76,000 shingles last Saturday...The steam dryer (kiln), which is capable of holding about 1,300,000 feet, is a new feature of that mill. The steam heat so far reaches about 150 degrees...

The 1880 Orange County Schedule IV census, Products of Industry, provided the following information about the Bancroft Shingle Mill:[107]

> ...Bancroft and Sons Shingle Mill, Orange, Texas. Capitalization: $15,000; employees: maximum, 35; average, 30 men; 3 boys under 16 (as shingle bundlers); work hours: 9 and a half winter; 10 and a half, summer; daily wages: skilled, $4.00; unskilled, $1.50; annual wages paid: $11,000; months in operation: 8, idle for 4; equipment: two 4-gang saws, 1 circular saw, 4 boilers, two 75-horsepower steam engines; raw materials and value: cypress logs worth $18,000; product: 14,000,000 shingles; product value: $35,000: origin of logs: Sabine River—mill did no logging.

By the early and middle 1880's, cypress logs were often becoming scarce, and often the mill had to be shut down until a river freshet and overflow allowed logs to be floated out of the swamps. Fortunately about 1883, the Texas and New Orleans Railroad offered Bancroft a lucrative contract. The railroad would furnish the Bancroft mill with cypress logs from railroad-owned lands to be cut into crossties by the shingle mill's circular saw; also they offered to pay the mill 20 cents for each solid heart crosstie the mill cut, and all the logs not containing wood of solid heart would be given to the mill to be cut into shingles.

In 1885, T. Bancroft and Sons discerned that they would soon run out of cypress timber for shingles anyway, and the owners decided to build a new sawmill to cut pine lumber. With $30,000 in cash in the bank, the sawmiller bought a new double-circular mill from the Filer and Stowell Company of Milwaukee; also a steam log turner, a 40-foot log carriage, with steam "shotgun feed;" an automatic saw sharpener; timber sizer; edgers and cut-off saws; and a complete new planing mill, with moulders and matching machines and a resaw. By February, 1886, the new sawmill was in operation, but lacking a dry kiln at first, rough lumber was stacked on the yard to season.[108]

By March, 1889, the Bancroft shingle mill was cutting 75,000 shingles daily, whereas the new Bancroft sawmill was cutting 75,000 feet of pine lumber each day.[109] At first, the Bancroft sawmill was logged by the Sabine Tram Company of Newton County, but eventually, E. W. Bancroft and L. T. Grubbs organized the "Tram Arthur Logging Company," with intent to supply logs to the mill from the new Bancroft pine timberlands in Louisiana.

On April 14, 1892, the stacked lumber on the yards of Bancroft's sawmill caught fire and burned for 48 hours, consuming 4,600,000 feet of lumber, worth $35,000, of which $17,000 was covered by insurance. The sawmill, planer, and 1,000,000 feet of lumber in dry sheds were saved. Two or three blocks of nearby dwellings also burned. The Bancroft blaze was one of Orange's most difficult fires ever to combat, and hundreds of volunteers, including a train load of men and equipment from Beaumont, fought the fire.[110]

In September, 1896, Thomas Bancroft died, and the Bancroft mills passed to his sons. In January, 1897, the Bancroft Lumber Company was incorporated, with a capital stock of $100,000, consisting of 1,000 shares with a par value of $100 each. Each brother was to receive 200 shares, and the new officers became A. J. Bancroft, president; E. W. Bancroft, vice president; and George Bancroft, secretary-treasurer. George Bancroft also acted as mill superintendent.[111]

In August, 1899, the brothers bought a 13,800-acre tract of prime long leaf pineries in north Calcasieu Parish, Louisiana, from Gay, Wasey and Company of Chicago for $12.50 an acre. The land contained an estimated 160,000,000 of mature (12 inch and larger) stumpage, sufficient to supply the mill for about ten years. The purchase also created an immediate need for rails and rolling stock. Bancroft contracted to buy a 28-ton locomotive, twenty logging cars, and six miles of steel rails from the J. I. Campbell Lumber Company, and agreed to pay for the purchases in dressed lumber shipped to Campbell's many retail lumber outlets in central Texas.

Hence, the Bancroft Lumber Company was soon staggering under a mountain of debt. The brothers borrowed $50,000 from a Beaumont bank. After paying a down payment of $43,612 to Gay, Wasey and Company, they also issued three annual notes for the same amount. Hence, they were soon committed to pay off $180,000 in notes in three years, as well as a percentage of their daily lumber cut to Campbell Lumber Company to pay for the rolling stock. To raise money to meet obligations, the brothers sold the 400 shares of uncommitted stock in Bancroft Lumber Company for $40,000 to George C. Vaughan Lumber Company of San Antonio. J. W. Link, an Orange attorney, negotiated the transaction with the Chicago firm.[112]

Link was soon to negotiate another transaction for the Bancroft firm. In the spring of 1901, the Kirby Lumber Corporation authorized Link to negotiate with Bancroft Lumber Company toward the purchase of all Bancroft company stock, real estate, and equipment. Both Bancroft and Vaughan Lumber Company agreed to accept $400 for each share of stock.

Since Link's fee of $10,000 for the transaction had to be paid out of the sale price, the old owners were paid a total of $410,000 for nearly all of the Bancroft assets, and as of September, 1901, Bancroft agreed to start sawing logs for Kirby, even though the company still owned much lumber on the yard that was committed to its former customers. A part of the agreement was that George Bancroft remain on his job as plant superintendent of "Mill D" at Orange (the new designation given to the old Bancroft mill in Kirby's "alphabet soup" nomenclature of its sawmills) at a salary of $250 a month.[113]

KIRBY LUMBER COMPANY'S "MILL D" AT ORANGE

Long before Kirby Lumber Company purchased its new "Mill D" at Orange, the old owners built "two big lumber-carrying barges" and purchased the steam tug *Margaret*. The barges were used to haul lumber for the Mexican export trade to Sabine Pass, where the lumber was loaded aboard ocean-going schooners, and to supply freight and animal fodder to the company-owned commissary and log camp at Whitman's Bluff, 100 miles north of Orange, where company-owned logs were skidded into the Sabine River. One thing that certainly made the Bancroft mill attractive to Kirby was its large export trade which, along with large railroad orders for crossties and timbers, had saved many an East Texas sawmill from bankruptcy during the economically-turbulent 1890's.[114]

In 1900, the Bancroft mill was rated at 90,000 feet daily. During 268 days the mill operated in 1900, it manufactured 23 million feet of lumber, averaging 85,800 feet for each work day. For the first year that Kirby operated the sawmill, George Bancroft raised the mill's output to more than 26 million feet in 345 operating days. The count must have included some night shifts since Sunday mill operation ordinarily was seldom conducted in East Texas. Bancroft resigned in October, 1902 (although he remained on the job a few weeks longer), and noted that Kirby had refused to increase his $250 monthly salary.[115] However, Bancroft had organized and become president of the Orange National Bank in 1901, and he probably wanted to work full-time as a banker.[116]

In 1901, the Bancroft mill had no dry kilns, but it had developed an elaborate system of air-drying. A system of rails carried the lumber dollies from the sawmill to the yard and from the yard to the planer. In the yard, "stationary bins" ten feet wide were built for air-drying the lumber, over which roofs were built to keep out the rain. The date that each "bin" was stacked full of lumber was then recorded, and after four or five months, each "bin" was emptied on schedule and sent to the planing mill for dressing.[117] By 1904, a number of "Porcupine" brick dry kilns had been added by Kirby with a daily capacity for steam-drying 30,000 feet.[118]

No company commissary was maintaind at Orange by either Bancroft or Kirby because a number of mercantile houses were located nearby. Hence, despite three major sources of information about the Bancroft or Kirby mill, no explanation was offered about how either company paid off its employees in an age when "mill checks," good only at the company store, was the norm in East Texas. In 1902, Kirby employed 85 men at "Mill D," and E. W. Bancroft, who continued to log "Mill D" until Kirby bought him out, worked fifty loggers on the 13,800-acre pinery at Whitman's Bluff. Very quickly, Kirby began logging "Mill D" over the Orange and Northwestern Railroad from Buna as well, in case the Sabine River water level fell too low to permit logging from Whitman's Bluff.[119]

As of 1904, Kirby had made no significant changes in mill machinery, except adding the dry kilns. The saws still consisted of a double-circular rig, edgers, trimmers, slashers and cut-off saws, but no band or gang saws had been added. The planing mill consisted on one Hoyt sizer, one Hoyt matcher, one Holms matcher, one Houston moulder, one resaw, one rip saw, and one picket machine. In March, 1904, a Santa Fe sawmill circular credited "Mill D" with a daily output of 150,000 feet and with one sawmill and two planers. However, the writer considers that source to be in error since Kirby's inventory of a month earlier does not verify it.[120]

It is unclear exactly why Kirby's "Mill D" failed so miserably to produce a profit after October, 1902, except that the company obviously suffered greatly after the loss of George Bancroft as superintendent. After Bancroft left, Kirby only owned the company for the next 33 months before selling out to Miller-Link Lumber Company on July 1, 1905.[121] In February, 1904, Kirby carried the "Mill D" saw and planing mills on its inventory at $90,569 and its dry kilns at $13,154, for a total of $103,723. However, Kirby Lumber Company sold its Orange plant to Miller-Link for only $60,000. One record of 1902 noted that George Bancroft was "the most successful (manager) in the bunch of mills owned by Kirby Lumber Company." Bancroft noted as well in his "Retrospective" that after he resigned, Kirby tried "several mill managers at their sawmill in Orange," inferring that Kirby replaced its mill superintendent there at the rate of one each year. Perhaps it would have been more profitable for Kirby Lumber Company had it given George Bancroft the raise that he sought.[122]

THE LEOPOLD MILLER LUMBER INTERESTS

Perhaps the reader, having already been introduced to J. W. Link, should become better acquainted with Leopold Miller before the writer continues with Miller-Link Lumber Company. Leopold Miller was an even more successful "Horatio Alger" immigrant-figure than Alexander Gilmer was because Miller had suffered no fire losses or ships sunk. Miller came to Orange from Germany in 1881 and quickly became successful as a commission merchant. He soon invested his profits by buying large tracts of timberlands in Jasper, Newton, and north Orange counties. In 1883, Sheriff G. W. Michael built a large shingle mill, which he sold out to Leopold Miller in 1887. One source noted that Miller:[123]

> ...In 1888 organized the L. Miller Lumber and Shingle Company...For the first few years, the mill was kept sawing shingles...The company erected a large lumber mill alongside the shingle mill, and put in new boilers, a double circular and a band mill, gang edgers (gang saw or "bull edger"), trimmers, steam...(racial slur-correct name is log turner), flippers, shotgun feed...This mill made a daily cut of 100,000 feet and...The Lutcher and Moore Lumber Company bought out the L. Miller Lumber and Shingle Company in December, 1901...
>
> ...Miller came here in 1881, opened a big department store, and made money fast...In 1887 he bought the shingle mill...and soon ran the daily cut from 75,000...to 250,000...He sold his mill...to the Lutcher and Moore Lumber Company for nearly a half million dollallrs. He still conducts a large mercantile business and has the trade of four large mills...He is president of the Orange and Northwestern Railroad, and vice president of First National Bank...

A New Orlans editor reported on Leopold Miller in a March, 1889, article, observing that:[124]

> ...It is said that L. Miller has the latest stock of general goods in the city. His last year's business was $125,000. Mr. Miller is also...one of the partners of Orange Shingle Company and is also interested in the large furniture house of L. Miller and Company...

The same article added that in 1889 the Orange Shingle Company was the largest shingle mill in Orange with a daily cut of 135,000.

Whereas for years, Miller had been limited to logs floated down the Sabine River to his mill, he dreamed of a railroad that would connect his mill with his large timber holdings in three counties. He soon teamed up in February, 1901, with W. H. Stark, vice president of the Lutcher and Moore Lumber Company, as the chief investors of the Orange and Northwestern Railroad from Orange to Buna (later extended to Newton). However, the same month that the last rails reach Buna, December, 1901, Miller sold out his lumber interests to his partner and others at Lutcher and Moore Lumber Company, which soon became the latter's "Lower Mill" until it burned down on August 1, 1915. After 1901, Miller served for years as president of the Orange and Northwestern, but with his expanse of timberlands still intact, he was only biding his time until he and J. W. Link could organize the Miller-Link Lumber Company in 1905 and buy out the Kirby Lumber Company's "Mill D."[125]

THE MILLER-LINK LUMBER COMPANY

Under Leopold Miller and J. W. Link, who bought "Mill D' in July, 1905, the old Kirby mill seemed to blossom in a way that it never did while Kirby Lumber Corporation owned it. By 1905, J. W. Link became actively caught up in sawmilling, and he certainly guided the new company's financial progress, even if he lacked the mechanical skills of many sawmillers. Although two histories of Miller-Link Lumber Company survive from 1910 and 1916, little space in them is devoted to the mechanical improvements at the mill. In 1909, however, the mill underwent a complete rebuilding to increase its cutting capacity to 125,000 feet daily. Since Miller was one of the earliest band sawmillers at Orange, the writer believes that the owners replaced the double-circular rig with a single-cutting band saw and perhaps a 36-gang saw. The machinery replacements occurred in May, 1909, at the time that Miller-Link added a No. 77 American Woodworking matcher, a high speed machine for dressing flooring, beaded ceiling, centermatch, shiplap, and siding. The machine produced the company's "Faultless" brand of products as follows:[126]

> ...A specialty is made of "Faultless" flooring, (beaded) ceiling, siding, and finish. Since the adoption of this trademark as applied to the high class of material shipped under this name ..."Faultless" is the product of a modern, high speed machine installed in this plant a little more than a year ago (1909)...

It became obvious that logging the Miller-Link sawmill was a trade-off with Kirby Lumber Company. Kirby logged its Call sawmill around 1906 from some of the L. Miller timberlands in Newton County, whereas Kirby provided the Miller-Link mill at Orange with logs skidded into the Sabine River at Whitman's Bluff.[126] Beginning in 1905, the Miller-Link sawmill at Orange also received 24 car loads of logs daily from the C. E. Slade logging camp at Quigley, Jasper County, on the Orange and Northwestern Railroad.[127] Miller-Link log trains on the Orange and Northwestern traveled at passenger train frequency and scheduling. Logging ended at Quigley in 1909, and it was soon transferred to Bunker Hill, Jasper County, also on the Orange and Northwestern, thirty miles north of Orange. At Bunker Hill, the Miller-Link loggers utilized both a steam log skidder and a steam log loader to facilitate the speed of handling, and their logging spurs crowfooted into the company's pineries on both sides of the railroad. At the height of logging activities at Quigley before 1909, loggers and their families numbered about 150 persons, and probably a like number lived at Bunker Hill.[128]

On September 1, 1908, Miller-Link tried to expand into Newton County, and the company bought the 50,000-foot Newton Sawmill Company plant at Newton. However, the year 1909 witnessed lumber overproduction and plummeting prices to the point that Miller had to close down the Newton plant in 1909 in order to keep the Orange plant open. A most grisly death at Newton in February, 1909, witnessed a young Orange man, Walter Bradford, being drawn into the sawmill machinery and was ground to bits; but since sawmill deaths occurred daily in that age, it probably had nothing to do with the mill closing. It is unclear to the writer exactly what happened to the Newton sawmill, but probably it was either sold or dismantled.[129]

Early in 1910, J. W. Link sold out his interests in the company to L. Miller, and Link moved his family to Houston. There was then considerable reorganization in the lumber company. Miller brought in Oscar S. Tam of Alabama as vice president and general manager over all mill operations. Two of Miller's sons were also active officers of Miller-Link Lumber Company. Joe Miller served as vice president and assistant general manager, and Maurice Miller was secretary- treasurer and purchasing agent for the company.[130]

Less is known about Miller-Link during the years after 1910, but it appeared the sawmill continued to prosper at least until the end of World War I. The *Directory of American Sawmills* recorded that in 1915, Miller-Link Lumber Company was still sawing 125,000 feet of yellow pine lumber daily. Some of its products were both for export and for railroad timbers. The company's sawmill and planer could saw and dress timbers up to 42 feet long, whereas the maximum length it could kiln-dry was thirty feet.[131]

Another history of the Miller-Link Lumber Company of 1916 indicated that nothing had changed drastically at the sawmill, which still was cutting 125,000 feet daily. Its logs were still coming from Jasper and Newton counties and from Calcasieu and Beauregard parishes in Louisiana. The Miller-Link mill owned seven miles of Sabine River frontage and river log booms, giving it ample space to store its logs, even to build up a surplus. Company officers had remained the same since 1910, and in 1916, Leopold Miller was still president of the Orange and Northwestern, the Orange Grocery Company, the Orange Light and Ice Company, and the Orange Rice Mill Company, and he was vice president of the First National Bank and the Yellow Pine Paper Mill Company.[132]

The advent of World War I was to add unprecedented demand for lumber in order to build wooden ships at the Orange shipyards. Some of the ships were as large as 318 feet long, weighed up to 4,800 tons, and consumed as much as 1.5 million feet of lumber. Once again, Miller-Link moved outside of Orange to purchase a sawmill and again disaster resulted. In 1917 the company bought the R. W. Weir Lumber Company sawmill at Texla, a 75,000-foot mill that employed 200 mill hands and loggers, and a few months later the mill burned in 1918. Although Miller-Link soon rebuilt it with another double-circular mill, the owners soon sold out to the Peavy-Moore Lumber Company of Deweyville.[133]

The history of Miller-Link Lumber Company after 1918 is even more unclear. Since the huge Orange Lumber Company sawmill at Orange also burned in 1918, Miller-Link Lumber Company sawmill was actually one of the last three or four great sawmills left in town. Although the years 1918 through 1928 were years of prosperty, George Bancroft noted that Miller-Link experienced financial reverses, year uncertain but probably about 1922, and went bankrupt. Bancroft added that around that date, a hurricane flattened the last remaining stand of Kirby Lumber Company timberland in Calcasieu Parish from which the Miller-Link sawmill was logged. The company was finally auctioned at a bankruptcy sale to Peavy-Moore Lumber Company, and the old Miller-Link sawmill was soon dismantled and moved.[134]

THE ORANGE LUMBER COMPANY

The Orange Lumber Company traced its origin to a combination saw and shingle mill built at Orange in 1876 by McKinnon, Jackson and Downs. Known at first as McKinnon, Jackson and Company, the firm became John McKinnon and Company in 1879 when McKinnon bought out his partners.[135] The first mill is believed to have had a capacity of about 30,000 feet of lumber and 25,000 shingles daily. When the 1880 Products of Industry census for Orange County was made, the following mill statistics were recorded:[136]

> ...John McKinnon and Company, Orange, Texas. Capitalization: $35,000; employees: 25 throughout year; daily work hours: 9 in winter, 11 in summer; daily wages paid: skilled, $3.00, unskilled, $1.50; annual wages paid: $10,000; months in operation: 12; equipment: one 4-gang saw, 2 circular saws, 2 boilers, one 80-horsepower steam engine; raw materials and value: cypress logs worth $31,000, mill supplies worth $500; product: 5,000,000 feet of lumber, 2,000,000 shingles; product value: $50,000; origin of logs: Sabine River swamps—mill did no logging.

About 1888, John McKinnon's health began to fail, and he sold a majority interest in his company to M. T. Jones Lumber Company of Houston. A Galveston editor observed in the same year that:[137]

> ...J. McKinnon and Co. own a fine mill and a planer of a little smaller capacity. Their machinery is modern and their Mr. John McKinnon is an active, energetic millman. They commenced with small mills and gradually acquired their present proportions...

By March, 1889, the name of the McKinnon mill had been changed to Orange Lumber Company. In that year the mill had a capacity of 60,000 feet, but all shingle-making by the company had already ceased.[138] About 1891, the controlling interest of a neighboring sawmill, Wingate and Company, also passed to M. T. Jones Lumber Company, and for a few years, John McKinnon was also active in the management of that firm.[139] Also in 1891, the Orange Lumber Company was converted into a joint stock company. In 1898 M. T. Jones died, and his nephew, Jesse H. Jones, the new head of the parent firm, M. T. Jones Lumber Company, promoted C. F. Pannewitz of Orange as general manager of both the Wingate and Orange Lumber Company sawmills. Under Pannewitz, the Orange Lumber Company shipped 25.4 million feet of lumber in 1901 and 30.2 million feet in 1902. Also major improvements were added to the Orange Lumber sawmill in 1902 when a huge Corliss engine and a gang saw were installed. By 1904 the Orange Lumber Company was reported as cutting 150,000 feet daily, with 225 mill hands on the payroll. But some of Pannewitz' best efforts came to naught as the Wingate sawmill at Orange burned down for the third time in April, 1902, and the Orange Lumber Company's plant burned down on March 28, 1905.[140]

The Orange Lumber Company's property was immediately rebuilt, however, and by September, 1906, the new mill's cutting capacity was reported once more as 100,000 feet daily.[141] Very little is known by the writer about the last years of that sawmill. It did manage to help supply the great demands for lumber at Orange during World War I. *Gateway To Orange* observed that the Orange Lumber Company plant burned for a second time about November, 1918, and since its stumpage reserve was about exhausted, the mill was not rebuilt. And the screech of its mighty band saw was silenced into the dustbin of history.[142]

SAWMILLS AND LOGGING ELSEWHERE IN ORANGE COUNTY

Vidor, Texas, became a principal logging camp of the Miller-Vidor Lumber Company about 1907 after that firm had purchased several thousand acres of prime timberland in western Orange County several years earlier. Officers of the Miller-Vidor firm included Charles H. Moore, chairman; A. W. Miller, president; C. S. Vidor, vice president; and B. I. Sparks, secretary-treasurer. All the men were from Galveston, and Moore was the veteran lumber dealer, who shared ownership with Eberle Swinford in two of the early shingle mills at Orange. The writer believes too that Moore had bought up the huge western Orange County timber reserve when he first came to Orange in 1876. The Miller-Vidor firm bought the United Lumber and Export sawmill at Beaumont in 1905, and quickly organized the Beaumont Sawmill Company to operate it. Miller-Vidor also organized the Orange Sawmill Company at Orange in 1907 and hired C. L. Hannah to manage their sawmill there.[143]

Miller-Vidor established its logging camp at Vidor, built logging tram railroads and spur lines all over western Orange County, and built its skidway for logs on the east bank of the Neches River. The Vidor logging camp employed 175 loggers and shipped 600,000 feet of logs to its mills daily. Logs were either dropped into the Neches River to be floated down to its mill plant near the Mobile Refinery in south Beaumont or logs were shipped on log trains over the Southern Pacific tracks to mills at either Orange or Beaumont. The Miller-Vidor sawmill at Orange (which will be dealt with separately) burned down in August, 1910, and after Miller-Vidor's Beaumont sawmill burned on January 5, 1918, the logging operations at Vidor gradually drew to a close.[144]

Texla, on the Orange and Northwestern Railroad, was first known as Bruce when the Harrell-Votaw Lumber Company built a sawmill there in 1905. According to the *Southern Industrial and Lumber Review* of 1906, the mill had a cutting capacity of 75,000 feet. In 1907, Harrell-Votaw sold out to the R. W. Weir Lumber Company of Houston, who changed the town's name to Texla. The mill employed 200 mill hands and loggers until Weir sold out to Miller-Link Lumber Company in 1917. In 1918 the sawmill burned, but Miller-Link soon replaced it with a double-circular mill of the same size in 1919. About 1920, Miller-Link sold out to the Peavy-Moore Lumber Company of Deweyville, who operated the mill until the timber was exhausted in 1929, the mill was dismantled, and the site was abandoned.[145]

Mauriceville, named for Maurice Miller of Miller-Link, was the junction point of the Orange and Northwestern Railroad and the Kansas City Southern Railroad. In 1949, the town had one sawmill, with a capacity of 50,000 feet daily, and the plant employed 125 mill hands and loggers. The mill was later dismantled.[146]

Terry was a longtime flagstop and rail employee hotel and transfer point, where Southern Pacific train crews exchanged trains as far back as 1885. Named for John W. Terry, a railroad official, a sawmill of the Terry Lumber Company was operated there from 1915 until the timber was cut out in 1917.[147]

Established on the Kansas City Southern Railroad in 1897, one small sawmill town was first known as Baer, then Stanford, and then as Adrian before it became Doty, Texas, in 1905. The Pearsall Lumber Company operated a sawmill there from 1902 until 1907 when the timber there was exhausted and the mill was dismantled. A section house remained at Doty, until 1935, when the rail facility there was abandoned.[148]

Oilla, north of Orange, was laid out as a townsite on the Texas and New

Orleans Railroad in 1913. The Russ Daniel Lumber Company operated a 35,000-foot sawmill there in 1915, which cut yellow pine, cypress, white oak, red oak, gum, and hickory logs. About fifty men worked for the mill, which cut both dimension and timbers, as well as laths, box shooks, crossties, and shingles. The Daniel sawmill is believed to have cut out and been dismantled around 1918, and after some years, the Oilla post office was discontinued as well.[149]

The writer has never located any logging or sawmill information for either Orangefield or Bridge City. In 1909, there was one sawmill in Orange County that the writer cannot connect with any named village. One editor noted that the:[149a]

> ...Alexander-Hardee Lumber Company are reported to have disposed of their mill and timber holdings in Orange County in January. They owned a sawmill and 3,000,000 feet of standing timber on the main line of the T. and N. O. (rail) road several miles west of Orange. The purchasers were C. G. Hooks of Kountze and a Mr. Crosswise of Dearborn, Texas...It is reported that the purchasers have arranged to secure a considerable amount of additional timber that can be hauled to the mill...

Only one other Orange County sawmilling site is known by the writer to have existed outside the city of Orange. Lemonville, was established as a townsite in western Orange County on the Kansas City Southern Railroad in 1897. In 1900 the Lemonville Lumber Company built a 30,000-foot sawmill there, which changed hands two or three times during the next four years.[150] In 1904, the Lemon Lumber Company operated a 60,000-foot sawmill and two planers there.[151] In October, 1905, the Lemon Lumber Company "made a shipment of kiln-dried saps to Port Arthur" to be trans-shipped by water to Europe.[152] It is not known exactly when Alexander Gilmer purchased the Lemonville sawmill, but when Gilmer died in July, 1906, his estate "owned a large stock in the Gilmer Lumber Company at Lemonville, of which he was president." Another source reported that the sawmill capacity there was increased to 125,000 feet daily and employed 200 mill hands.[153] Later, the estate sold the sawmill to the Talbert-Dewey Lumber Company, who operated the mill until 1928, when the timber was exhausted, the mill was dismantled, and the site was abandoned.[154]

THE A. E. SMITH MILLS AND THE MILLER-VIDOR LUMBER COMPANY

In 1884, A. E. Smith built a shingle mill at Orange, and in 1889 his mill was cutting 60,000 shingles daily.[155] As the cypress timber grew ever more scarce, Smith added a sawmill to his shingle mill in 1902, with a daily cut of 50,000. A year or two later, Smith increased the cutting capacity of his sawmill until by 1906, the A. E. Smith sawmill was cutting 125,000 feet daily.[156]

In 1907, the Miller-Vidor Lumber Company of Galveston bought out the A. E. Smith mills and began manufacturing both lumber and shingles at Orange. The owners organized the Orange Sawmilling Company in January, 1907, with A. W. Miller as president; C. S. Vidor, secretary-treasurer; and C. L. Hannah as mill superintendent at Orange. In January, 1910, the owners discontinued the Orange Sawmilling Company, and operations at the company's four mills at Orange, Beaumont, Milvid, and Timber were reorganized directly under the parent company, Miller-Vidor Lumber Company.[157] A company advertisement of 1910 stated the four mills produced a total of 100,000,000 feet of lumber and 50,000,000 shingles annually. The writer believes that all shingle-making activities were consolidated at Beaumont and Orange, rather than at the two yellow pine sawmills at Timber and Milvid, and that Miller-Vidor made pine shingles as well as cypress.[158] The Orange mills were logged from Miller-Vidor cypress swamps and pine timberlands near Vidor in Western Orange County.[159]

A fire at Orange at 6 P. M. on August 3, 1910, completely destoyed the Miller-Vidor shingle, saw and planing mill, and the Orange fire company was unable to combat the blaze because of its remote locality at Cypress Bayou. The loss was only partially covered by insurance and exceeded $100,000. About 150 jobs were lost at Orange as well. All that remained were the dry kilns, called "Moran" kilns, which were 30-feet by 50-feet in size, held 25,000 feet, and could be steam-heated to 210 degrees by two boilers installed for that purpose. The Orange plant also saved one rough dry shed and two dressed sheds. Most timber, however, was shipped by water. Export stock was stored in Cypress Bayou until the rafts were ready to be towed to Sabine Pass. "Export saps" were towed in barges to Sabine Pass to be loaded on ocean-going ships.[160] C. L. Hanna believed that the plant would probably be rebuilt because of Miller-Vidor's huge reserve of uncut timber.

The Miller-Vidor sawmill at Orange owned one mile of railroad track that connected the mill with the Southern Pacific tracks, and it had switching arrangements with two other railroads. The key plant personnel at Orange included C. L. Hannah, superintendent; Oscar Smith, bookkeeper; Alfred M. McKinley, planing mill foreman; C. F. Balter, sawmill foreman; T. L. Cour, shipping clerk, and T. G. S. Goodwin, river foreman. Although there had been early plans to rebuild at Orange, they never materialized, and thereafter all log trains at Vidor were shipped to the Miller-Vidor sawmill at Beaumont, until it burned on January 6, 1918.[161]

A SUMMARY OF ORANGE COUNTY SAWMILLING

The history of sawmilling in Orange County was so unique that it had only one counterpart in all of East Texas, and that was at Beaumont. By 1920, there were only three large sawmills left at Orange and three smaller ones elsewhere in the county. The Miller-Link mill went bankrupt and was dismantled. The sawmilling era there ended on December 16, 1930, when the last logs floated down the Sabine River was used up and the two big Lutcher and Moore sawmills were shut down for all time, silenced by the Great Depression. A way of life at Orange had really sung its swan song and died, leaving only memories in its place.[162]

Another sawmilling casualty when the supply of sawdust and shavings expired was the Orange Pulp and Paper Mill. Known variously as the Orange Papermill Company and Yellow Pine Papermill Company, the mill and its methods had been a failure at Mobile, Alabama, before the mill was moved to Orange about 1905, where new paper-making processes were tried and proven successful. The paper mill used up all the sawdust and shavings that in Beaumont were usually stacked one hundred feet or higher on every vacant lot. Yet, despite Orange's ridding itself of all the sawdust hazards to mill fires, the rate of sawmills that burned down was not reduced. It was never the writer's intent to add paper mills, and particle board and veneer mills to these chapters, but the Orange paper plant, the first of its kind in East Texas, did add a unique chapter to Orange's history.[163]

Orange was likewise a perfect example of a community's over-dependence on a "one-product" economy, and the problems entailed when there was no economic diversity to absorb financial shocks created by the sawmills that were either dismantled or burned down or the depletion of the sawmills' timberlands. A good source of information about early Orange's financial "ups-and-downs" can be found in either J. Eddie Johnson's MA thesis at Lamar University or his published article, "The Economic Development of Orange County, Texas, 1900-1920," published in 1977.[164]

In 1879, when most East Texas counties could still not boast of a single "large" sawmill, the city of Orange manufactured and shipped 52,000,000 shingles, 40,000,000 feet of lumber, and 7,500,000 wood lathes.[165] By 1905, the Lutcher and Moore Company alone could manufacture 100,000,000 feet of lumber a year, and not strain an engine or shear a pin. By 1879, Orange also had two sash and door factories attached to its mills. In June, 1878, Capt. G. B. Burr (master of the steamboat *Era No. 8*) and Boyd built a planing mill at Orange, that was variously referred to as a "portable house factory" as well. In 1880 Hart and Goodman added another planing mill at Orange, but generally the independent planing mills at Orange were unsuccessful because each sawmill eventually added a planing mill to its inventory.[166]

Certainly no other city or county in East Texas witnessed the rate of and great number of sawmill fires that the city of Orange did, beginning with the burning of the Empire Mills on May 31, 1856, and ending with the burning of the Orange Lumber Company sawmill in 1918. Two sawmill fires were intentionally set due to outlaws hiding out in the area; others were probably arson as well. As far back as when the shingle bundlers struck for higher wages at Orange about 1884, one of the "dirty tricks" endured by sawmillers was railroad spikes driven into company timber, intended to shear teeth off the band, circular, and gang saws. Two Orange sawmillers, A. Gilmer and D. Wingate, lost four or five sawmills to fires in each one's lifetime. The writer estimates that between 18 and 20 sawmills burned at Orange over seven decades of time, and each fire plowed a path of financial misery for each family whose livelihood was lost. It was always sad whenever a sawmill cut out and shut down, but employees could always see that day approaching. Fires robbed hundreds of their jobs without even a day of warning.

It has been said that "oil built Beaumont," recognizing no part of the role that sawmilling played there, but "old King Lumber" certainly built Orange without fear of intrusion from any other industry. Oh yes, there was an early rice mill and a large foundry, but no other sizeable industry reached Orange until the shipyards of World War I arrived. And even now, six decades after the last of the big band saws quit its screeching, Orange still reaps the benefits of lumber money that eyes can see all over the town. Today the beautiful Lutcher Theatre and the Lutcher Art Museum are only two of the lovely institutions that owe their origins to the Stark Foundation; another lumber-era landmark, the beautiful Edgar Brown home, was willed to Lamar University.

Polk County

Cabbage-stacked locomotives such as this were used to haul logs to the W.T. Carter & Bro. sawmill at Camden in Polk County.

POLK COUNTY

THE GEOGRAPHIC DESCRIPTION

Polk County, a civil entity containing 1,095 square miles, with its county seat located at Livingston, is situated in the southeastern extremity of Texas. It is bordered on the east by Tyler County; on the south by Hardin and Liberty counties; on the west by San Jacinto and Trinity counties, as well as the Trinity River; and on the north by Angelina County and the Neches River. Originally about ninety percent or more of the county was heavily forested with virgin long leaf pine trees, which often grew to four feet in diameter and a hundred feet tall. There were probably about three billion feet of pine and hardwood forests there in 1880, which was about the first year of large-scale sawmilling, and by 1900 only about two billion feet remained. The terrain and soil structure of Polk County vary considerably — from swamp lands along the Trinity and Neches river (as well as numerous creek) bottoms, where cypress and hardwoods thrive; some level lands in the southern sectors; which become gently and moderately undulating terrain in the northern sectors, a series of ridges, interspersed with valleys and creek bottoms. Polk County is home to a great variety of trees. Three varieties of pines - long leaf, short leaf, and loblolly - are native to the county, as are white oak, red oak, pin oak, cypress, cottonwood, ash, hickory, beech, birch, willow, and tupelo gum trees. Elevation in the county varies between seventy and 260 feet.[1]

Polk County has had four or five railroads operating within its boundaries, some of them being short line or sawmill tram roads, and some of them being narrow gauge, rather than standard, railroads. The first was the Houston East and West Texas line (later the Texas and New Orleans or Southern Pacific), built between 1875 and 1885 from Houston to Shreveport. The Waco, Beaumont, Trinity, and Sabine Railroad (the "Trinity Tap" or "Orphan Katy"), an east-west line, was built between Trinity and Colmesneil, Texas, between 1881 and 1882, and that railroad passed through some of the most beautiful virgin timber in the county. The Trinity and Sabine was abandoned in 1936.[2] The Moscow, Camden and San Augustine was a 7-mile, short-line tram road between Moscow and Camden, operated by the W. T. Carter Lumber Company before World War II. Other short line railroads or tram roads were the Livingston and Southeastern, the Knox Lumber Company's tram from Soda to Livingston, and the Beaumont and Great Northern, the Carlisle Lumber Company's trunk line from Trinity to Livingston. As state-chartered railroads, these were somewhat apart from the many narrow-gauge trams operated by numerous lumber companies to log their mills.[3]

POLK COUNTY'S ANTEBELLUM SAWMILLS

In a time period exceeding 150 years, Polk County has probably seen more sawmills erected within its geographic confines than any other East Texas county. Certainly one of the earliest steam sawmills in East Texas, that of M. B. Menard and Thomas F. McKinney, was built on the Trinity River at a place known as Menard's Bayou or Menard's Creek. At one time, the Republic of Texas authorized a post office at that point to be known as Menard's Mill. Although some sources have indicated that this mill was begun in Liberty County in 1833, it seems most likely that the writers meant the Mexican Municipality of Liberty, in what is today a part of Polk County. Menard's sawmill was of the "muly" or upright sash saw variety, probably having a daily output of between 2,500 and 3,000 feet. Beginning in 1837, M. B. Menard obtained a league of land on Galveston Island and laid out the townsite of Galveston, where soon much of the mill's output found ready sale. The other partner, T. F. McKinney, was building steamboats at Lufkin's Shipyard at Galveston for his large cotton business, McKinney, Williams and Company. In February, 1842, the boiler at Menard and McKinney's mill exploded, killing two slaves, but it was soon repaired and back in business. Menard's Mill lasted for about fifteen years, from 1833 until 1848.[4] Although the products of Menard's mill were principally exported to Galveston Island because of the owners' association with the Island City, most of the antebellum sawmillers of Polk County cut lumber mainly for local consumption, their intent being that buyers should purchase the lumber at the mill and haul it away in their wagons.

Another of the early Polk County sawmillers was John F. Carr of Smithfield, Polk County's principal antebellum river port on the Trinity. One source observed that:[5]

> ...The years between the organization of Polk County and the Civil War were also a period of development of the lumbering industry there. Several water-powered mills had been put up in the county as early as the late fifties (1850's), marking the beginning of the manufacture of lumber for the outside world. One of the first sawmills was built by John Carr and was located on Menard Creek...

John F. Carr was the most important antebellum steamboat builder on the Trinity River, and one reason he cut his own lumber was to provide himself with suitable boat timbers. In fact, one source noted that he built "three sawmills, a grist mill, and a cotton gin," all located at or near Smithfield. Others who operated sawmills in that vicinity were "R. Williamson, A. B. Carr, Philip Sublett, and John S. Ada."[6] Although Carr probably built about five steamboats at Smithfield, his best-known sternwheeler was the *John F. Carr*, which served the Confederacy at the Battle of Galveston on January 1, 1863, and afterward was a cottonclad gunboat in Matagorda Bay for the remainder of the Civil War.[7]

Another sawmill of Polk County was built by Monroe Drew at Mariana or Drew's Landing on the Trinity. No details about that sawmill survive.[8] Tom Hickman of Corrigan claimed in 1940 that the first sawmill in Polk County was that of Moses Meekins, built at Moscow in 1855. However, with the information that survives about Menard's Mill, that statement seems unlikely. Meekins, who owned large tracts of timber lands and who probably operated his mill with slave labor, built his sawmill on the east side of town.[9] There were probably several other early, water-powered lumber mills, often combined with grist mills and cotton gins, in Polk County, about whom time and history books have forgotten about.

THE SAWMILL TOWN OF BOWERS

The new rail line in Polk County, which gave many sawmillers a new outlet to market, was the Waco, Beaumont, Trinity and Sabine Railroad (also known as the Trinity Tap or the M. K. T.'s "Orphan Katy"), built in 1881 and 1882 between Trinity and Colmesneil. In 1885, a man named A. W. Morris dismantled a sawmill near Montgomery, and relocated it on the Trinity Tap at a point 21 miles west of Colmesneil and ten miles east of Corrigan. The town of Bowers took its name from the man who supervised the relocation and erection of the mill.[10] The sawmill burned in 1887, but was quickly rebuilt in a more elaborate fashion as the following quote reveals:[11]

> ...Bowers is a sawmill town two miles west of Barnum...It has an estimated population (in 1889) of about 300 people. The mill at this place is one of the best along the Trinity and Sabine Railway and is owned and operated by Mr. A. W. Morris, a gentleman of large and varied experience in the lumber business. Its capacity is 65,000 feet (daily) from the saw and 50,000 feet from the planer. The old mill burned down two years ago, and a larger mill has been rebuilt, and it is elegantly fitted up with modern, labor-saving machinery. It is logged by two dinkeys, or small locomotives, that run over a steel railroad extending some five miles into the pineries. There are 3,000,000 feet of lumber in the yard, about one-fourth of which is dressed. A mercantile business (commissary), carrying a $5,000 stock, is run in connection with the mill...
>
> ...J. H. Broome is foreman of the mill; J. T. Hufham is in charge of the planer and dry kilns; Mr. Tobe Pounds is chief of the store; J. T. Bostick is bookkeeper, while the energetic Guy H. Morris takes a personal supervision of the entire business, which under his watchful eye is successful.

Two articles of April, 1893, confirmed that the Bowers sawmill had already been sold to W. T. Carter and Brother of Barnum, and that A. W. Morris had been retained by Carter to superintend the Bowers facility. A list of sawmills on the Trinity and Sabine Railroad, published in 1893, revealed lthat the Bowers sawmill still cut 60,000 feet daily and belonged to W. T. Carter and Brother.[12] Another article, published at the same time, disclosed that:[13]

> ...The Bowers mill of W. T. Carter and Brother, under the efficient management of Colonel A. W. Morris, has not lost a week of time since last August, making an average

> of 55,000 feet per day, and their planer is turning out 35,000 feet (daily) of dressed lumber.
> ...Their lodging department (boarding house) is about eight miles north of the mill and connected by a good tram road, with Mr. Fuller (as) engineer. Eight of their (Bowers') Arkansas (dry) kilns were burned down on the 29th of last month (March 29, 1893), and new ones are almost completed in their place, which, when completed, will make twelve kilns in operation...

The stand of virgin long leaf pine available to the Bowers mill was cut out by 1898, and the sawmill was no longer listed as sawing lumber on the Trinity Tap Railroad in January, 1900.[14] The post office at Bowers was established in 1887 and was discontinued in 1898. The Bowers mill was dismantled either in 1897 or 1898, but the author is uncertain whether it was removed to Barnum, two miles to the east, or to the new sawmill town of Camden, which was under construction in 1898 and to which place W. T. Carter and Brother were moving their base of operations. The Barnum sawmill burned down during the summer of 1897, and there was no sawmill listed at Barnum among the listing of Trinity Tap sawmills published in 1900 either.[15]

THE SAWMILL AT BARNUM

The sawmill town of Barnum, located in northeast Polk County near the Tyler County line, was situated only two miles east of Bowers and twelve miles east of Corrigan. It had its origin in 1881 when the Trinity and Sabine Railroad built to that point, and W. T. Carter of Trinity bought some huge tracts of pine timber near the Tyler County line in 1881. There are two sources for the town's name, one source noting that it was named for P. T. Barnum, the circus showman, and the other stating that Carter named the town for a friend of his, who owned a sawmill at Groveton.[16]

In 1882 Carter built the sawmill at Barnum, and he brought his brother, E. A. Carter, as a partner into the firm, to be known as W. T. Carter and Brother. According to one writer, the first tram road at Barnum was a "pine pole railroad" (wooden rails), over which log cars were pulled by mules.[17] The following description of Barnum was published in 1889, as follows:[18]

> ... Barnum, one of the prettiest locations along this road

> (the Trinity Tap), is nineteen miles west of Colmesneil, has a post office, a general store (commissary), a neat two-story hotel with fifty boarders, operated by W. L. Owen, proprietor; a public hall, a school house, and church, and a population of about 350. The splendid mill at this place is an ably-managed enterprise, and is the property of those popular mill men, Carter and Bro., whose principal employees are Jack Thomas, office clerk...; Capt. A. B. Caton, woods foreman...M. A. Moses, planer foreman; L. C. Carter, mill foreman; and Julius Patzchke, yard foreman.
> ...Mr. W. T. Carter, senior member of the firm, was inducted into the mill business many years ago while assisting his father in running a mill at Trinity. His father's mill and store were destroyed by fire some twelve years ago (1877), the loss amounting to about $3,000. It was soon rebuilt by Mr. W. T. Carter, who removed to Barnum when the T. and S. Railroad was first extended east to that point. His mill and planer at Barnum were burned in 1887, ...and were rebuilt with new and latest improved machinery within 60 days after the fire. Their loss aggregated about $10,000...
> ...The mill and planer are now being run regularly, the planer putting in extra time every night...There is fully 6,500,000 feet of lumber now in the yard in Barnum. Of this, about 2,500,000 feet are dressed lumber, all snugly stored in neat lumber bins, secure from the weather. From Barnum is shipped monthly about 100 box car loads of lumber, most of this going to Missouri, Kansas and Nebraska...The firm has recently established an office in Kansas City, under the management of E. A. Carter. The firm owns 12,000 acres of yellow pine lands near the mill, and has an iron rail tram extending seven miles into the forests, over which two locomotives log the mill.
> ...The moral tone of the community is excellent. Preaching and Sabbath school are had every Sunday, and a good school is had eight months each year. There is seldom as much refinement and intelligence found in a town the size of this as exists here...

During 1891, Carter and Brother at Barnum remained busy "building a new tram road, putting in waterworks, a Wickes gang saw, and placing their mill in splendied repair."[19] In April, 1893, the mill at Barnum was "again running after the blow-up (boiler explosion?) of several months ago, cutting 50,000 (feet daily), and when the gang saw is finished, will considerably increase their cut."[20]

The Barnum mill owners also experimented with agriculture, especially hog-raising, to a degree not engaged in by any other East Texas sawmill, as the following quote discloses:[21]

> ..About three years ago (1893),...W. T. Carter and Brother... decided to do a little experimental farming. Accordingly, they cleared, fenced, and cultivated about 200 acres of land from which their mills had used all the merchantable timber. They fertilized the soil and grew on it a diversity of crops — corn, rye, oats, barley, potatoes, peanuts, melons, pumpkins, squash, etc. Two years since (1895), they stocked the place with hogs of varied grades, ranging from the razorback to Berkshires and Red Jerseys. This season they will market 40,000 pounds of choice pork. Chief among the...articles of hog diet was squash. Four acres were planted early last year, turning out a most fabulous yield from May till October, two plantings being necessary...This year about three acres...began yielding in May, and daily for three months an average of three dray loads, or sixty bushels, were...fed to the hogs raw...

Another article of 1896 indicated that Carter's agricultural and sawmilling activities were industries which complimented one another, as an account by a Galveston editor revealed:[22]

> ...W. T. Carter's gang and circular mill at Barnum keeps pegging away and is doing as well as their hog ranch, which indicates for the sawmill a prosperous condition. These gentlemen are putting in another large dry kiln...The work of removing the old Fleming mill from Bowers is progressing rapidly, and it will not be long until W. T. Carter and Bro. will have another large plant in operation. They are combining with their sawmill a hog ranch, on which they have a very large number of valuable porkers...

Destruction, however, was lurking around the corner from Barnum, and in the summer of 1897, the huge Barnum lumber installation went up in smoke, as one news account so vividly recalled:[23]

> ...Messrs. W. T. Carter and Bro. of Barnum, whose plant was totally destroyed by fire some time ago, are going ahead, preparing to rebuild. They have not as yet decided on all the details...Their net loss will considerably exceed $100,000...In a letter...written shortly after the conflagration, they said, "We are perfectly solvent, and while our loss was a heavy one, we hope to be on our feet again soon...

By 1898, however, the fate of Barnum, destined to become a ghost town, had been sealed, for one account soon observed that: "...The large sawmill plant being erected by W. T. Carter and Bro. at Camden is about completed."[24] Within a short time, housing and anything else that had survived the fire were removed to Camden, and nothing but the ashes remained behind to remind one of the former mill town that had turned to ghost town. Most of the employees were transferred to Camden, and the post office at Barnum was discontinued in 1902. A detailed account of the Camden sawmill and its facilites will appear later.

THE SAWMILL AT BENFORD

The first sawmill at Benford, located a few miles east of Corrigan, was established in 1889 when Bennett and Stanford Lumber Company built its sawmill there. According to one source, the town was first called Silver Hill because the early mill owners paid their employees with silver currency. Eventually, the name of Benford was conceived, derived from the first syllable of Bennett's name and the last syllable of Stanford's name.[25]

At some time around 1900, the mill owners sold out to the Burkett and Barnes Lumber Company, who in turn sold out to a Trinity County firm in January, 1904, as one news account later recalled:[26]

> ...Benford Mill Purchased—The (J. M.) West Lumber Company of Westville, Texas, has consumated a deal with Burkett and Barnes of Benford, whereby the former comes into possession of the big sawmill at the latter place. This is one of the largest mills in East Texas, having a daily cutting capacity of 60,000 feet of lumber. The new owners will put in a new planer, dry kilns, etc...The West Lumber Company...owns a number of mills at various places...

A year later, the only known history of the sawmill at Benford was published, and a Beaumont editor revealed about the new West Lumber Company sawmill that:[27]

> ...This is one of two mills owned by this firm, the other being at Westville in this state. The old mill formerly belonged to Burkett and Barnes, who bought it from George W. Stanford. They at once began extensive improvements, in fact, remodeling the old plant and making it an up-to-date one in every respect...
>
> ...The owners...have surrounded themselves with experienced men...The new planing mill is completed and the machinery will be moved in at once. The Standard dry kiln will be finished in a few days. The frame of a large dry shed is up...An artificial (log) pond will be made (dug out) at once. The logging is done close in...A tram road is being constructed, which will be in operation as soon as the stumpage close in is exhausted.
>
> ...Officers: J. M. West-president; Lynch Davidson-vice president; W. West-secretary and treasurer; C. B. Granbury-bookkeeper and shipping clerk; S. H. Prossey-mill foreman and filer; R. P. Weams-woods foreman; Frank Josserand-sawyer; Abe Ryman-planer foreman; R. H. Moody-yard foreman; Estus Starling-mill engineer and foreman; Phil Ramsey-locomotive engineer; and Dr. M. B. Canon-physician...
>
> ...Mill statistics: Capacity-65,000 feet daily; feet in yard-1,500,000; stumpage reserve-15 years; miles of tram-one-half mile; locomotives-one; log cars-ten; water tank-elevated 30 feet high; saws-one circular; houses-30; commissary-to be built soon; mill engines-one 14x20, one 12x18 (inches)...
>
> ...Planer: Matchers-3; resaw-1; planer engine-14x18 (inches); men on payroll-110; wages paid per month-$5,000; dry kiln-1 Standard; teams-16 mules...

After a few years, West sold out to his vice-president, Lynch Davidson, who organized the Continental Lumber Company. Davidson's successors included the Mardez Lumber Company, the Glynn Lumber Company, and the Ragley Lumber Company. By 1922, when all operations at Benford had ceased, the installation consisted of a pine sawmill, a hardwood sawmill, a commissary, and a drug store. All facilities were dismantled in 1936, and the site of the old mill town became a ghost town among the second-growth forests it had formerly helped to level.[28]

THE SAWMILLS OF FANT-PETERSVILLE-POTOMAC-WAKEFIELD

The Polk County sawmill town that was first known as Fant or Fant's Mill, and was later known as Wakefield, was founded in 1881 by John Napoleon Fant, according to *The Handbook of Texas*,[29] although the actual founding date could have been two or three years later. A Galveston editor of 1891 reported: "...N. Fant is getting his new machinery in, and when all his repairs are done, he will have one of the best small mills in Texas."[30]

In April, 1893, a list of sawmills on the Houston East and West Texas Railroad reported that the Napoleon Fant sawmill had a daily cutting capacity of 28,000 feet daily. By 1900 the capacity had been increased to 50,000 feet daily.[31]

Fant may have experienced some trouble shipping his lumber prior to 1894. Located ten miles north of Corrigan, the mill shipped lumber in the Houston East and West Texas box cars which were narrow-gauge and hence, could not be transferred to the trackage of a standard-gauge railroad. Fant sold his mill to Peter Edward Josserand, J. W. West, and Josserand's cousin, George Johnson, all of Trinity County, in 1902. The Fant post office became Petersville in 1903, but the name was changed again to Potomac in 1907 when the mill was sold to Ed Ingram.[32]

One source noted that Petersville was also called Josserand after 1902, but it was Peter Josserand that gave Fant the name of Petersville, perhaps coined from his own given name. Peter Josserand was a part of and from the sawmill town and post office of Josserand, Trinity County (2 miles east of Groveton), and he knew the post office would not approve of a second town named Josserand in the state of Texas. Although Peter Josserand was owner or part-owner of five sawmills, he lived at Petersville and managed the sawmill there from 1902 until 1904.[33] After lumbering ceased there about 1920, Wakefield or Potomac quickly became a ghost town and only twenty people resided there in 1940.

THE SAWMILLS AT ASIA AND MULVEY MILLS

During the early 1880's, soon after the Trinity and Sabine (the Trinity Tap or "Orphan Katy") Railroad was built across Polk County, Sam Allen and Company built four sawmills along the Trinity Tap, two respectively at Asia and Mulvey Mills in Polk County, one at Mobile in Tyler County, and the fourth as Allen and Williams Lumber Company at Corrigan. Some early Polk County historians, including Aline Rothe's research paper and article on "Asia" in the *Handbook of Texas*, connect the Asia sawmill owners (Sam Allen, John Kirby Allen, Augustus C. Allen) with being the founders of the City of Houston. While the author concedes that the mill owners were sons, brothers, or perhaps nephews of the founders of Houston, he considers it necessary to point out that the John Kirby Allen and August Chapman Allen who founded Houston died respectively in 1838 and 1864, decades before the Trinity Tap Railroad was ever built.

Mulvey Mills, located two miles east of Barnum, was situated very near to the Tyler County line. The author believes that Sam Allen had purchased a small tract of pine land, perhaps 2,500 acres, in the midst of a huge pinery owned by W. T. Carter and Brother, and that Allen built his small sawmill at Mulvey Mills in order to cut out his timber since he otherwise would be unable to tram the logs to his sawmill at Mobile. Only one news article about Mulvey (that the author has knowledge of) appeared in print while the mill town actually existed, as follows:[34]

> ...Mulvey Mills, two miles west of Fleming (Tyler County), is the property of Sam Allen and Company. This is a mill of about 20,000 feet daily capacity. It stacks the lumber direct from the saw on to the (box) cars (of the Trinity Tap Railroad), which is shipped thence to Mr. Allen's mill at Asia, west of this place, to be dressed...

Nevertheless, Mulvey Mills did not appear and then disappear overnight as one might perhaps think. In a list of the Trinity Tap sawmills published in April, 1893, the Mulvey sawmill of McDougald and Allen was still cutting 25,000 feet daily.[35] The author believes the mill cut out its timber and was dismantled about 1896 or 1897 since Mulvey was no longer listed at a Trinity Tap sawmill in a list published in 1900. Mr. — McDougald had earlier been a mill supervisor at the Allen mill at Asia.

The Asia sawmill was built about 1882, but unlike Mulvey, where available timber was in short supply, the mill at Asia lasted much longer, even though *The Handbook of Texas* states that the timber there had been cut out by 1890. The Asia sawmill was still cutting 25,000 feet daily in 1893 and 50,000 feet daily in 1900.[36] The following article about Asia appeared in print in 1889, as follows:[37]

> ...Asia is located only a few (3) miles west of Corrigan, on

> the Trinity and Sabine Railroad...Years ago Messrs. John K. Allen and John Donovan bought teams and contracted to log one of Mr. Sam Allen's several East Texas mills. Their success was such as to elicit from Mr. (Sam) Allen a proposition to take them into partnership, and each of them became one-fourth owners in the Asia and Mobile (Tyler County) mills, Mr. John K. Allen becoming manager of the former (Asia), and Mr. Donovan of the latter mill, their subsequent success being equal, if not superior, to that formerly enjoyed by them in the logging business.
> ...Asia has a population of about 200 souls, the country being...surrounded by vast pineries, from which the mill is logged by means of a tram road, over which the "Donovan," a splendid steam locomotive, hauls the timber from the woods. The sawing capacity of the mill is 30,000 feet daily; that of the planer is (greater), sufficient to dress the lumber from the mill at Asia and the one at Mulvey. The Asia mill will ship 1,000,000 feet or more this month, and expects to ship 10,000,000 feet during the present year.
> ...The mill here, like most others, runs a commissary. Messrs. Earl and Simmons superintend this branch. Mr. John K. Allen is general manager of the Asia mill. His love for the chase is only equalled by his success as a nimrod (probably meaning bear hunter)...His family reside with him at Asia in a home widely-known for the generous hospitality of its residents.
> ...Mr. Sam C. Allen is foreman of the planer (mill), the sawmill foreman being Mr. Benjamin Sanford. The yard, in which is neatly stacked a large stock of lumber, is presided over by Mr. J. W. Dennard. The jovial Judge E. J. Streeter, whose wounds received in the "late unpleasantness" (Civil War) attest to his patriotism, has charge of the repair shops...Your correspondent feels under special obligations to the courteous bookkeeper of this concern, who hails from the beautiful Alamo city...

Early in 1891, Allen and Company built a complete new, two-story sawmill at Asia.[38] In April, 1893, a Galveston editor published an account of the Asia sawmill as follows:[39]

> ...The mills of Allen and Company at Asia and Mobile

> must not be forgotten as they rank as two of the best upon this (Trinity Tap) road. The Asia mill is cutting 40,000 (feet daily), and when Manager McDougald finishes the proposed improvements, the capacity will be 55,000 feet per day. Their planer cuts about 40,000 feet per day of dressed lumber. The bookkeeper, Mr. Earl, and foreman, Mr. McCarty, are quite an addition to any mill's force...

In 1895 Sam Allen had to close down the Asia sawmill for two years due to lumber overproduction and a general stagnation of lumber demand, but by September, 1897, the Asia sawmill was back in full-time production.[40]

The author has no information about when the Asia sawmill was cut out of timber and dismantled, but assumes it was after 1910. He agrees with the editors of *Pictorial History of Polk County, Texas* that "by 1900 the sawmill (at Asia) was expanding."[41] Nevertheless, as nearly all sawmill towns of that era experienced, the day finally arrived when the last log was sawed, the mill was dismantled, the employees moved away, and the once-thriving mill town became only one more of the ghost towns of the East Texas forests they had helped to level.

THE MILL TOWN OF STRYKER OR PLUCK

Another of the early sawmills of Polk County was built at Stryker, a mill town about five miles east of Corrigan. The mill was built by D. M. Angle and G. H. Stryker, who were cousins from New Jersey, and the following year a post office was established there.[42] In 1889 a Galveston editor published the description of the mill town of Stryker, as follows:[43]

> ...The Angle Lumber Company, situated at Stryker, five miles from Corrigan, on the Trinity and Sabine Railway, has a sawmill capacity of 45,000 feet per day - steam-fed (log carriage), live rollers (to convey lumber), slab conveyor, new tram engine (locomotive), and generally the most modern improvements and timber land to hold it down for fifteen years.
>
> ...Here are the Huyett dry kilns with a capacity of 80,000 feet per day, which only stops from Saturday at midnight to Sunday at midnight. The planer is most conveniently arranged and fitted up with two 63-inch-flue, steel boilers and an Atlas (steam) engine. The woodworking machinery is new, and (is) therefore the latest and best, and like the kilns, never

> loses any working time, but makes the full fifty-two weeks in the year. The capacity is from 30,000 to 50,000 (feet) daily.
>
> ...The company was burned out some two years ago - planer, kilns, lumber, everything - but with the energy that characterizes this company, everything was rebuilt and put in still better shape. And now with the mill, kilns, and planer operated at full capacity, the orders are still way behind, and the stock of rough lumber on hand is low.
>
> ...The general office of the company is in Houston, where are the headquarters of D. M. Angle, the president of the company. G. H. Stryker is superintendent and general manager here, while Mike Cassidy makes the "wheels go round" (mill foreman) at the sawmill, trying to bury Joe Dunbar, the yard foreman, in lumber. Joe is generally on top, though, sometimes only the rim of his hat is visible. Eb Smith, the shipping clerk, worries the life out of the planer boss for lumber, and the railroad for cars.
>
> ...Mr. Angle and Mr. Stryker are from New Jersey, a tract of land formerly owned by the Camden and Amboy Railroad (?)...Mike Cassidy, the mill foreman, is a cosmopolitan; Joe Dunbar, a Pennsylvania Dutchman by adoption; Eb Smith, a native Texan; and the planer foreman, — Munger, a "Yank" from Yankeeland. He ought to be a good lumberman, that is, so long as he is not a stone cutter, for he says Vermont has two things in abundance, wood and rock, with rock in the lead.
>
> ...The commissary is A-1, with the genial Mr. Atkinson in charge —"Bolly," as he is called. This is a term of affection when a man comes into the commissary after something, and has got nothing coming to him. I am indebted to Mr. Wm. Martin, the genial hotel man of Stryker, for many courtesies during my sojourn at this place.

In July, 1891, another article reported that: "...the new mill of the Angle Lumber Company at Stryker, on the Trinity and Sabine (Railroad), is running smoothly, turning out between 50,000 and 60,000 feet of lumber daily."[44] In April, 1893, however, a list of sawmills on the Trinity Tap reported a mill capacity of only 40,000 feet daily for the Angle Lumber Company at Stryker.[45]

Apparently the Angle Lumber Company cut out its available timber about 1898, and the sawmill was dismantled soon afterward. The Stryker post office was discontinued in 1899. No Stryker lumber production was reported in a list of the Trinity Tap sawmills published in 1900. One source reported that Mrs. G. H. Stryker died in 1887, followed by her husband in 1892, and were buried in the local cemetery. For many years afterward, their descendants returned every four years to visit their graves.

About 1902 the area around the ghost town of Stryker became known as Pluck and only 25 residents were still living there in 1940. Also, sometime after 1900 the Taylor and Echols Lumber Company built a small sawmill there and cut logs at Pluck for a few years. Since then, the old ghost town of Stryker or Pluck has had plenty of time to return to the forests its sawmills had helped to level.

SOME EARLY POLK COUNTY LOGGING CAMPS

If a sawmill town was by its very nature a "temporary" location, then the logging camp, or log Front as it was known, was even more so. Often the Front was constantly in motion as the "steel gang" laid or removed spur tracks into new stands of pine forest or from cutover areas. Nevertheless, some log camps enjoyed some semblance of permanence over a long span of years, even to living quarters provided, a commissary, with church and school facilities for the families of the loggers or "flatheads."

One of the best-known logging camps in Polk County was Camp Ruby, located fourteen miles east of Livingston. W. T. Carter and Brother of Camden operated 16 miles of main line tram road to Camp Ruby, over which several log trains moved daily to Camden.[46] (This was in addition to Carter's rail line from Moscow to Camden.) An early Polk County historian described log camp life at Camp Ruby in 1937 when she observed that Carter and Brother at Camden:[47]

> ...will continue to cut lumber indefinitely because of the perpetual system that they use. They cut in a circuit (circle?) so as to keep timber growing. Their main camp is at Camp Ruby; smaller camps called 'mule corrals' are located convenient to the timber being cut. A few men care for the mules, and other men go to and from Camp Ruby on a log train. This company gives employment to 800

> men. They maintain a Class B high school at Camp Ruby and a three-teacher grade school at Camden...

A mill such as Carter's at Camden, cutting 100,000 or more feet of lumber daily, might require 150,000 feet of log stumpage daily in order to survive. An average log car carried from 3,000 to 4,000 feet of logs, which might tally up to fifty or more log cars, comprising three or four trains, arriving at the sawmill's log pond daily. Although no manpower statistics for Camp Ruby were located, it seems probable that 250 men may have worked as "flatheads" for Carter and Brother at Camp Ruby. The camp was named for Ruby Caton, the daughter of A. C. Caton, Carter and Brother's first woods foreman.[48] (Note elsewhere that Captain A. B. Caton was Carter's woods foreman at Barnum as early as 1889.)

Another Polk County logging camp was operated by the Kirby Lumber Company at Camp Seale, located near Segno in southeast Polk County. Camp Seale flourished between 1934 and 1948, and about twenty miles of tram road, over which several log trains ran daily, carried about 150,000 feet daily of log stumpage to the big Kirby sawmill at Honey Island, northwest of Kountze in Hardin County.

A former Kirby employee, Jim Whiteside, recently explained to the author that some loggers rode the log train daily from Honey Island to Camp Seale, whereas many loggers lived permanently (until 1948) at Camp Seale in about fifty of Kirby's portable houses. There were also church, commissary and school facilities there. When logging ceased at Camp Seale, the portable buildings were returned to Honey Island aboard tram cars. Each small room was so constructed by Kirby carpenters that it could be bolted on to another small room (or two) in order to make a "shotgun house" of two or three rooms.[49]

Around 1905, the J. I. Campbell Lumber Company of Warren, Tyler County, maintained its log camp at Stutts, located in eastern Polk County, somewhere south of the Alabama Indian reservation. In fact that company hired Indians to gather pine knots to fire its locomotives with. To provide its stumpage, the Campbell Company chartered a 22-mile regular-gauge tram road, the Warren and Corsicana Pacific Railroad, which ran from Warren in Tyler County to Stutts and Camp Wood in Polk County. That railroad and the Stutts log camp may have established a record distance for log tramming. Not only did the Stutts log camp provide 100,000 feet of stumpage daily for Campbell's Tyler County sawmill, it also provided an equal amount of logs that were trammed a total of 54 miles daily (22 miles from Stutts to Warren, and 32 miles over the Texas and New Orleans main line from Warren to Beaumont) from Stutts to the J. F. Keith Lumber Company sawmill at Voth, near the mouth of Pine Island Bayou north of Beaumont.[50]

THE SAWMILLS OF CARMONA

The once thriving sawmill town of Carmona, located on the Trinity Tap Railroad, eight miles west of Corrigan and two miles from the Trinity County line, was founded about 1885 by the Carmona Lumber Company and in 1890 was sold to William Cameron Lumber Company of Waco. By 1890 that huge retail lumber concern owned four Southeast Texas sawmills, at Carmona, Saron, Rockland, and Nona; sixty retail yards scattered out from Waco to St. Louis; and several flour mills. The town derived its name from a Mexican land grant, issued to Juan Carmona, upon which the mill was situated.[51] The first newspaper article about Carmona in the author's possession was published in 1890 by a Galveston reporter who remained in Carmona for only one night and left the following description:[52]

> ...I remained one night with Colonel Richard Hayward, general manager of the Cameron Lumber Mills Company. This is a new concern with a capital of $250,000. Their mill capacity is 60,000 feet per day. New machinery is being added to the sawmill and planing mill, and their dry kiln capacity is being doubled. C. M. McWilliams is head bookkeeper, and Mr. Lockridge has charge of the large mercantile interests of the firm. The head office is at Waco, with W. B. Brazelton at the head...

A list of the sawmills on the Trinity and Sabine or Trinity Tap Railroad in April, 1893, listed the Carmona sawmill capacity as being only 40,000 feet,[53] but another Carmona news article of the same month and year recorded that:[54]

> ...The Cameron lumber mill company of Waco operates a fine mill at Carmona, cutting 60,000 feet, and with the addition of their planer, handling 40,000 feet (daily) into various assortments of dressed lumber. The financial department (bookkeeping) is under Mr. McWilliams and the commissary is under Mr. Singleton...

The third newspaper article about the Cameron mill at Carmona that the author owns was published in 1897, as follows:[55]

> ...The Cameron lumber mill company's mill and planer at Carmona is putting in full-time. This is one of the few companies that believes in using more labor and less labor-saving machinery. And in this way does its part to give employment to idle labor and help alleviate the condition of the laborer. Manager T. W. Brazelton has just returned from a trip to Beaumont and Houston...

Another list of Trinity Tap sawmills published in 1900 showed the Carmona sawmill as cutting 60,000 feet daily. Also in 1900, the *American Lumberman* published an excellent history of the Cameron mill at Carmona, which will follow almost verbatim:[56]

> ...The official title of the Cameron interests at Carmona is the Cameron Lumber Mills Company, and the manager is C. Lipscomb. Capt. Thomas Waties, the general manager of all the yellow pine mills of the Cameron interests, also lives in Carmona. This mill is No. 3 in the line of properties acquired by Mr. Cameron. He bought it in 1890, succeeding the Carmona Lumber Company. The equipment is an Allis circular with a capacity of 18,000,000 feet annually (about 75,000 feet daily)...
>
> ...The company has eighteen miles of narrow gauge track with three engines (locomotives) and thirty five (tram) cars, and has a timber supply sufficient for 20 years to come. It carries about 8,000,000 feet of dry yard stock on hand at all times...The dry stock is carried in a large dry shed...In this connection, attention is called to the Andrews dry kiln, which has just been completed. The shed is a big one and will handle a large amount of lumber...This dry kiln is connected with the mill by a tramway, also with the planer and with the yard...
>
> ...The plant at Carmona consists of the sawmill and an excellent planing mill, well-equipped and conveniently modeled, and so situated that they are easy of access to the dry kilns and also to the railroad for shipping facilities...
>
> ...There is one building in connection with the plant at Carmona, which is not usual in a sawmill plant; that is a roomy and well-equipped machine shop where work is done in repairing the locomotives of the nearby Cameron plants, along the Trinity Valley railway. This machine shop is as big

> as shops in large towns, and the mechanics can practically rebuild an engine, except for the castings, no matter how badly it may be wrecked or broken up by any accident which is likely to occur to a logging engine. It is rather a departure to do this kind of work in a sawmill town...
>
> ...The commissary at Carmona is not so large as it is at some other places, but it has a well-assorted stock and does a large business with the employees and the people of the surrounding country. The office is located in a separate building, or rather an addition to the commissary itself, and is very convenient to the whole plant.
>
> ...All of the logs are hauled in from the woods on the company's own track and switched to the log skidders and from there they are rolled into a car, which is pulled up to the saw deck by a wire cable. There is no water or mill pond at the Carmona plant, so everything has to be handled dry. A new set of boilers has just been put in and equipped for handling the sawdust and refuse, and the carriers (conveyers) are the usual kind that carries the refuse away to the burners of the boilers...

Nothing more is known about the Cameron mill there before it burned down in 1914. In the same year, the Saner-Ragley Lumber Company built its sawmill at Carmona - on "the ashes of the Cameron mill," one might say - except that Saner-Ragley built one mile south of the old Cameron mill site and called their new location Ragley.[57] In 1931, W. T. Chambers visited Carmona and left a long, sometimes charming, sometimes very critical, account of sawmill and community life in Carmona, that is quite unsurpassed, as follows:[58]

> ... Carmona is a small sawmill community...property of the Saner-Ragley Lumber Company, which purchased a 10,000-acre tract of virgin pine forest here about twenty years ago, erected a mill to cut timber, and built a village to house employees engaged in the work...The tram railroad by which logs are brought to the mill has been lengthened as the logging "front" receded, and is now about eight miles long.
>
> ....Carmona is situated in an expanse of typical southeastern Texas cut-over and fire-swept forest land. It has an open stand of young second-growth pine and hardwood timber...

> .and farmers use the land as a free range for cattle, hogs, goats and sheep. The village landscape is dominated by the sawmill with its large low buildings, high smoke stacks, rows of lumber ricks (stacks), and large mill ponds. The residential area is beside the mill. The rather small, old, unpainted wooden buildings have paper and galvanized iron roofs. They are arranged in rows along sandy streets...A company store or commissary retails general merchandise, including such staple commodities as flour, corn meal, bacon, beans, potatoes, canned vegetables, work clothing, shoes, and gasoline. Neatly painted, relatively-new community church and public school buildings present a striking contrast to the characteristically weather-beaten structures of the village. The entire population, about half of which is Colored, depends directly or indirectly upon the sawmill payroll for its livelihood...

Chambers described the Carmona sawmill as cutting 40,000 feet daily (which did not make it one of the giants of the industry), and added that the turpentine industry had been closed due to the low prices offered for "naval stores" (rosin and turpentine). He wrote of the loggers or "flatheads," who felled the trees; of the "swampers," who carted and dragged the logs with mule teams and oxen to the nearest tram track. No skidder was used by the company in that work, but a steam log loader placed the logs on the tram cars, which were then hauled to the mill and dumped into the mill pond. The "log pondsman" then guided the logs onto the endless chain conveyer, which carted logs up an inclined trough and on to the "shotgun" carriage. The mill sawyer then manipulated the log on the carriage into the band saw which squared the timber, but left some of its bark-covered log slabs to be carried away to the boiler furnace to be used for fuel. Chambers then described home life in Carmona as he viewed it, as follows:[59]

> ...Rather heavy continuous labor and modest earning power are dominating factors in the life of the mill hand. He is tired and hungry when he comes home from work at the close of the day...Hence relatively little time is spent working about the house, reading, or engaging in social activities. Cultural attainments tend to be dwarfed, and interests are narrowly-centered upon the satisfaction of such basic and immediate needs as food, shelter, and clothing. The mill people...(express) loyalty

to political leadership which secured semi-monthly payment in cash in lieu of monthly payment in company "checks," which are accepted at the commissary in payment for merchandise...

...Women participate freely in the work and cares of mill town life. They prepare and serve the family food regularly, and many of them produce food stuffs by tending a vegetable garden, milking a cow, and raising chickens...The women of the community spend much time keeping house, caring for children, washing and ironing clothes. Limited income...discourages the employment of help or the purchase of labor-saving devices which lighten household tasks.

...The people save little or none of their income...This is said to be about as true of the well-paid men... Apparently this is an adjustment to sawmill town environment, which thwarts the development of a stable, home-owning society by causing all the people to live in a rented house and depend upon a single set of officials for continued employment there. As a result families are unprepared for sickness, accident, or loss of employment due to occasional depression of the lumber industry. In such cases the company grants employees limited credit at the commissary and discontinues the collection of rent...

...The home and home life at Carmona are fairly typical of Southern sawmill towns...Little is done to increase the attractiveness and beauty of the home and community. All buildings except the church and school are owned by the lumber company. None of them are painted... Employees move so frequently that they have small incentive to improve their rented houses...

...Salaried employees, skilled workmen, and their families compose the local aristocracy. The superintendent of the mill is the highest paid...and heads this group. It includes the managers of the different departments, the commissary manager, saw filer, blacksmith, mechanics, engineer, log scaler or grader, sawyer, and physician. The ministers and teachers of

> the community are also admitted. This group has a relatively high standard of living...
> ...Community life is well-organized at Carmona. The commissary is a community store. There is a community or mill physician. The company deducts $1.50 a month from the earnings of married men and $1.00 from those of single men to use in employing a company physician, and his medical services are available to the people without additional charge. The company purchases group accident insurance for its employees...
> ...The church is a community institution. Methodist and Baptist ministers conduct services on alternate Sundays ...Services are attended by the religious people of the community irrespective of denominational membership. The public school has grades from one to six inclusively, and pupils who wish to continue their work farther are transferred to the school in Groveton. The white and Negro sections of Carmona are very distinct. They are situated some distance apart on opposite sides of the railroad track, and each has its own church, school, and self-contained social life...
> ...In the meantime the fate of Carmona is uncertain. Will it successfully develop another economic base; or must it become, as many other Southern sawmill towns have, merely a few rows of rotting abandoned buildings along silent, deserted streets in an expanse of sparsely-settled woodland? End of Chambers' article.

And so ended one of the most insightly descriptions of the East Texas sawmill town ever written. Soon after Chambers visited there, the Saner-Ragley sawmill was greatly enlarged (even though its very existence was threatened by vanishing log stumpage), and the Carmona sawmill was still producing 1,000,000 feet of lumber a month in 1945. What was perhaps the most remarkable feat was that Saner-Ragley Lumber Company managed to prolong the existence of the mill and town as long as it did. But the inevitable day that all sawmillers feared finally arrived; Carmona cut its last log and then lined up to take its place among the ghost towns that have dotted the second-growth forests of East Texas.[60]

THE SAWMILL TOWN OF ONALASKA

W. T. Chambers' word picture of sawmill life in Carmona was so utterly typical of life in almost any nearby sawmill community that his account could as easily have have been applied to Groveton, or Woodville, Jasper, or Silsbee, and hardly anyone would have noticed the difference. Chambers wrote of a way of life that was as foreign to him as Mexico, but he made no attempt to discolor it with biases, prejudices, or partiality. On the other hand, a diatribe, entitled "The Feudal Towns of Texas," published in *Harper's Weekly* by George Creel in 1915 about sawmill life in Kirbyville and Jasper, is so discolored by Creel's hates and biases that one questions its authenticity immediately.[61] Emma Haynes' description of sawmill life in Onalaska is so marked by its exuberance, impartiality, and child-like innocence that one has no cause to question its accuracy either, although her memory for detail may have varied. After all, she was writing about her home town and perhaps had no recollections of other towns of her childhood with which to compare Onalaska. Obviously Haynes was writing about the approximate same time period (1931) when Chambers was writing about Carmona. The author will not be the first person to quote Haynes' writings about Onalaska verbatim, for as far back in time as 1947, Dr. H. P. Easton quoted her on an entire page, four long paragraphs, recognizing the value of her story to Texas sawmill history in his two-volume doctoral dissertation, entitled "A History of The Texas Lumbering Industry."[62]

Onalaska, in extreme western Polk County, near the Trinity County line, became a sawmill town in 1902 with the arrival of William Carlisle's sawmill and the Beaumont and Great Northern Railroad from Trinity to Livingston. The name Onalaska was to come from Carlisle's other mills, for he was destined to own sawmill towns named Onalaska in four states — Wisconsin, Arkansas, Texas, and Washington.[63] Although some of the dates in Emma Haynes' account of Onalaska will conflict with those furnished by Aline Rothe to *The Handbook of Texas*, her story is nevertheless the best and follows in part:[64]

> ...The largest and most modern sawmill and sawmill town ...was built by William Carlisle in 1905 at Onalaska...He scouted around and found the great pines in Polk County and bought up about 150,000 acres. (Aline Rothe's article in *Handbook of Texas* gives the Carlisle sawmill dates in

Onalaska as being from 1902 until 1908.)

...He sent L. O. Jackson...to Polk County in 1905 to superintend the building of a mill...He had to go farther west (into the river bottoms west of Livingston), and there he encountered mosquitoes...that work on the day and night shift...so Mr. Jackson...screened the houses...

...Jackson and Carlisle had some problems building the mill. There was no railroad so they built one, then known as the B. and G. N. (Beaumont and Great Northern). Starting from Trinity, Texas, they cut a right-of-way at the outer edge of the Trinity River bottom and built a track. They used a small ground-hog mill for sawing the trees from the right-of-way into lumber for the first middle-sized mill and to carry lumber to market...

...As soon as a few livable houses were built, Jackson sent for Mrs. Jackson to come...The middle-size mill cut lumber for the big mill and some houses; then the big mill was completed in 1907 (date seems doubtful), and the railroad was extended to Livingston. The cutting capacity of the big mill was 125,000 board feet per ten hour day. It employed eleven to twelve hundred men; sometimes a night and day shift were used. (Note: Normally a 125,000 board feet mill, including woods crew and two mill shifts, worked no more than 600 men.)

...The town was laid off by streets. The houses for the white people, except for the superintendent and bookkeeper, which were more elaborate, were built by the same (floor) plan in different sizes to accommodate small, medium, and large families. All were painted. Jackson built a large ice plant, the first modern one in Polk County, and a cold storage room that took care of a car load of meat, apples, etc. The commissary was a large, up-to-date general merchandise store, with dry goods, a grocery, a hardware, and a drug department. It employed a dozen men and women and supplied the people with a variety of good merchandise. Other buildings were a good electric light plant, a warehouse, a post office, a railroad station, two well-furnished churches, a three-teacher school for the white people; a one-teacher school for the Negroes; a

> general office, and a theatre equipped for picture shows (movies) and concerts. Pictures were shown several nights a week. A block was set aside for a park. The first concrete sidewalk laid in Polk County was laid in this sawmill town, from the house of the superintendent to the general office. Several others of lumber were laid convenient to the offices, store, depot, and other places...
>
> ...Mr. and Mrs. Jackson were devoted Baptists, but they took as much interest in the Methodist Church and encouraged employees to attend church. Jackson arranged for services to be held at the Baptist Church on two Sundays each month and at the Methodist Church the other two Sundays in order that people might attend both churches. For the same reason prayer meetings were held at one church on Wednesday night and at the other church on Thursday evening. The churches have not been kept in repair since the mill was moved...
>
> ...The women of this sawmill town, and of most sawmill towns, were fortunate in that they had little work to do. (Note: This statement remains highly doubtful for Onalaska or any other mill town.) They did nothing more than house work, and many had Negro cooks and maids. The women visited, attended Eastern Star meetings, rode horseback, went for buggy rides, had parties, went to the picture show (movies), and went to church. The children played anti-over (Annie over?), townball, hide-and-sick. The girls had dolls; the boys went swimming.
>
> ...The men worked from seven till six, went home for supper at six-thirty, and sat in the easy chair and smoked. On Sunday they went to church and visited. This was a quiet, peace-loving, church-going town due to the Christian influence of the churches and the superintendent of the mill...

In some respects, one is almost led to believe that Ms. or Mrs. Haynes was the mill superintendent's daughter or otherwise lived some kind of sheltered existence. At any rate, it's interesting to note that Onalaska's houses were painted, whereas those of Carmona and most other mill towns were of drab, unpainted clapboard construction. Mill houses at Kirbyville was described as "gray, dingy boxes," and those at Browndell (both sawmills in Jasper County) as "rotten shacks." It appeared that the women of Onalaska lived lives of leisure, enjoying the help of servants, whereas those at Carmona lived drab, laborious existences scattered between household chores, raising children, gardening, and milking cows. However, Haynes' writings about Onalaska and Polk County sawmilling are priceless for the information they contain.

In 1908 Carlisle sold out to J. M. West of Westville, Trinity County, and the mill name was changed to West Lumber Company. "Jim" West was slated to become the largest and wealthiest of the Polk County sawmillers, one might say "the prince of the Polk County pines," because he would eventually own the mills at Onalaska, Potomac, and Benford, as well as two mills, including Westville, in Trinity County. The sale of William Carlisle and Company at Onalaska was perhaps the second largest sawmill and stumpage transfer in East Texas, totalling almost $5,000,000 for Onalaska, 143,000 acres of virgin forest, and the Beaumont and Great Northern Railroad. No date was given by Haynes about when the mill closed, but it was apparently between 1925 and 1930. Evidently, agriculture already existed in Onalaska and soon replaced timber-processing to some degree since Haynes wrote of the farmers living in town, as farmers in Europe did, and going back and forth daily to their farms. As of 1940 Onalaska still had a population of 650 people, but since then it has largely regressed to ghost town status.

In addition to the West sawmill, there was also a large hardwood mill at Onalaska in 1915. The Philip A. Ryan Lumber Company mill cut 50,000 feet daily of white oak, red oak, gum, elm, walnut, cottonwood and magnolia, which was sold to the Midwestern and European furniture and barrel makers.[65]

SOME LITTLE KNOWN SAWMILLS OF POLK COUNTY

Laurelia, an early sawmill town on Laurelia Creek in north central Polk County as well as on the Houston East and West Texas Railroad, was known as Holshausen prior to 1900. Judge Claiborne Holshausen built the first sawmill at Holshausen about 1880. About 1890 he sold out to C. A. Bender and Sons Lumber Company.[66] In April, 1893, the C. A. Bender sawmill was cutting 30,000 feet of lumber daily. A 1900 list of sawmills along that railroad revealed that the Holshausen mill was cutting 60,000 feet daily, but did not show the name of the owner. Charles Adolph Bender, a native of Germany, made and lost two fortunes in Missouri and

Texas before he began sawmilling, the first because he chose to fight for the Confederacy. About 1875, he bought a 26-mile block of land, stretching from Humble to Cleveland, for 25 cents an acre. The owner of three sawmills at Holshausen, Humble, and Spring, he may have been the first Polk County sawmiller to use "commissary checks," or tokens made of both metal and cardboard, and since he was also a successful rancher, his tokens were unique in that Bender's cattle brand was printed on them. Like G. B. Moore, the millionaire sawmiller that Bender sold out to in 1902, Bender acquired a third large fortune, which he and his sons managed to hold on to.[67]

Holshausen apparently became Laurelia around 1902. G. Bedell Moore of Orange bought the Bender mill and soon became president and treasurer of the newly-organized Hilgard Lumber Company of Laurelia. Moore was the founding partner along with H. J. Lutcher in the two huge Lutcher-Moore sawmills at Orange in 1877. Moore died in October, 1908, and Mason Williams, superintendent of the Hilgard Lumber Company, shut down the sawmill immediately.[68] The sawmill was soon purchased by the Davidson-Ingram Lumber Company in 1909. The mill burned down in March, 1910, and was never rebuilt. Since the available timber there was already cut out, Laurelia quickly died and became the ghost town that it remains today.[69]

Lima was the site of an early sawmill on the Houston East and West Texas Railroad, built by M. D. Tackaberry about 1895. In 1900, the Lima sawmill was cutting 25,000 feet of lumber daily, but the name of the owner was omitted. No other information about that mill has become available to the author, and the old mill site is now only a ghost town in the forest.[70]

Kurth Station was a small sawmill town between Corrigan and Stryker, founded in 1882 by Joseph H. Kurth, who had left his native Germany only four years earlier. In May, 1887, Kurth bought the C. L. Keltys sawmill in Keltys, Angelina County, and in 1888 Kurth dismantled his Polk Couty sawmill. His son, Ernest L. Kurth, was born at Kurth Station in 1886. During that period, Kurth formed a partnership with Simon Henderson and Sam Weiner, both of Corrigan, and the trio built the Angelina County Lumber Company of Keltys into one of East Texas' largest timber industries.[71]

Bering was a sawmill town, seven miles south of Moscow, on the Houston East and West Texas Railway, beginning about 1882. Emma Haynes recorded that "the first large mill with a planer in Polk County was built at Bering" by the Bering Lumber Company. The owners were Conrad, W. G. and E. J. Bering. Of necessity, the Bering mill must have installed a planing mill as soon as it was built, for by 1889 a large number of Polk County sawmills owned planing mills. In April, 1893, the W. G. Bering sawmill at Bering was cutting 27,000 feet of lumber daily. A large commissary was operated at Bering until all milling activities ceased there in 1920, during a part of which time W. G. Winston of Moscow was the mill superintendent.[73] A Galveston editor of 1896 observed that:[74]

> ...Mr. W. G. Bering's mill, near Valda, Texas, is running at present, but is not making full time. They are short of water, and Mr. Bering is one of those sensible men who believes it is better to run the mill only when they have orders. Mr. Bering is a thorough gold standard man, and while living in a community of "free silverites," is doing his best to convert the erring...

"Free silver" was the dominant political issue during William Jennings Bryan's candidacy for president in 1896, and resulted in the famous "Cross of Gold" speech. While the years 1895-1897 were not considered 'depressions years,' many East Texas sawmills shut down for long periods due to lumber overproduction and minimal demand for lumber by any one other than the nation's railroads.

Valda became a very important sawmill location in early-day Polk County when Tom L. Hackney built his mill there and named the site after his daugther. In 1893, Valda, with a daily cutting capacity of 55,000 feet, was second in size in the county only to the W. T. Carter sawmill at Bowers, which had a 60,000 feet daily capacity. In 1900 the Valda sawmill was cutting 60,000 feet daily, and in 1904 the Hackney mill there sawed 75,000 feet daily and employed 150 men. Having apparently exhausted his timber reserve in 1906, Tom Hackney dismantled and moved his Valda mill, probably to one of the other three sites in Polk County where he would engage in sawing timber. Valda gradually became a ghost town on the cutover landscape it had so severely scarred.[75]

THE SAWMILLS OF CORRIGAN

Corrigan received its name from Pat Corrigan, an official of the first Houston East and West Texas Railroad train to pass through the town in 1882. Very quickly Corrigan became the most important rail junction in Polk County when the Trinity and Sabine Railroad reached there as well in 1882 en route from Trinity to Colmesneil. About the same year, Sam Allen and T. H. Williams founded the first sawmill at Corrigan with a cutting capacity of 25,000 feet of lumber daily. A Galveston editor of 1889 observed that:[76]

> ...Corrigan is at the intersection of the Trinity and Sabine and the Houston East and West Texas Railroads...twenty-two miles (north) from Livingston...It is a thriving town of some 1,000 population, and well-supplied with schools for both white and Colored scholars. It shipping facilities are excellent and the soil is well-adapted for raising fruits of all kinds...
>
> ...Allen and Williams have a saw and planing mill here; the sawmill with a capacity of 25,000 feet per day. The mill is not the largest, but is well appointed and makes full time. The planer is likewise a good mill - capacity, 50,000 feet and continuous in its work (planing the output of other mills as well). Mr. (T. H.) Williams is a thoroughly experienced mill man and has been in the business with Mr. Allen for many years. He was formerly at Luling and Kingsbury in charge of the Allen lumber yards at those points. W. P. Ryan, the planer foreman, is the right man in the right place.
>
> ...Corrigan is well-supplied with hotels and good ones too. Taking everything into consideration, especially the splendid shipping facilities, it should become a fine town. There are quite a number of substantial business men here. J. L. Henderson and Co. and Allen and Co. carry large stocks and both do an immense business. In the mills at this point, there is at present a scarcity of hands, and employment could easily be found for forty or fifty more...

It is indeed interesting that the Simon Henderson, referred to earlier as a partner in the Angelina County Lumber Company at Keltys, grew up in the Henderson store in Corrigan that his father owned. Another interesting, but much shorter article about Corrigan appeared in a Galveston newspaper only a year later, as follows:[77]

> ...Corrigan is lively and everybody is busy. T. H. Williams of Allen and Williams reports more orders than he can attend to. R. S. Cawthon has charge of the merchandise department of this firm. He has been with them for ten years. Colonel Jim Henderson is shooting birds and is

> going into the corn patches of his neighbors at a lively pace...

In Aline Rothe's article about Corrigan in *Handbook of Texas*, she referred to an earlier Allen and Williams sawmill at a point two miles west of Corrigan known as Gaul.[77] Strangely, the location of Gaul seemed identical or very close to that of Sam Allen and Company's sawmill at Asia. It is unknown to the author how long the Allen and Williams sawmill remained at Corrigan or what happened to it. In April, 1893, the Allen and Williams mill was still cutting 25,000 feet daily, but no sawmill at Corrigan appeared on the mill lists of either railroad in 1900.[78]

According to the *Handbook of Texas*, Corrigan eventually became the location of a second large sawmill, as follows:[79]

> ...Lumbering has always been one of the principal industries. The town was in 1946 the site of a large sawmill, established by the Edens-Birch Lumber Company in 1923 and enlarged in 1927. The company also operates several small mills in the vicinity of Corrigan, and has a large store and several other business interests in the town...

Over the years, Corrigan had so built up its timber processing, farming and livestock industries, as well as other industry, until there was never any risk of its becoming a ghost town. In fact, in 1946 Corrigan was still the second largest town in Polk County, with a population in 1950 of 1,420 persons.[80]

SAWMILLING AT CAMDEN

Earlier the author related the facts that were available about the W. T.Carter and Brother sawmills at Bowers and Barnum. Carter was born in 1856 in Cherokee County and cut his eye teeth on long leaf sawmilling in Trinity County. After the big mill at Barnum burned in September, 1897, W. T. and E. A. Carter decided to move their new base of operations several miles to the southwest to a new sawmill, and the townsite which sprang up around it, which Carter's wife named Camden. Early in 1898 Carter had received a court judgment in the sum of $100,000 against the Trinity and Sabine Railroad, a locomotive of which had started the fire which destroyed the Barnum mill.[81]

Emma Haynes observed that W. T. Carter and Brother planned to build perpetuity into their sawmill operations with their circuit logging, and not use a "cut and run" technique, a proven method that in a few years would leave only a ghost town to mark their lumbering trail through the forest. Since the story of the tram road to and from the logging camp of Camp Ruby appeared elsewhere, there is no need for repetition here.[82]

Robert Maxwell of Nacogdoches provided the following description of Camden, as told to him by Carter S. Caton and Nora Havard, as follows:[83]

> ... A typical, if somewhat more attractive than average, company town was Camden...According to veteran workers, when the town was built in 1898, no definite plans were made for development and growth. The streets were not laid out in blocks or squares; the principal streets curved around a hill, two streets forming an entire circle, commonly called the "Loop." The streets were unimproved and were largely sand. By 1914 about 450 houses had been built for the employees. A row of houses for the managerial personnel and Carter's house, standing somewhat apart on the hill, were also completed about the same time. In the business section of town were the general store or commissary, the hotel-boarding house, the company offices, and the depot for the Moscow, Camden and San Augustine Railroad. Water was supplied by wells, each serving one to four houses. Water was hand-carried, usually by the children, from the well to the kitchen. Sanitary facilities were simple, and baths were infrequent. The kerosene lamp was the standard light in the house until electricity became available in 1926...

Employees' houses were of typical clapboard construction, built of upright one-by-twelve boxing boards. Wives kept garden plots, milked cows and churned butter, and kept chickens and livestock which ran everywhere in the streets.The only transportation to the outside world was the train to Moscow, which served both as a logging road and a common carrier. Passengers on the Moscow 'cracker-barrel' typically rode "in a bright-red combine with a green roof and rattan seats."[84] One article described the Camden railroad as being:[85]

> ...perhaps the most famous of the Texas short lines. . .This lively little 'cracker-barrel' was the property of W. T. Carter and Brother Lumber Company, chartered in 1898... For many years the "mixed train daily" of the M. C. and S. A. (Moscow, Camden and San Auguntine Railroad) made

> the run, pulled by a 2-6-0 Mogul locomotive, that had earlier seen service on the Panama Canal. Passengers rode in a bright-red combine with a green roof and rattan seats, that had carried commuters on Long Island...From all over the world fans and nostalgic visitors came to ride behind the venerable iron horse on the shortest main line track in Texas...The M. C. and S. A. developed into a considerable tourist attraction, discussed wherever railroad enthusiasts gathered...

Another source of the history of Camden and the Carter sawmill observed that:[86]

> ...W. T. Carter and Brother operated this new timber mill at Camden until the summer of 1910 when it burned, resulting in the total loss of the sawmill proper. Having suffered the two fires of 1898 (actually Sept., 1897) and 1910, Carter made it his decision to erect a steel and concrete mill...Erection on the steel commenced on Christmas Eve of 1911, with the new steel, double-band sawmill starting operations in 1911. The mill, with several modifications, has been in continuous operation since that time...In 1919 W. T. Carter retired from active management of the company...It was two years later, in 1921...that Carter died at age 65 after...45 years in the lumber and timber business...
>
> ...During its long history, W. T. Carter and Brother has provided employment for many hundreds of people and has assured the continuation of that employment through its program of forestry management which provides a continuing source of raw material...

In her pictorial history of the county, Ruth Peebles recalled of Camden, certainly one of the great sawmill ventures of Polk County, that:[87]

> ...the once-picturesque town in the piney-woods setting with its mills, office buildings, log pond, commissary, drug store, boarding house, homes, churches, hospital, school, and engines and tram road attracted many tourists. ...Often referred to as a story book town, Camden became another sawmill ghost town when it was sold to the U. S. Plywood Corporation...

THE SAWMILLS OF SODA AND KNOXVILLE

One of the early sawmill families of Polk County, whose names would become involved in a notorious murder case at Hemphill in 1922, was the Knox family, consisting of William Hiram Knox, Sr.; his son, Hiram Knox, Jr.; and the latter's wife, Lillian Miriam Marshall Knox. The elder Knox had already earned a sizeable fortune, while sawmilling in the exhausted Great Lakes forests, prior to 1897 when he resettled his family in Texarkana and operated retail lumber businesses in Texarkana and Mission. A very attractive nurse, Lillian Marshall first attached herself to the Knox family when the elder Knox became ill at Texarkana, and Hiram Knox, Jr. soon married her following the death of his first wife. One source credited the Knox men with being quite violent, noting that one son had been killed in a dance hall brawl, whereas the elder Knox had killed a white man for abusing one of his black employees.[88]

Around 1900, Hiram Knox, Sr. purchased a large tract of virgin pine forests, believed to have totaled about 10,000 acres, about ten miles east of Livingston. Two sources noted that the Knox Lumber Company's first milling venture in Polk County was at Soda in 1902, where they employed forty men, but the author is not aware whether that occurred before or after the Callahan Lumber Company operated its mill at Soda.[89] In 1903, Hiram Knox, Sr. chartered the Livingston and Southeastern Railroad, which was a ten-mile combination tram road and common carrier that Knox built to connect the Houston East and West Texas rails at Livingston with his new mill town of Knoxville.[90] Late in 1905, a Beaumont editor published the following notation, to the effect that:[91]

> ...The big sawmill at Knoxville is about ready to commence making lumber. Not Knoxville, Tenn., but Knoxville, Polk County, Texas, seven miles east of Livingston. Mr. Knox is the owner of the mill and also the tram or railroad that runs to Knoxville, which is in the pine woods at the terminus of the railroad. He owns several thousand acres of fine pine land, all paid for, and if he doesn't get rich, it's his own fault...
>
> ...The new mill will have a daily cutting capacity of 80,000 feet of lumber and will give employment to 250 men all the time. A large village has been built up at the terminus of the road, and as stated, is called Knoxville...

After seven years of successful lumbering at Knoxville, the Knox Lumber Company's Polk County stumpage reserve was exhausted. Hiram Knox, Sr. purchased a 25,000-acre tract of virgin pine forests near the Sabine River in Sabine County in 1912, and began building a large sawmill there when he suddenly died. Hiram Knox, Jr. dismantled the Knoxville sawmill and moved everything to a new town he founded named East Mayfield, which was on the outskirts of Hemphill. In 1920, Knox experienced sudden financial reverses, and he sold everything at East Mayfield to the Southern Pine Lumber Company. Following a period when Hiram Knox, Jr. was suffering considerable mental depression, his lifeless body was found in his home with a bullet in his head, a pistol in his hand, and letters in his pocket addressed to his mother and his attorney. The Sabine County sheriff, however, ruled that Knox' death was murder, and East Texas was soon electrified to learn that Mrs. Lillian Knox had been charged with his murder. Although the case went before a grand jury, a lengthy investigation revealed that there was insufficient evidence to indict her, and Lillian Knox was 'no-billed' of the murder charge.[92]

THE SAWMILLS AT LEGGETT, FREEMAN, AND NEW WILLARD

Ralph Leggett was probably the first Caucasian person to live in the vicinity of Leggett, located about eight miles north of Livingston. The first sawmill known to have been erected there was the Freeman Lumber Company, owned jointly by James R. Freeman and Ralph Leggett, which sawed lumber there from 1880 until 1882. In 1883 the Young Lumber Company operated a sawmill briefly at Leggett.[93] In 1889 the Polk County Lumber Company built its sawmill at Leggett, and in 1891, a Galveston editor observed that: "...the Polk County Lumber Company has its mill running in a first class way."[94]

In 1892 Freeman bought out the Polk County Lumber Company at Leggett, and in April, 1893, a Galveston newspaper noted that the Freeman Lumber Company at Leggett was cutting 50,000 feet daily, which made it the third largest sawmill in Polk County as of that year.[95] About August of 1896, the Leggett sawmill burned, as the following account revealed:

> ...Mr. James Sinclair, of Sinclair and McArdle of New Orleans, and Mr. W. E. Stalnaker, of the Filer-Stowell

> Company of Milwaukee, have been down at Leggett, Texas, during the present week, trying to put in a new mill for the Freeman Company, whose mill was recently burned....[96]

Apparently the new sawmill was completed at Leggett. Peebles' *Pictorial History of Polk County* confirmed that the Freeman Lumber Company lasted at Leggett only until 1896, but the author has no information as to whom the proprietor sold out to. Peebles likewise noted that "a disastrous fire in 1897 burned some buildings (in Leggett). . .The fire also destroyed the mill, planer, and two million feet of lumber. . ."[97] A record of the sawmills on the Houston East and West Texas Railroad in 1900 did not list any sawmill as being existent at Leggett, but one subsequent sawmill, Thompson and Tucker, built a mill there in 1909.[98] Leggett remained a sawmilling town until 1922.[99]

The sawmill town of Freeman, later New Willard, seven miles north of Livingston, was the site of several sawmills, the first one built by James R. Freeman in 1886, who promptly named the village there after himself. After a year, Freeman sold out to the Jefferson Lumber Company, who operated the mill until 1889, probably until the company's timber reserve was exhausted. As had long been the policy of the Thompson and Tucker Lumber Company of Trinity County to buy up huge tracts of timber for their stumpage reserve, the author believes that that company had long before purchased most of the forests in the New Willard area or else Thompson and Tucker would not have relocated there.

According to Aline Rothe's article in *Handbook of Texas*, the Thompson Brothers Lumber Company closed down and dismantled their mill at Willard (now Woodlake), Trinity County in 1910, and reassembled it at Freeman, Polk County, which they soon renamed New Willard. A year earlier, Thompson and Tucker had started up a mill at Leggett.[100] Both companies were from one and the same family. John Martin Thompson, the father, began sawmilling in Rusk County in 1852, and later, teamed up with Henry Tucker in 1882 to found the Thompson and Tucker Lumber Company sawmill at Willard, Trinity County. Four of J. Martin Thompson's six sons (Hoxie, J. Lewis, James A., and Liggett) later organized the Thompson Brothers Lumber Company.[101]

The Thompson brothers and their employees were baseball enthusiasts during the years before World War I, some of the brothers playing on the team themselves. The sawmillers provided an "excellent diamond" at New Willard, as well as baseballs, bats, uniforms, and other sports paraphernalia. The Thompson players, known as the "Trimmers," were apparently an outstanding and winning team among the smaller sawmill towns of Southeast Texas.[102]

In July, 1922, the Thompson brothers sold their three mills - at New Willard, Trinity, and Oakhurst - to the Texas Long Leaf Lumber Company, and soon afterward Paul T. Sanderson of Zwolle, Louisiana, arrived to take charge of all three mills. The proprietors operated a commissary and drug store, as well as housing, churches, schools, and a physician for their 300 employees. During the 1920's, Sanderson bought up all of the cutover timberlands belonging to the Thompsons, William Cameron, and other lumbermen until he acquired 252,000 acres, and the Texas Long Leaf Lumber Company became one of the first sawmill concerns to practice reforestration on its cutover stump lands. New Willard still had a population of about 600 in 1940, and sawmilling ended there about 1955. Another lumber concern, L. T. and Arthur Sloan, also operated a sawmill at one time at New Willard, but the author has no information about the time span involved.[103]

THE SAWMILL AT BUCK

According to information appearing in *Pictorial History of Polk County*, the sawmill towns of Stanley and Buck seem to be one and the same, the town and post office being changed to Buck after the arrival of the new mill owner, H. D. "Buck" Reynolds, about 1904. The town had its origin about 1885 when the first mill owner, David Zimmerman, built the Livingston Lumber Company sawmill, two miles north of Livingston. Page 121 of Peebles' book gives the location of that mill as being at Stanley. Oddly, no sawmills named Buck or Stanley were recorded on the lists of the Houston East and West Texas Railroad's sawmills for either 1893 or 1900. Zimmerman sold out to John W. Cochran, who in turn sold out to Waterman and Die.

After Buck Reynolds bought the Livingston Lumber Company from Waterman and Die, Reynolds dropped the word "Manufacturing" from the firm's name, and the Buck post office was established in 1904. Reynolds also transferred many of his former Arkansas employees to Buck, where a church, school, commissary, hotel, and company housing were built. No sawmill statistics survive for the Reynolds mill. In 1912, Reynolds' son, Bill Reynolds, bought a large locomotive in Pittsburgh, on which he served as engineer while steering the locomotive, the "Old One Hundred," over several privately-owned tracks to Polk County. After Reynolds sold out in 1918, the locomotive was owned by and carried logs to the New Willard sawmill. Both Ruth Peebles and Emma Haynes noted that the Reynolds sawmill was moved to Honey Island, Hardin County, in 1918, but Aline Rothe's article in *Handbook of Texas* reported that the mill was dismantled and sent to Liberty County instead.[104]

The author has only one other record of Stanley, Texas, if indeed Stanley and Buck were one and the same, as follows:[105]

> ...The Russell mill at Stanley in Polk County has cut nearly all its timber, and will move to another place sometime soon...

OTHER EARLY POLK COUNTY SAWMILLS

Another early sawmill location was Seven Oaks, situated about five miles south of Moscow and halfway between Corrigan and Livingston. The first sawmillers there soon after 1880 were the Hackney brothers, Bob, George, Dave, and Thomas, and the latter had seven oak trees enclosed within his yard fence that had been planted there by an earlier settler. By April, 1893, Tom L. Hackney had either sold out or moved his 55,000-foot mill to Valda or elsewhere, and the new sawmiller at Seven Oaks in 1893 was Lafayette Mason, whose small mill cut only 30,000 feet of lumber daily. The two earliest sawmillers at Seven Oaks also operated a commissary and owned company housing. No sawmill location was recorded at Seven Oaks in the railroad's sawmill list of 1900.[106]

After the last of the available timber at Seven Oaks was exhausted in 1905, the last of the early sawmillers there moved his mill to Tyler County, as a Beaumont editor revealed:[107]

> ...Mr. C. D. Hanna, who for several years has been running a large sawmill at Seven Oaks, in Polk County, has about used up all his timber, and will move his mill to a point half a dozen miles east of Colmesneil, where he has bought several large tracts of pine land...

The Rock Island Tie and Lumber Company, built during the 1890's by Tom Manry, Sol Bergman, and Jim Lewis, was located on a rocky gulley, four miles east of Moscow. The Rock Island sawmill was not mentioned in either railroad list of sawmills for 1893 or 1900. The name may have been derived from the local rock formations, or of course, Rock Island, Illinois, may have been the home town of one of the owners. The Rock

Island sawmill was logged by Sam Gay and J. G. Bruce, and no other information is available.[108] At a place known only as Rocky Creek, two sawmills are known to have been in existence, the first by Jim Baird during the 1870's, and the second by W. H. Faircloth during the 1880's.[109]

Lamont, four miles from Livingston, was the location of at least three sawmills, the first owned by Middleton Tackaberry during the 1890's. The name supposedly came from the hometown (Lamont, Ohio) of Tackaberry's father. Another sawmill operator there was the Ward Lumber Company, date unknown, and the last known sawmiller there was the Lamont Lumber Company in 1906. Tackaberry also had sawmills or sawmill property at Baldwin and Lima on the Houston East and West Texas Railroad, although the former may have been only a rail switch to connect the 25,000-foot Lima sawmill site with the railroad siding at Baldwin.[110]

Moscow was the site of two early sawmills mentioned by the Polk County historians. The Moses Meekins mill was an antebellum, water-powered mill, of the upright sash saw variety, established there about 1855. The other mill was Tom Hackney's Lone Star mill, operated there in 1885. It seems implausible that subsequent sawmills did not operate there at future dates as well, and of course, Moscow was the rail junction or switch through which the W. T. Carter and Brother sawmill shipped their lumber to market.[111]

Many of the early sawmills of Polk County are so ill-documented that only the name of the mill owner and the location are recorded, but dates and mill statistics are unknown. Such a place was Bluewater, where Perry Connally once owned a sawmill and Jim Fowler owned a combination cotton gin, grist mill, and sawmill.[112] William Faircloth owned a gin, grist mill, and sawmill at Magnolia, near Onalaska, and Jesse Leggett, the postmaster at Marston, four or five miles north of Livingston, owned a sawmill that was turned at first by water power, but was subsequently converted to steam power.[113] It is interesting that Abe Peebles owned sawmills at both Chink and Valda in 1886, and later the Hackney Lumber Company owned mills at both places, implying perhaps that Peebles had sold out to the Hackney interests.[114] East Tempe was recorded as the site of a sawmill location on the Houston East and West Texas Railroad in 1880, and L. T. Sloan and Son owned a sawmill at Elmwood, as well as a tram road to connect it with the railroad at Livingston.[115] Jack Williams owned a sawmill at Goodrich in 1898, and Eggland and Groce operated a barrel stave mill there.[116] The Hackney Lumber Company owned sawmills at various times at Lone Star (near Corrigan), Valda (in 1893), Phoenix, Seven Oaks and Chink.[117] James Harrison owned a sawmill at Midway, between Livingston and Woodville, and George Harrison operated a mill at Segno.[118]

In 1909, the Texas Long Leaf Lumber Company owned a sawmill at Oakdale, four miles north of Livingston. That firm, based in Louisiana, would eventually buy out the Thompson mills around New Willard.[119] John O. Selby owned a water-powered, sash sawmill on Choate's Creek as early as 1854; Coker Brothers operated a mill in the Williams Community, date unknown; and Schutze Wright owned a mill at Kickapoo about 1880.[120]

THE SAWMILLS AT OR NEAR LIVINGSTON

From research sources available to the author, it does not appear that Livingston, despite its special location on the Houston East and West Texas Railroad, experienced a sawmill history of any great extent. As the junction point of the Livingston and Southeastern Railroad to Knoxville and the Beaumont and Great Northern Railroad to Onalaska, Carlisle, and Trinity, thousands of box cars of lumber must have passed through Livingston en route to market. In lists of the Houston East and West Texas sawmills published in 1893 and 1900, no sawmill was recorded at Livingston during each of those years.[121]

Certainly the first water-powered, sash sawmill there was that of Moses Choate, built near Livingston in 1851. The next sawmill was that of Daniel M. Fleming and Charles N. Fisher, who sawed timber at Livingston during the 1870's. By 1888, Fisher was building the big Liberty Lumber Company mill at Liberty.[122] Others who operated early-day mills at Livingston, dates unknown, were L. T. and Arthur Sloan, C. B. Kelly, and A. E. Simmon. Thomas McDonald built a water-powered, sash sawmill east of Livingston in 1846; Harvey P. Haynes built a sawmill north of Livingston, date unknown; and the Witherspoon Brothers built a sawmill west of Livingston, date also unknown.[123] The Ogletree Lumber Company built a sawmill at Livingston about 1930, and it was perhaps the Ogletree mill that was still there in 1948.[124] On a visit to Livingston in 1984 the author observed a small sawmill, of perhaps 25,000 feet daily capacity, situated on the east side of Highway 146 a few miles southeast of Livingston, near Schwab City, but he does not recall its name.

A SUMMARY OF POLK COUNTY SAWMILLING

Over a period of 160 years (1833-1993), it is no wild guess to surmise that at least one hundred sawmills sawed timber at one time or another in Polk County, and even that figure may be too low. Sometimes the same sawmill was erected and dismantled from at least three or four locations within the county. If a sawmill operated at one location for as long as ten years or more, it seemed almost a certainty that one mill would be lost to fire, and perhaps two, since sawmill fires all over the long leaf belt were so commonplace. And of all the sawmill fires in East Texas between 1900 and 1920, perhaps as many as thirty percent involved arson.

Like all of its neighboring counties, Polk County enjoyed a fabulous sawmill history, which probably provided employment for 4,000 people at any given moment between 1890 and 1930. But along with the economic and employment advantages that sawmilling produced, it also bequeathed two distinct disadvantages, the first being the countless acres of unsightly, cutover stump lands that scarred the county's sylvan beauty for decades; and secondly, the numerous ghost towns, that resulted from the "cut out and run" logging techniques of yesteryear.

The author feels he cannot praise two Polk County lumber companies - W. T. Carter and Brother of Camden and the Texas Long Leaf Lumber Company of New Willard and Trinity - too highly for their devotion to reforestration at a time when no one else cared and their efforts to preserve the forests in order to assure a perpetual supply of logs.

Unlike its neighbors on the Sabine and Neches Rivers, whose mills were almost solely devoted to pine and cypress-milling, Polk County had a great variety of hardwoods, many logs of which were floated down the Trinity River to mills at Liberty and Wallisville, the lumber of which was exported to France, Germany, and England for use in the furniture industry, tool handle, and wine barrel factories. Perhaps a quarter-billion feet of oak, ash, elm, cottonwood, hickory, beech and walnut were so harvested. Certainly another two billion feet of pine and cypress were sawed down by 1925 until all the great stands of virgin timber were depleted. And on any still, clear day at sunset, provided the reader cups his or her ear in the right direction, one might still discern the faint echoes of the big band saws, each of them screeching its way through a log and carving its own niche into the history of beautiful Polk County.

Tyler County

A bust of legendary lumberman John Henry Kirby stands in a chapel at Peach Tree Village in Tyler County.

TYLER COUNTY

TYLER COUNTY'S PHYSICAL DESCRIPTION

Tyler County is a geographic entity, containing 908 square miles in Southeast Texas. The county was created by an act of the Texas Legislature in 1846. It is bounded on the east by both the Neches River and Jasper County. The county's western neighbor is Polk County, whereas it is bounded on the north and south respectively by Angelina and Hardin counties. The East Texas Railroad, now the Texas and New Orleans, first bisected the county from north to south in 1881, making it possible for about ten sawmills to open in the county during the next ten years.

Tyler County is one of the most heavily-timbered counties of East Texas, and it must have been a locality of immense beauty and sylvan grandeur before its virgin forests were cut away, leaving only the unsightly cutover stump lands instead. The county is crisscrossed by ten or twelve creeks, namely, Hickory, Wolf, Caney, Billums, Cypress, Turkey, and Horsepen and a few other creeks. Most of the county is gently undulating, a series of ridges and valleys, many of the latter being creek bottoms. Agriculture probably never involved as much as ten per cent of the land area, and except for that devoted to cattle-raising, is much less than that today.[1] For a view of how large the Tyler County saw logs of 1902 were, the reader should consult the Kirby Lumber Company's hardback book of that year, *Timber Resources of East Texas*, page 164. The photograph on that page shows Woodville saw logs that appear to be from forty to fifty-five inches in diameter.

THE SAWMILL AT HYATT

About three miles south of Warren, where Highway 69 crosses Hickory Creek, is the ghost town of Hyatt, where once about 1,000 people resided, and where a large sawmill cut out all the surrounding timber at the rate of 125,000 feet daily. The community was founded in 1882 by two Houstonians, Jonas Shearn Rice and William M. Rice, and an associate, Captain F. A. Hyatt of Beaumont.[2] Hyatt, however, soon pulled out of the partnership upon winning the Federal contract to build the first Sabine River jetties at Sabine Pass that same year.

The first references to Hyatt in newspapers occurred in October and November, 1884. A Galveston reporter was evidentally staying there when he date-lined two of his published letters from Hyatt, but he left no record at all of the sawmill or the town that it spawned.[3] Two records of 1886, however, indicated that the sawmill there was cutting 45,000 feet of lumber daily,[4] and that for the year ending September 1, 1886, the Hyatt proprietors shipped 1,046 box cars of lumber products.[5] By 1888 the Rice Brothers mill was cutting 55,000 feet daily.[6]

The first detailed account of the Rice mill at Hyatt was published by a Galveston editor in 1889, as follows:[7]

> ...Hyatt is the seat of the mills of Messrs J. S. and W. M. Rice, which is one of the finest in East Texas, and is located about 39 miles from Beaumont, in Tyler County. Messrs. Rice have recently rebuilt their mills, with increased capacity, and are turning out from 60,000 to 70,000 feet per day, with all the requisite drying kilns, planing mills and extensive iron (tram) ways.
>
> ...Their lumber is long leaf pine, and is claimed to be equal in quality to the Calcasieu (lumber) of Louisiana. About 100 men are employed in these mills. There are from 75 to 100 houses belonging to the mills, and a neat country hotel, all for the accommodations of the employees. Many of the employees have fine gardens, furnishing an abundant supply of vegetables. A fine clear spring is contiguous to the mills, which furnishes an abundant supply of fish and bathing facilities.
>
> ...The village of Hyatt contains about 800 inhabitants, and all its surroundings indicate a thrifty and healthy place. Your correspondent is indebted to Mr. Joe Rice for his

> entertainment at his elegant residence convenient to the mill...Mr. Rice is experimenting in grape culture, and hopes ere long to set before his guests the generous wine from the pure juice of the grapes of his own production...

A year later, another Galveston reporter visited Hyatt, and he too praised the sawmill, its owners, and employees, with the same ornate descriptions of the previous year. A Galveston editor observed in 1890 that:[8]

> ...I halted for a few hours at Hyatt, where is located one of the finest sawmills on the Sabine and East Texas Railway. This valuable property is the plant of the Rice Brothers (J. S. and W. M.), old Houstonians, who are favorably known throughout the state and highly esteemed, both as business men and gentlemen of culture...Notwithstanding the cut of this mill is large, more orders are received than can be filled. The property was greatly enlarged and improved last year, and is now amply supplied with all first class machinery. A large commissary is run in connection with this mill, which commands the trade of the town and surrounding country.
>
> ...While here I met Mr. E. C. Parsons, who will probably enter the county sweepstakes races (election?) this fall for one of its offices. He thinks the air of Woodville will suit him to a T.
>
> ...At this place I enjoyed the hospitality of Colonel Joe (Jonas) Rice and his gifted lady, who occupy a large and spacious home, a beautiful type of all that is noblest and best in man's journey from the cradle to the land of shadows. May they long live to enjoy it!...

Jonas Rice did hold a colonel's commission in the state militia. His mill is believed to have consisted of a single 84-inch circular saw, cutoff saw, resaw, edgers and trimmers, live rollers, and probably a 10 or 12-gang saw, with a shotgun-feed log carriage. In 1891 another newspaper's report indicated that the proprietors at Hyatt were still increasing cutting capacity and enjoying record lumber cuts as follows:[9]

> ...Messrs. J. S. and W. M. Rice have gone over 80,000 feet per day for the past two weeks, one day cutting 91,000 feet of lumber. This gives them a record on the Sabine and East Texas (railroad that) is hard to beat...

> At the same time, the Hyatt sawmill had 2,750,000 feet of lumber on the yard, cut, drying, and awaiting planing and shipment.

In April, 1892, the Rice Brothers had to face the total loss of the sawmill facility at Hyatt by fire and the inevitable result that only a fraction of the loss was covered by fire insurance because sawmill insurance rates were so high. Nevertheless the owners soon replaced the loss with another mill of the same size. A Galveston reporter of that year reported that:[10]

> ...Messrs. J. S. and W. M. Rice went to Houston on the 7 o'clock train last evening. They estimate the loss of their mill, which burned, to be about $30,000, with about $10,000 insurance. Their many friends here deeply sympathize with them in their misfortune, and hope to see them ere long rebuild. The destroying of this plant throws about 150 men out of employment...

The Hyatt sawmill was quickly rebuilt, with perhaps a few improvements made over the old mill. Destruction had been total, sawmill, planer mill, dry kilns and lumber on the yard. In March, 1896, the proprietors added a 1,000,000-foot dry shed to the other equipment in the mill yard.[11] In May, 1896, they added another facility that was certainly unique to the Tyler County of that year - an ice plant - as one editor soon recorded:[12]

> ...J. S. and W. M. Rice have started their ice plant, which is being run in connection with their sawmill, and are turning out enough cooling material to ward off to a certain degree the intense heat of the summer months...

In the fall of 1896, at a time when many area mills were running on half or three-quarter time, the Hyatt sawmill was reported as still running regular hours, and even cutting some timber for export.[13]

In March, 1897, the Hyatt mill had to shut down for a while "while the construction force is building five miles of railroad into an unusually fine body of timber. The route...is slow, having to build sixteen bridges, one of which is 1,000 feet long." This was a standard guage tram road because, two months later, the owners bought a locomotive from the Houston East and West Texas Railroad.[14]

A few months later, another article explained some of the difficulties the Hyatt employees encountered while moving portions of their tram road from one forest to another, as follows:[15]

> ...The sawmill...at Hyatt has lost several parts of days during the past fortnight (14 days) because of a scarcity of

> logs. They had about cleared up where their tramming crew had been working, and the delay in the log supply was caused by the usual picking up of scattered logs, and the getting in shape of track to the new sections on which they would cut...Their planer is about the busiest thing in the lumber district. It is being operated by two distinct crews, and is putting in twenty-two hours out of each twenty-four...

An article of 1904 indicated that the owners of the Hyatt sawmill were already trying to convert a part of their cutover stump lands to agriculture, but reforestration as a daily practice of Texas lumber producers was still decades away. A Galveston editor noted in 1904 that:[16]

> ...Hyatt...is considered one of the largest mills in East Texas. The land around Hyatt is well-adapted to the raising of fruit and truck (vegetables), and the Rice Brothers have a large peach orchard half a mile south of Hyatt, containing several thousand fine peach trees. The capacity of the mill is approximately 80,000 feet...There are at present 3,500,000 feet of lumber on the yard...

At some time in 1904, it seems apparent that the owners overhauled the Hyatt mill and increased its capacity to 100,000 feet daily. And the increase came at a time when the mill's timber reserves were practically used up. One account observed that:[17]

> ...Hyatt logs (trams) from the east side (of the railroad), and the tram road now is sixteen miles long, extended into and up the Neches River bottom. It was reported quite recently that the company has made arrangement to buy a large stumpage (feet of uncut timber) in the Neches River bottom...

The author doubts that the above-mentioned purchase of timber land ever took place since only one other news article about the Hyatt sawmill ever appeared in print. It added that the cutting capacity at the Hyatt mill had climbed to 123,000 feet daily, and that logs arriving at the mill by tram train were being dumped into Hickory Creek, where they were being used as needed. A portion of the last article of 1904 owned by the writer reads as follows:[18]

> ...Hyatt is situated in an ideal fruit and truck-growing country, and many of the sawmill people raise enough vegetables for their own use and have a surplus to sell. In

> common with every other mill in East Texas, the proprietors have erected a first-class school and church house at Hyatt. Public school is taught on an average of nine months in the year, and there is a scholastic population of 100 pupils...
>
> ...The mill at Hyatt has 15 miles of tram (rails) that run from the mill to within a short distance of Spurger, a small village near the Neches River...The flatheads are usually paid about 35 cents per 1,000 feet, and two good flatheads can cut 20,000 feet per day of good timber...When the logs are cut several miles from the mill, a boarding house is put up in the woods for the employees, and sometimes quite a village thus springs up in the depths of the forest. It is called a "log front."...
>
> ...Following is the list of officers in charge of the business at Hyatt: F. A. Rice, superintendent; W. P. Cleveland, bookkeeper; E. F. Dunlap, shipping clerk; Tom Tucker, commissary manager; Dan Nevills, mill foreman; S. P. Chappell, planer foreman; J. R. Glover, woods foreman; E. Everitts, master mechanic; and Mr. Wells, millwright...
>
> ...Hyatt has a splendid baseball team, composed of mill employees, and an effort will be made to organize the sawmill baseball league, to be composed of teams from all the mills on the Texas and New Orleans Railroad between Nacogdoches and Beaumont...

Although the author has found no article about Hyatt's disentegration to ghost town status, he believes the last timber there was cut out in 1906, after which the mill site was abandoned and was reassembled at Ward, Louisiana, where the Rice Brothers cut lumber for the next eleven years. After the town was abandoned and the mill moved away, the Hyatt post office was discontinued on June 15, 1907.[19]

It might be added that Jonas S. and William M. Rice were born in Houston before the Civil War, and they were Texans of rather patrician ancestry. J. S. Rice was graduated from Texas Military Institute, and during the 1870s, he was in the Houston Light Guards, and colonel and adjutant general of the First Brigade of Texas Militia. The brothers were nephews of William Marsh Rice (the Houston millionaire who was murdered in New York), the benefactor of Rice University, and William M. Rice was a trustee of the college for years thereafter. Jonas Shearn Rice was financial agent of the state prison and superintendent of the Texas Penal System from 1895 to 1902, years that he was residing both at Hyatt and Houston. He was also president and chairman of the board of Union National Bank of Houston from 1905 until his death in 1931.[20]

THE SAWMILL AT MOBILE

Mobile, a sawmill ghost town in northwest Tyler County, was a mill community about which very little knowledge has survived. Located six miles west of Colmesneil, the Mobile sawmill was situated on the Trinity and Sabine Railroad, with access to markets via the East Texas Railroad, Colmesneil, and Beaumont, and on the west, via Trinity and the International and Great Northern Railroad. The first sawmill was established there by Sam Allen in 1883, and according to one source, the community received its name from a Negro handyman, who was from Mobile, Alabama.[21]

One early reference to Mobile was published in 1889, as follows:[22]

> ...Several days ago, Messrs. John K. Allen and John Donovan bought teams and contracted to log one of Mr. Sam Allen's several East Texas sawmills. Their success was such as to elicit from Mr. Allen a proposition to take them into partnership, and each of them became one-fourth owners in the Asia and Mobile mills,...and Mr. Donovan of the latter (Mobile) mill...

As of 1889, Sam Allen and Company owned four small mills along the Trinity and Sabine Railroad, those at Corrigan, Asia, and Mulvey Mills, all in Polk County, and Mobile in Tyler County. Donovan, who was one-quarter owner of the Mobile mill, served as the first and long-time superintendent of the Mobile sawmill.

A Galveston reporter of 1890 left the following record of his first visit to Mobile, as follows:[23]

> ...I may add that among the other points of interest visited on the Trinity and Sabine Railroad may be mentioned Mobile, in many respects one of the more desirable points along the whole line. The water here is most excellent, many cold springs abound in this vicinity, while game is plentiful.
>
> ...Here is located another large Allen mill. Colonel John Donovan, the veteran sawmill man and managing partner

> being in charge. The November shipments of this mill were 1,000,000 feet. The large commissary is in charge of two competent and courteous gentlemen, Messrs. Allen and Hunter, who are not only fine business men, but good entertainers as well...

By 1893, the Mobile facility was still grinding out lumber at a lively pace, but the sawmill's management team was periodically changing, as one record revealed:[24]

> ...Mobile sawmill is running at the rate of 40,000 feet per day, with the planer cutting its proportion of dressed lumber. Mr. B. Allen will leave in a few days for Houston and will be succeeded by Mr. Little as general manager...

The chronicles of early Tyler County also recorded some history of the early sawmill town, that vanished to ghost town status, as follows:[25]

> ...Another town which has now passed from the scene was Mobile, where Sam Allen operated a sawmill in the 80's and 90's. The post office at Mobile was established on September 18, 1883, with F. K. Wise as the first postmaster. Sam Allen later sold the mill and land to W. E. Trotti, who operated the mill until 1900, at which time it was sold to John H. Kirby, who continued the operation for some time thereafter. Mobile was a substantial town of its day, but with the passing of the supply of timber, its reason for existence passed also, and the town disappeared...

Actually, a letterhead of the Southwestern Lumber Company, Mobile, Texas, dated June 16, 1902, revealed that as of that date, the Mobile sawmill still belonged to W. E. Trotti and associates. The letter, addressed to F. M. Aldridge at Aldridge, Jasper County, indicated that the Mobile sawmill had a daily cutting capacity of 75,000 feet, and the company officers were D. R. Swift, president; B. A. Trotti, vice president; and Dr. W. E.Trotti, secretary, treasurer, and general manager.

In 1902 the Mobile sawmill became "Mill N" in the Kirby Lumber Company's "alphabet soup" of sawmill nomenclature, and the company's publication of that year observed that Mobile's:[26]

> ...Mill N is located on the Missouri, Kansas, Texas Railroad (successor to T. and S. R. R.) a short distance from the point where that road crosses the Texas and New Orleans railway (formerly E. T. R. R.). The mill has had a checkered career, and previous to its purchase by the

> Kirby Lumber Company, had been operated by a number of firms and individuals.
>
> . . . The mill was first owned by Sam Allen of Houston, and afterward passed into the hands of Dr. W. E. Trotti and associates. After the second owners assumed charge of the mill, the plant was given a thorough overhauling, and the new machinery set in motion in November, 1897. Later a stock company was organized, and under the name of the Southwestern Lumber Company, the mill was operated until it passed into the hands of the present owners. The (Kirby) company has rebuilt the plant, adding many improvements in labor-saving machinery, and greatly increasing its capacity, which is now about 85,000 feet of lumber a day...

In 1904, the "big planer at Mobile" was reported as "running on three-quarter time and reports abundant orders."[27] A report, however, observed that Mill N at Mobile was "being dismantled," which proved to be quite incorrect.[28] In 1907, the annual output of the Mobile plant for 1907 amounted to 14,277,000 feet, about equal to eight other Kirby mills, but considerably less than the four largest mills.[29] Although the Mobile sawmill was still in operation as of September, 1910,[30] that year is believed to have been the final year it was in operation before the last log was cut, and the sawmill site was abandoned. Fifteen years later, the Mobile post office was discontinued on October 31, 1925, when the handful of farm residents still remaining were placed on rural delivery service. Thus, the Mobile sawmill's steam whistle was silenced for the last time, and the mill site joined the ranks of East Texas ghost towns currently covered with pine needles.[31]

THE SAWMILLS AT ROCKLAND

For about twenty years following 1882, the northern terminus of the East Texas Railroad, later the Texas and New Orleans, was at Rockland, Texas. At first, the sawmill town was given the name of Rockingham, but the name was changed the same year to Rockland. On July 18, 1883, a post office was established there, with John W. Delaney as the first postmaster.[32]

There was no sawmill in Rockland in June, 1886,[33] as well as two months later, when a record of Rockland's annual rail shipments revealed that 265 bales of cotton had been shipped from there that year, but no lumber.[34] By

June, 1888, the Rockland Mills were cutting 45,000 feet daily.[35] During the next few months, R. D. Davis, the hotel keeper, built a second sawmill at Rockland, about the same time, although another source of the same period referred to it as "William Davis" mill and gave its cutting capacity as 25,000 feet. One editor left a quote about the Rockland mills as follows:[36]

> ...Your correspondent pursued his migrations on the S. and E. T. railway (East Texas Railroad) to its terminus, the village of Rockland, situated on the banks of the Neches River. The scenery is rather rough and broken, but is picturesque and pleasant to the eye. Here is located the mills of the Rockland Lumber Company, which is composed of Messrs. J.(oseph) A. Carroll, J. W. Delaney (the postmaster), and A. D. Carroll, who own large bodies of timbered land in Tyler, Angelina, and Jasper counties. The capacity of this mill is 50,000 feet of lumber per day, works 75 men, has on hand 3,000,000 feet of dry lumber, has a large Texas trade, and receives orders from Kansas and Mexico.
>
> ...The morals of the village and of the employees ought to be first class owing to the fact that two Baptist ministers are employed—Rev. L. E. Howard as foreman of the mill, and Rev. S. C. Johnson as foreman of the woods camp. Mr. A. D. Carroll is general manager of the mill, and he is able and experienced in his business. Mr. Carroll's residence is a neat and substantial cottage, just completed, that would be an ornament to any town. Mr. R. B. Davis also has a mill at Rockland; also a two-story hotel that furnishes refreshing nutrition to the weary traveler, who finds himself at Rockland.

In 1890, the proprietors of the Rockland Mills sold out to W. H. Aldridge, who owned the Aldridge Lumber Company until 1898, when he sold out to William Cameron and Company of Waco. In October, 1890, the Aldridge Lumber Company had 2,250,000 feet of unsold lumber on its yard, and in March, 1891, there were 2,600,000 feet of unsold lumber drying in the Aldridge yards at Rockland.[37] Also in 1891, one editor observed that: ". . .The Aldridge Lumber Company at Rockland is bridging the Neches River and will be using timber from the Angelina side before long."[38]

In 1892, the same Galveston newspaper added the following comments:[39]

> ...Rockland's Growth—The big (Aldridge) mill at this place now gets its timber supply from Angelina County, the (Neches) river now being substantially bridged, and the tram road extending into the pineries beyond. The mill is running on full time, and the lumber trade is apparently getting better...Several new cottages dot the suburbs of Rockland, and talk of extending the road from here to Lufkin...is causing enterprising homeseekers to locate here...
>
> ...The rock quarries here, from which the jetty walls at Sabine Pass are being built, are sending out ten cars of rock daily....(These were the jetties being built by Capt. F. A. Hyatt, mentioned earlier, and formerly associated with the Hyatt sawmill.)

At some time late in 1894 or early in 1895, fire destroyed some or all of the Aldridge Lumber Company's mills at Rockland. No description of the fire loss was ever found by the author, but one short article of August, 1895, observed that: "...The first lumber from the Aldridge Lumber Company's mill at Rockland, since the disastrous fire at that place, came down this evening."[40]

An article of January, 1897, also gives some information about the overhauling of the Aldridge mill at Rockland, as follows:[41]

> ...W. H. Aldridge has just completed general repairs on his mill at Rockland, including the changing of three miles of tram track, overhauling of his locomotives, the erection of a new water tank, slab elevator, the repairing of the big mill's boilers, increase of steam feed capacity, and various minor repairs, making his plant almost equivalent to a new one. The tram now enters virgin pineries, and the mill will run on full time, after a shut-down of about four weeks, pending the above changes...

In 1898, Aldridge sold out to William Cameron of Waco, who was perhaps Texas' first lumber millionaire. For one of the best account of the Rockland mill under Cameron management, *American Lumberman* published the following description in 1900, as follows:[42]

> ...The Rockland plant is one of the finest of the Cameron interests...and has recently been rebuilt, as it burned down

> on November 4, 1898. There is some very good timber at Rockland, enough to last about 25 years...Charles Ball was found at the office in Rockland upon the arrival of the writer and showed him over the plant. There is plenty of room at Rockland and the whole property lies in a long, irregular parallelogram alongside the railroad track...
>
> ...It will not take long to stock this yard, as the timber supply is very good and accessible. The railroad equipment consists of fifteen miles of narrow gauge track, four engines and forty (tram) cars...The equipment there is a McDonough friction power mill with one circular and one McDonough resawing rig. The total capacity is 125,000 feet a day or 30,000,000 feet annually. The plant proper consists of the mill, planer, Andrews dry kiln, dry sheds, etc...
>
> ...The Rockland mill site at one time was unhealthy and there was more or less sickness among the men until recently. But the company has built a large and roomy boarding house on the high land, so that at present it has no trouble whatever with sickness. This hotel or boarding house is much better than one usually finds at a sawmill...
>
> ...After inspecting the plant thoroughly, a ride on the logging locomotive was taken to look at the log cutting work at the camp. The camp at this point is portable; the houses resemble those that were built for making box cars, and were taken off the wheels and sidetracked there in the country. They are good houses, however, because they are light and comfortable and easily-handled and transported from one place to another. They save time in establishing camp and afford better quarters than the ordinary house which is built in the woods...Under the direct management of Mr. J. B. Hubbard, assisted by Mr. Ball, there is no reason why Rockland (sawmill) should not continue to enjoy the reputation it has always held as a yellow pine saw mill...

In October, 1904, the Rockland tram was eight miles long, and a Beaumont editor recorded that:[43]

> ...The Rockland mill has pulled up all its tram road west of the Texas and New Orleans railroad and moved it to the

> east side. It now penetrates a large area of the finest pinery in East Texas...

A month later, the same editor observed during an extensive East Texas drought that:[44]

> ...The Neches River is reported so low at Rockland that it can be forded with ease in a number of places. One of William Cameron and Co.'s large 100,000 (feet) daily capacity mills is located at Rockland, and logs are now being trammed from a find body of pine eight miles east of Zavalla...

In June, 1905, a newspaper article reported that the Cameron Company was negotiating with Kirby Lumber Company to buy the Rockland sawmill. It added that "the body of timber owned by the Kirby people east and north of Rockland is not surpassed by any other body of lumber in the South, and it is a vast mine of almost inexhaustible wealth."[45] At the same moment, another report noted that Kirby was dismantling its Mobile sawmill, so perhaps Kirby was indeed considering to move its Mobile operations to Rockland. If so, however, both plans were abandoned. In the published records of all Kirby mills both in 1907 and 1910, there was no Kirby mill located at Rockland, but Kirby Lumber Company was still sawing at its Mobile mill in both 1907 and 1910.[46]

The most extensive history of the Rockland sawmill ever found by the author was published in 1905, as follows:[47]

> ...This lumber proposition has been in existence many years. The mill was first built at this place by Carroll and Son, and afterward passed into the possession of W. H. Aldridge, who held it until 1898, when it became the property of Cameron and Co. It has always been an excellent producer, and today it is one of the largest mills in the South. It has the sawmill, planer, and (dry) kilns at great distances from each other, and in case of fire in one, the others could easily be saved. As an additional fire protection, its water tank is at an elevation 150 feet above the level of the plant and gives pressure of 40 pounds to the square inch. Connections are made by pipes with all parts of the mill, and hose stations are so arranged that four streams could be used at any point. The tank is 20x24 feet. There is an abundance of water at all times, a good stream running into the pond, and in case of a necessity, by means

of a canal to the Neches River, a supply could be obtained from that source.

...The machinery is of the latest improved (style) and so installed as to insure the most economical handling of the lumber. A large two-room Andrews dry kiln turns out 50,000 feet per day. The mill and offices are brilliantly lighted by electricity, as is also the commissary. This mill is managed by a body of men, who have spent a large portion of their lives in the lumber business and understand the details perfectly. Superintendent J. E. West is a man of great executive ability, and has associated with him as heads of the various departments, men who are eminently fitted for the positions they hold.

...Mill officers: superintendent-J. E. West; bookkeeper-W. H. Jones; time keeper-C. N. McCarty; commissary manager-J. M. Weaver; mill foreman-L. A. Whatley; filer-C. H. Barclay; sawyer-Edgar Vandenberg; mill engineer-J. M. Hodges; yard foreman and shipping clerk-C. W. Richards; planer foreman-P. F. Dooley; planer engineer-N. M. Hardy; electrician-Hardy Donohue; machinist-Arthur Wilkerson; locomotive engineer-W. A. Bartrum; woods foreman-W. T. Allen; physicians-Dr. P. M. Bledsoe, Dr. W. T. Barron; teachers-Prof. Sam Mann, Mary A. Enloe.

...Mill statistics: daily capacity-100,000 feet; stumpage-8 years; miles of tram-14; tram cars-26; locomotives-3; feet of lumber in yard-8,000,000; saw-one circular; sawmill engine-350 H. P.; 1 large dry shed; planer engine-225 H. P.; 1 large dynamo; men employed-300; monthly payroll-$15,000; commissary stock-$15,000; monthly shipments-2,000,000 feet; shipments for 1904-25,000,000 feet; number houses-130; 1 school house; 1 church; planer capacity-80,000 feet; log trains daily-4; hotels-2, the Weaver hotel and Jackson hotel...

The author has no subsequent information about the Rockland sawmill. Since it owned stumpage sufficient for eight years and may have purchased additional timber afterward, it is assumed that the log supply lasted until somewhere between 1915 and 1918. Doubtless, other sawmills were built there in later years. At its peak around 1910, an estimated 800 persons resided at Rockland, making it one of the largest of Tyler County's early mill towns, but the number had decreased to only one hundred by 1940. And when the main highway between Woodville and Lufkin was paved and passed only one-half mile to the west, about fifty residents of Old Rockland moved to a new location on the highway, leaving only a handful of people to populate the ghost town of Old Rockland on the shores of the Neches River.[48]

THE SAWMILL TOWN OF COLMESNEIL

One writer noted that "Colmesneil jumped into existence with a rush with the coming of the sawmill,"[49] and the author might add, the coming of two railroads, the East Texas road and the Trinity and Sabine Railroad. With 350,000,000 feet of stumpage in all directions within easy reach of the town's tram roads, and being so economically-situated at the junction of two railroads, no town in East Texas was located any more favorably than was Colmesneil. And it would seem as well that the history of Colmesneil's sawmills would be the great economic success story of Tyler County. Instead, it was a history of failure, and probably of mismanagement as well, beyond belief.

Colmesneil received its name from Captain W. T. Colmesneil, who was the first and most popular conductor on the East Texas Railroad. The town was sub-divided into town lots about 1882, and its post office was established on October 24, 1882.[50] Although two sources of 1886 and 1888 do not show any lumber as being shipped from Colmesneil,[51] this much is certain. Colmesneil and the Yellow Pine Lumber Company had grown to such an extent by 1889, in only seven years time, that the sawmill had to have been in operation by no later than 1883. At first, the location consisted of two towns, Ogden and Colmesneil, but the latter quickly absorbed its neighboring city.

A Galveston editor published the first and most detailed record of early-day Colmesneil of 1889 that survives, as follows:[52]

> ...This town contains about 2,200 inhabitants, white and black, the former being largely in the majority...The former town of Ogden is a part of Colmesneil now, the twin villages having consolidated some eighteen months ago. This is the present junction of the Trinity and Sabine Railroad (Trinity Tap), but the line is being extended eastward...The Sabine and East Texas crosses the Trinity

and Sabine here and extends to Rockland...

...This town is well-supplied with schools and churches. The Baptist and Methodist churches are well-fitted up... Builders are now at work erecting a Catholic Church... The high school building is fitted up with patent car seats with attached desks, a good supply of charts, blackboards, and an organ. There are 137 pupils assigned to this school. There is another school building in North Colmesneil to which 47 little pupils have been assigned...

...The manufacture of lumber plays an important part in the business of Colmesneil. The Yellow Pine Lumber Company's mills at this place are perhaps the neatest and most completely-equipped mills in the entire South. This magnificent plant is fully-supplied with all the modern, labor-saving applicances...has an energetic, faithful crew of men, and puts in full-time almost the year round. Its daily output is 75,000 feet from the rotary (circular) saw and 15,000 feet from the gang (saw), making an average daily cut of 90,000. Its drying rooms are splendidly arranged. The daily drying capacity is 50,000, the Stuyvesant blower process being used.

...There are six planing machines at full blast, dressing about 100,000 feet of lumber daily, a large portion being shipped here by the adjacent mills...The latest... improvement is the steam lumber stacker, being operated by two men on deck and two more to feed lumber into it, stacks ready for the drying room 40,000 feet a day. This is the only steam lumber stacker in Texas as yet. It is said to do the work of twelve men.

...The mill is logged by its own tram road, extending five miles eastward into the pineries. It is a steel rail, standard gauge road...This gigantic enterprise gives employment to about 300 men, the average monthly payroll being about $6,000.

...Mr. E. P. Cowen of Dallas is the controlling financial light...Since W. H. Carson became general manager of the business, some eighteen months ago, the mill has flourished ...Mr. Carson is now assisted by J. H. Pollitt...Mr. John F. Stunkel...has charge of the yard. Mr. J. M. Mitterer...is

> planer foreman...H. M. Fleming...mill foreman...Frank Benjamin is now in charge of the engine...Mr. J. Valentine... is bookkeeper...Captain J. C. Haralson...is foreman of all forces operating in the woods...J. B. Simmons has charge of the mercantile department, assisted by W. J. Colvin, the storekeeper. The stock averages about $15,000.
>
> ...There are here 23 business houses, seven hotels, two livery stables...ten mercantile firms...The Colmesneil *Times*, a neat, spicy paper, all home print, is published here every Wednesday...

A year later, another Galveston reporter wrote another glowing account of Colmesneil, although not as lengthy as the one of 1889, as follows:[53]

> ...The Yellow Pine Lumber Company's mills at this place are perhaps the neatest and most completely-equipped plant of its kind in the entire South. Its daily output of lumber is 75,000 feet. Its drying rooms are splendidly-arranged, and have a capacity of 50,000 feet daily. There are some six planing machines in full blast, dressing about 100,000 feet of lumber daily. A steam stacker is in use here, the only one west of the Mississippi, and two men do the work with ease that usually requires fifteen men to perform. The mill is logged by its own tram road, extending fourteen miles into the pineries. It is a steel rail, standard gauge, on which are used two standard locomotives and 24 flat cars...
>
> ...Some 300 men are employed making a grand total of about 1,000 people maintained by this company. The mill is covered up with orders. I saw one today from Denver, Colo., for 100,000 feet of first and second-class flooring...
>
> ...Mr. W. H. Carson is the general manager of the company and under his faithful stewardship, the Yellow Pine has flourished...Mr. Jack Mitterer is still planer mill foreman, while H. M. Fleming is general foreman of the mill...W. E. Johnson, J. Brady, and others hold important positions...Mr. W. A. Walker...is chief bookkeeper, assisted by Mr. C. M. Valentine...
>
> ...The town is prosperous...The Yellow Pine Masonic Lodge, named for the mill here, whose members contributed the lumber with which the edifice, 50x100 feet, was built,

> has a large membership...The Colmesneil *Times*, under the efficient management of F. H. Robinson, is giving its people a fine local paper...

These flowery write-ups about Colmesneil were soon to end, however. No, it was not a case of exhausted raw materials, such as created so many East Texas ghost towns. Instead, as late as 1906, there was not a single sawmill located between Rockland and Doucette, even though an estimated 250,000,000 feet of stumpage still stood within close tramming range of Colmesneil.[54] With the aid of "hindsight," though, one can readily see where the owners of Yellow Pine Lumber Company made two mistakes. First, they were trying to produce and grow too rapidly. The owners were Dallas residents, rather than local, who seemed primarily concerned with "milking" the Yellow Pine Lumber Company and the neighboring pineries of quick money profits. Secondly, as one can readily detect from the write-ups, the owners had learned nothing from the many sawmill fires that had so often destroyed the neighboring mills, perhaps because the owners had not had to live next door to a sawmill themselves. Where write-ups of other mills bragged of the water supply, water pressure, sprinkler systems, hose carts, overhead tanks or mill ponds, or whatever else might be used to fight fires, all of that was obviously missing from the early newspaper accounts of the Yellow Pine Lumber Company. It was as if the owners considered the town and their pocket books to be "fire proof."

The 'day of reckoning' for the Yellow Pine Lumber Company came on April 17, 1893. Fire consumed the entire mill on that date, millions of feet of drying and planed lumber, company houses, etc., and it was only with great heroic effort on the part of many fire fighters that the entire town was not consumed. One account of the conflagration observed that:[55]

> ...As a result of the fire, over two hundred men lost their jobs. Most moved to find new jobs. Six weeks before the fire, a man named Butler took over the operation of the mill. The mill had $54,000 in liabilities and only $46,000 in available assets. The newspaper report was unsure if the mill would be rebuilt. It was not rebuilt....

The author's criticisms, however, should be tempered by some of the events that were occurring at Colmesneil before the fire, as well as financial difficulties that always seemed to overwhelm a sawmill at times. For instance, the mill remained closed for many months of 1891, while the Yellow Pine Lumber Company was in receivership. In April, 1891, it was sold at the court house steps in Woodville to Royal A. Ferris of Dallas for his high bid of $40,000.[56]

Also at Colmesneil, there was a Colored Sawmill Hospital Association to provide health care to black employees at a cost of fifty cents monthly. The employees utilized an outdated commissary building for the hospital that had been offered by the owners.[57] Colmesneil also had a young people's dramatic society which "will next month present "The Lady of Lyons" at the Masonic Hall Opera House."[58] In March, 1892, the residents of Colmesneil were petitioning the County Court for an election to incorporate,[59] which had been part of an overall effort to wrest the county seat and the new court house away from Woodville.[60]

For three years after 1893, Colmesneil had no sawmill. Early in 1896, "work on the new mill at Colmesneil" was "being pushed quickly ahead under the direction of H. E. Craig, who will manage the mill when completed."[61] For several years, H. E. and James L. Craig, who were former Galveston News employees, had operated a 30,000 foot sawmill at Spring Creek, seven miles north of Colmesneil. The mill, though, had been operated by H. E. Craig when his timber was cut away, and he moved the mill to Colmesneil.[62] Craig and his partner, S. H. Meadows, planned to operate the new sawmill under the firm name of Colmesneil Lumber Company. One article of May, 1896. revealed that:[63]

> ...The new mill being erected at Colmesneil by Craig and Meadows is nearly completed. This plant will only manufacture high grades, and while a small one, the owners expect to do well by cutting only that which will find a ready market at good prices...

A few months later, another article indicated the Colmesneil Lumber Company was prospering (although it was actually in financial straits) as one editor observed:[64]

> ...The lumber mills here are running at full blast. The Colmesneil Lumber Company has orders ahead to keep the mills active for several months. They now are making all kinds of rough and dressed lumber, including moulding of various designs. They expect to manufacture a large lot of hardwood lumber...

By the end of 1896, though, the Colmesneil Lumber Company was in receivership, with one of its partners, S. H. Meadows, appointed by the court as trustee. Meadows kept the sawmill running at full capacity, but, its liabilities exceeding its assets, the court ordered the sawmill's sale to the highest bidder on February 2, 1897. The new owner, Robert I.

Douglass, immediately closed down the mill for repairs and overhauling, with intent to restart the mill in late April of 1897.[65] No other records of the Colmesneil Lumber Company were ever found after May, 1897, leading the author to believe that the mill either burned or went bankrupt during that same year.

A Galveston newspaper article of 1897 seemed to be making a plea for the renewal of the sawmilling industry at Colmesneil by reporting how much long leaf pine stumpage was available there, in all directions from the town, as follows:[66]

> ...A very reasonable estimate shows that within a radius of 23 miles of the town (Colmesneil), 50,000,000 feet can be secured by tramming northeasterly to Jordan's Ferry...a half mile haul from each side of the tram would give 40,000,000 feet more...the usual tram spurs covering a radius of five miles from the center point between Colmesneil and the river, 112,000,000 feet more...

The article estimated that 227,000,000 feet of stumpage was available to any Colmesneil sawmill between the town and the Neches River.

By 1905, it appeared that the sawmill history of Colmesneil had already ended. One editor noted that: "Colmesneil, which was once a sawmill village, is now a thriving town, maintained by its fruit and truck industry, and is the only town between Nacogdoches and Beaumont that has a fruit and truck grower's organization..."[67]

Perhaps for decades to come, sawmillers remembered the misfortunes that the earliest mill owners of the town had experienced. Nevertheless, the author has one more early record of a sawmill at Colmesneil, or rather on the outskirts, although no detailed information survived, as follows:[68]

> ...The putting in of a new mill in the vicinity of Colmesneil by Mr. (C. D.) Hanna of Seven Oaks, in Polk County, will make the lumber business in that particular section much more interesting than it is at present...

Hanna had used up all the timber he owned in Polk County, and before moving to Tyler County, he bought several square miles of the finest pine lands east of Colmesneil. He moved his old Polk County sawmill to a point six miles east of Colmesneil.

Brenda Goode, in her published article, "The History of Colmesneil," said it best when she stated that, "as the winds whisper through the pines, one can hear sounds that hint of the coming of the railroads, the Yellow Pine Mill, a New Year's Eve ball at the Opera House..."[69] Colmesneil, however, never quite regressed to ghost town status, as other mill towns did, but thereafter it retained only a small fraction of its former size. The town's population remains at about 500 persons today.

OTHER MILL TOWNS AND LOGGING CAMPS

There are in Tyler County many ghost towns and logging camps, whose names and locations would not be readily known or recognizable by a majority of the county's citizens. One such early sawmill town was Seneca, located five miles southeast of Woodville on the East Texas Railroad, now the Texas and New Orleans. The first postmaster at Seneca was Robert W. Snelling, the mill owner, on August 18, 1887, and sometime afterward, the location became a logging camp around 1910, with the Seneca name changed to Milton.[70] Up until October 12, 1886 when a hurricane destroyed the R. P. Snelling sawmill at Kountze, there had been two Snelling sawmills on the East Texas Railroad.[71]

In 1886 the Seneca sawmill had a daily mill capacity of 25,000 feet, and during the year ending September 1st, the mill had shipped 787 box cars of lumber.[72] By June, 1888, the capacity of the sawmill was reported to be 30,000 feet a day. A Galveston editor left the following account of the Seneca sawmill of the same year, as follows:[73]

> ...At the Seneca sawmill, fifty miles north of Beaumont, we called for one hour. This mill is owned by R. W. Snelling and Company. Its capacity is 40,000 feet per day. To Mr. F. P. Gagne...we are indebted for many courtesies...

Two years later, another Galveston news article furnished the greatest amount of information known about the Seneca sawmill, as follows:[74]

> ...Five miles south of Woodville, on the Sabine and East Texas Railroad, the Seneca Mills are run by R. W. Snelling and Company; has a good plant, with a capacity of 40,000 feet a day. The mill is filled with orders and at a price which commands $2 a thousand more than the figures of six months ago. The mill has been overhauled and repaired, the capacity being increased by 10,000 feet per day, and all the modern improvements are at work in the mill to manufacture first class lumber. The logs are trammed to the mills about two miles, and it is one of the most healthy places on the road, being well-supplied with fine freestone water. Thie firm is composed of R.(obert) W. Snelling and F. P. Gagne,

two most obliging and thorough business men...

Only two other references have ever been found by the author to the R. W. Snelling and Company sawmill at Seneca. In October, 1890, the mill was reported as having 1,000,000 feet of unsold lumber on the yard, and six months later, the yard inventory of unsold lumber had dropped considerably, to only 350,000 feet.[75]

The author presumes that the Snelling sawmill at Seneca either burned or was dismantled by 1893 or 1894. However, at least one, and perhaps two, sawmills were erected there in subsequent years. About 1903, another sawmill was erected there by C. S. Vidor of Galveston. In June, 1905, one editor recounted that "the old Vidor sawmill at Seneca is being rebuilt." And a month later, the same newspaper added that: "The new Vidor mill at Seneca...is to be started up again in a short time."[76]

No sawmilling has taken place at Seneca or Milton, at least since 1940, but logging for utility poles, bark chipping, etc., has been carried on in that area almost continuously. In fact, one of the largest logging operations in Tyler County is still in progress nearby on the west side of Highway 69.

Another of the forgotten ghost towns of Tyler County was Summit Station, north of Colmesneil, built in 1887 by David J. Williams, Sr., one of the pioneer sawmillers of Southeast Texas. Williams was operating the sawmill of D. J. Williams and Co. at Kountze when the hurricane of October 12, 1886, destroyed his mill. Instead of rebuilding on the site, Williams gathered up whatever machinery he could salvage and rebuilt his sawmill at Summit. A newspaper account of 1889 recorded that:[77]

> ...Three miles north of Colmesneil, at Summit Station, are the Summit Mills, owned and operated by D. J. Williams and Co. This mill was built about eighteen months ago, and has proved to be a good investment, due to the long experience and good management of the proprietor...

In June, 1888, the Summit Mills were reported as cutting 30,000 feet daily.[78] As small mills, the Summit and Spring Creek mills would have consisted of a single circular saw, plus corresponding cut-off, edger, and trimmer saws, and each mill would have needed from sixty to eighty exployees to operate and log the mill. No information was found for Summit Mills in 1891, so Williams may have moved again by then. In 1902 Williams was back in Hardin County, sawing lumber at Batson. In 1904 he moved his sawmill to Thicket. The author has no information on the Summit post office.

Another nearby ghost town was Spring Creek, between Summit and

Rockland. The Spring Creek sawmill was also built in 1887, and by 1888 its daily cutting capacity was 35,000 feet.[79] An editor recounted in 1889 that:[80]

> ...The Spring Creek mills are situated seven miles north of Colmesneil, on the Sabine and East Texas Railroad. This place has a post office, general store, town hall, etc., is a delightful location, and has good spring water. H. E. Craig and his brother, Jim (J. L.), formerly connected with the Galveston News, are proprietors of this mill. The capacity of the mill is about 30,000 feet daily, and its planer has a capacity of about 20,000 feet. The Craig Brothers, if push and energy are worth anything, will win...

About 1895, the Spring Creek sawmill was moved to Colmesneil, in an effort to obtain a timber supply and fill the void left by the destruction by fire of the Yellow Pine sawmill. However, the subsequent history of Craig and Meadows' Colmesneil Lumber Company was one of financial woes, and it was soon auctioned at a receiver's sale to repay debts.

One Tyler County historian wrote of a sawmill built at Maydell, six miles west of Warren, about 1900. In order to get lumber from Maydell to the main line of the East Texas road, a new rail line was chartered in 1899, the Warren and Corsicana Pacific Railroad. Eventually, the tram road's seventeen miles of trackage were laid from Warren to Camp Wood (and later to Stutts in Polk County), but the railroad was abandoned in 1908 when it failed to receive common carrier status.[81]

By 1890 there were nineteen sawmills operating in Tyler County, and it's very doubtful if the author can pinpoint all of them accurately.[82] One of the earliest sawmill sites in Tyler County was that built at Smith Ferry, on the Neches River, but the mill could only supply local needs due to lack of a railroad (steamboats carried lumber only when no cotton was available). The Smith Ferry sawmill was built and owned by Randolph R. Smith. Built about 1870, it was a water-powered mill, which turned an ancient upright or sash saw to cut about 2,000 feet daily, and Smith also operated a grist mill and cotton gin. Smith Ferry, which had about 250 people in 1880, had dwindled to fifty people by 1900, and its post office was discontinued in 1926.[83]

Acol, some four miles from Colmesneil, was the headquarter logging camp in Tyler County for the Angelina County Lumber Company. The log camp may have lasted as little as four years since its post office was established in 1934 and was discontinued in 1938.[84]

Hicksbaugh, on Cypress Creek in south central Tyler County, was the location of the Lodwick Lumber Company sawmill between 1918 and 1930. S. B. Hicks and J. T. Wurtsbaugh owned the sawmill, and the name of the town waas derived from parts of each of their names. The post office existed at Hicksbaugh from February, 1918, until September, 1939. In 1940 Hicksbaugh still had about 250 residents, but the site has been a ghost town for many decades. No mill statistics or history of the Hicksbaugh sawmill has survived.[85]

Griffin, on the Texas and New Orleans Railroad in north central Tyler County, was a flag stop and switch for loading logs and other timber resources before World War II. At one time the log camp was known as Steger, but the source of both names is unknown. By 1948 no trace of the facility had survived, except for one store.[86]

In 1891 Alsobrooks, located somewhere between Rockland and Colmesneil, appeared to have been an alternative or successor name for Spring Creek. The sawmill there belonged to J. L. Craig and had 1,200,000 feet of unsold lumber in its yard. The author found a reference to Alsobrooks on only that one occasion.[87]

Spurger, in southeast Tyler County, was the site of small sawmills for many years, the last one thought to have been operating there around 1950. Unfortunately, no history of the mills is known to survive. The town's name was attributed to a local drunk's attempts to pronounce 'Spurgeon Whiskey' at the local saloon.

There are probably some more forest locations, where small sawmills operated at one time or another, but their steam whistles blew and then were silenced, leaving them with no permanent niche in recorded history. Although many tram roads crisscrossed the county, Tyler County never had but two railroads, even at the height of steam sawmilling there. After the Trinity and Sabine Railroad (the Trinity Tap) was shut down in 1923, the county has been left with no east-west railroad, and only one north-south rail outlet, the Texas and New Orleans Railroad.

THE SAWMILL AT HAMPTON OR FLEMING STATION

The sawmill town of Fleming Station, later known as Hampton, was located almost on the Polk County line, two miles west of Chester and thirteen miles west of Colmesneil. Fleming was built in 1882, soon after the trackage of the Trinity and Sabine Railroad (the Trinity Tap, later the

"Katie") was completed from Trinity to Colmesneil. The first record of Fleming and its sawmill appeared in print in 1889, when a Galveston editor published the following excerpt:[88]

> ... At Fleming Station, two miles west of Chester, one of the early towns of Tyler County, is situated the sawmill of D. H. Fleming and Son, formerly of Livingston. Its capacity is about 20,000 feet daily. There are about 1,500,000 feet of lumber on the yard. The mill will undergo repairs during this month, and will start up for active business again the first of September...

That first sawmill at Fleming is presumed to have been a small single circular mill, of perhaps 72-inch in size, with the usual cut-off, edger and trimmer saws. Throughout all the years of its operation, which certainly exceeded thirty years, the output of the Fleming plant was purposely kept low, never exceeding 30,000 feet, even after Fleming became the first band sawmill on the Trinity Tap road. A record of 1891 noted that "Fleming and Son have their new mill going." Whether or not the new mill at Fleming replaced a burned-out sawmill or not is unknown, but the Fleming mill burned out three times between 1889 and 1905. A record of 1893 also noted that the sawmill of Fleming and Son at Fleming was still producing only 20,000 feet daily.[89]

In the summer and fall of 1896, D. Hampton Fleming, the owner, decided to convert the Fleming sawmill to a single band saw as the following quote reveals:[90] (This may have occurred after a second burn-out.)

> ...The last installment of machinery has arrived for the big mill at Fleming Station. The tank was raised this week, and finishing touch is being put on all parts of the plant. It is the only band sawmill on the Trinity and Sabine road, and as soon as the new saw rigging can be put in, it will be ready for operation...

In January, 1897, another news excerpt of the Galveston editor recounted that:[91]

> ...The new band sawmill at Fleming Station on that (Trinity Tap) line, steamed up Tuesday, giving its machinery a trial. It is to begin running regularly in a few days, and enjoys the distinction of being the only band sawmill on that road. It is one of the best-equipped mills in the state...

It is to D. Hampton Fleming's total lack of greed that he chose to keep the cut of his mill purposely low, even though his band saw had a capacity to cut 75,000 to 80,000 feet daily, if desired. Obviously, Fleming was content to cut up his timber reserves slowly, in order to hold his employees to between seventy and eighty, and to keep the mill intact as long as possible by keeping his daily cut low. For reasons unknown, the town and post office's name was changed from Fleming to Hampton (Fleming's middle name) in July, 1896.[92]

By 1905, the Hampton sawmill was one of the oldest on the Trinity and Sabine Railroad, as the only history of that mill ever written reveals:[93]

> ...Hampton received its name from Mr. Daniel Hampton Fleming, who established a mill at this place 23 years ago. Mr. Fleming has been engaged in the lumber business in Texas for a number of years, at one time operating a sawmill at Lynchburg. Today the style of the firm is D. H. Fleming and Son. The mill is located on the Missouri, Kansas, Texas Railroad (the "Orphan Katie" replaced the Trinity Tap), 13 miles west of Colmesneil.
>
> ...The firm has lost three mills by fire in the last few years, the present mill having a capacity of 30,000. (The foregoing statement almost confirms that Fleming lost his first mill to fire in 1890 and probably the second in 1895 or 1896.) It is conveniently arranged and is being constantly improved. The firm has recently purchased a new locomotive, which will greatly facilitate the movement of logs.
>
> ...Mill statistics: Daily capacity-30,000 feet; feet of lumber on yard-1,250,000; output in 1904—6,000,000 feet; log stumpage-5 years; capacity of dry kilns-20,000 feet; number of houses-70; commissary stock-$10,000; 1 dispensary; saws-one band; mill engine-100 H. P. steam; planer engine-50 H. P.; 3 planers; one resaw; locomotive-1 Shay, 1 new engine; dry kilns-6 Arkansas completed, 3 buildings; miles of tram-8; tram cars-12; 1 natural log pond; men employed-100; wages paid per month-$3,000.
>
> ...Mill supervisors: superintendent-D. H. Fleming; assistant superintendent- John B. Fleming; commissary manager-W. M. Fleming; clerk-Ed Pope; mill foreman and filer-S. P. Russell; planer foreman-Joe Waldrop; woods foreman-J. V. Bumstead; sawyer-T. K. Matthews; yard foreman-J.

> A. Angell; locomotive engineer-Ira D. Holland; blacksmith-John Williams; head carpenter-M. J. Hay; physician-Dr. T. T. Stewart.
> ... Notes: Dr. Stewart was a graduate of the Memphis Hospital Medical College. A Presbyterian Church was organized last week...elder-D. H. Fleming; deacon-J. B. Fleming. Regular services are provided for; there is a membership of ten. The Royal Fraternity has an organization at this place. The mill is sawing considerable hard wood lumber...

The Hampton sawmill and town probably reached its zenith around 1905, when the mill had 100 employees, and the town population was probably between 400 and 500 residents. Although the author has no additional information about the Hampton sawmill, the last log was probably sawed there between 1915 and 1918. The last train ran on that line in 1923, and the Hampton post office was discontinued in 1926. Long after the mill was dismantled and the community had reverted to a ghost town, the last traces of its foundations, tram road and mill pond lay buried beneath a sea of pine needles ages ago.[94]

THE SAWMILL AT DOUCETTE

Doucette appears to have been the last location on the East Texas Railroad, where a sawmill was built in 1893. In the railroad records of sawmills and lumber shipments, dated 1886, 1888, 1890 and 1891, there is no indication of a sawmill existing at Doucette. However, in a list of such mills published in April, 1893, the sawmill name, "Nebraska Lumber Company, Carroll, Texas, capacity-25,000 feet," suddenly appeared. While the name "Carroll, Texas," may not confirm the town to be Doucette, the name "Nebraska Lumber Company" does.[95] Obviously, the Doucette mill proprietors had attempted what the Nona Mills owners had tried in 1881, when they named the Hardin County mill town "Carroll" instead of Nona, and the Post Office Department forced them to change the name to Nona to prevent missending of mail to Carrolltown, (there were no ZIP codes prior to 1960).

P. A. "Pete" Doucette, a Canadian, and William McCready built the Nebraska Lumber Company sawmill at Doucette in 1893, and a post office was established there the same year. While the author knows nothing about McCready, Doucette at various times owned sawmills at

Doucette, Woodville, and Sour Lake, and he was a logging contractor on Village Creek and the Neches River. One source described him as a "jolly, open-hearted man, who knows more about lumber and logging than most people can remember."[96] At one time around 1882, Doucette was a minor partner as well as contract logger for the Olive, Sternenberg and Co. mill at Olive, north of Kountze.[97]

The Nebraska Lumber Company seemed to have undergone a rather checkered experience under Doucette and McCready. One account noted that "the Nebraska Lumber Company's plant at Doucette is closed down on account of no water."[98] Another editor observed that:[99]

> ...The Nebraska Lumber Company's plant at Doucette is making full time. This is the only mill in Texas that can boast of a lady secretary and treasurer, Mrs. A. L. Doucette filling that position. President J. A. Hargis reports business good...

A Galveston newspaper ran several articles about the Doucette sawmill in 1897. In one of them, the editor revealed that:[100]

> ...The Nebraska Lumber Company has about completed its repairs and improvements and expects to begin running regularly Monday, with its capacity greatly increased for making more and better lumber. Among its new improvements is that of a new carriage with steam feed action...

Another Galveston news account of the same year observed as well that:[101]

> ...The Nebraska Lumber Company mills at Doucette have lately put in water works, getting a supply from springs nearby a mile west of the mill. Their new locomotive has arrived and they find it an easy matter to log the mill with it...The Nebraska Lumber Company at Doucette...has extended its tram into new pineries, whereby it is getting an excellent grade of timber...

However, misfortune seemed to be stalking that firm constantly, and a few months later, another editor recounted that:[102]

> ...Several days ago, the Nebraska Lumber Company of Doucette caught fire, burning the mill, planer, and a lot of lumber...the main lumber sheds and about 1.5 million feet of lumber...estimated loss $20,000...will be rebuilt...

However, a later follow-up predicted the end of the Doucette sawmill when it noted that "a receiver has been recommended for the Nebraska Lumber Company since that mill has no insurance."[103] Thus, Doucette ended his partnership with McCready at Doucette, and he soon built another small mill at Woodville in 1899 with L. J. Chapman.[104]

The next sawmill at Doucette was that of the Emporia Lumber Company, which in 1900 moved its 75,000-foot mill from Emporia, located a few miles north of Corrigan in Angelin County. The Emporia mill was founded in 1893 by S. F. and J. P. Carter and M. T. Jones, all of Houston.[105] Jones, a wealthy retail lumberman, was also chief stockholder of the Orange Lumber Company and D. R. Wingate Lumber Co. sawmills at Orange. Samuel Fain Carter became bookkeeper of the Long and Company shingle mill at Beumont in 1881, before purchasing an interest in the Texas Tram and Lumber Company sawmill at Beaumont and the Village Mills Company in Hardin County in 1883. With Jones, the two Carters had operated the Emporia sawmill until their available timber was cut out, but by 1900, the year that they moved their mill to Doucette, the Carters had bought out Jones' interests in the company.[106]

Under the Carter influence, the new Doucette mill was built into one of the most modern and improved sawmills in Southeast Texas. The company officers (S. F. Carter, president; J. P. Carter, vice president; E.L.Crooke, secretary) all resided in Houston, whereas E. S. Hickman was superintendent of the Doucette sawmill, by then renamed the Sunset Lumber Company, although it appeared that that name was used interchangeably with Emporia Lumber Company. Another editor observed that:[107]

> ...There will be no question about extensive improvements being made whether the present owners dispose of their holdings or run the mill themselves. One feature of this mill that commends itself...is the convenient arrangement of the dollyways. The inclination is such that the loaded trucks have only to be guided, disposing almost entirely with mules for the purpose...

Another Beaumont editor recounted in 1904 some of the changes that S. F. Carter planned to accomplish at Doucette, as follows:[108]

> ...The Emporia Lumber Company will in a short time commence important improvements to their Doucette mill...in Tyler County. The mill, which now has a daily cutting capacity of 75,000 feet, has only one circular saw. The circular is to be taken out, a double side, short and

> long, put in, together with two band saws of the latest improved pattern. The capacity of the mill will be increased ...placing it in the list of the largest Texas mills...
> ...The mill is at present being logged from a large body of fine timber in the direction of Town Bluff on the Neches River, and in addition the Emporia Company owns a vast quantity of timber a few miles east of Colmesneil—enough in fact to last a number of years. The enlargement of the mill's cutting capacity is made necessary by the present heavy demand for lumber...

In January, 1905, S. F. Carter attempted to sell out his Doucette mill and timber interests, and at least for a while, the W. A. Pickering Lumber Company of Kansas City operated the Doucette mill, as well as several sawmills in Western Louisiana. Some weeks after the sale was announced and after Pickering was already operating the mill, the Beaumont *Journal* reported that, "owing to a hitch somewhere after the trade was closed, the matter fell through and the mill still belongs to the Emporia people."[109] S.F. Carter's biography states that he sold his lumber interests in 1906, and a history of Tyler County states that Carter sold out to the Thompson Brothers lumber interests of New Willard and Houston.[110] Whatever the truth, the Pickering Company was operating the Doucette mill when the sawmill and town's first history was published, as follows:[111]

> ...This is one of the oldest mills (which it wasn't) in Eastern Texas. It has born the name, among others, of Nebraska Lumber Company. It passed through several hands (untrue) before it was organized as the Sunset Lumber Company, property of Emporia Lumber Company of Houston.
> ...January 14, 1905, it was purchased by the Pickering Lumber Company, whose headquarters are in Kansas City, and who own several other plants. They (Pickering) have given it the name of Wolf River Lumber Company. Hundreds of acres of denuded pine forests attest the work that it has done in past years. The company owning it has always been associated with large enterprises, and its possession by them is a guarantee that it will soon be one of the most complete and largest producers in the South. Continued improvements have been made and much greater (improvements) are anticipated.

> ...The men in command are thoroughly-educated in the lumber business...Three new Berlin planing machines have just ben installed, and a pumping station three-fourths of a mile distant, pump daily from a 400-foot well—11,000 gallons of water, which furnishes an ample supply of water for all purposes. The mill does a large export business in kiln-dried saps. The company builds its own (log) cars and repairs its own machinery.
>
> ...The mill in detail is as follows: capacity-75,000 feet; feet lumber in yard-1,000,000; monthly shipments-1,750,000 feet; output for 1904-10,000,000 feet; stumpage-15 years; miles of tram-16; tram cars-35; locomotives-2, and three Shay engines; houses-98; churches-1 white, 2 colored; school house for colored; commissary stock-$10,000; 1 dispensary; men on payroll-225; monthly payroll-$8,000; 1 circular saw; mill engine-125 H. P. steam; planer engine-100 H. P.; 3 new Berlin planers; one 14-30 sizer; one 11-15 matcher; one 10-inch matcher; two Hall and Brown matchers; one picket machine; planer capacity-75,000 feet; 1 two-room porenpine (?) kiln; one 400-foot well, with 11,000 gallon capacity daily; one elevated tank, 100 feet above grade of mill; town population-500.
>
> ...Mill officers: superintendent-E. S. Hickman; bookkeeper and cashier-W. C. Stockroy; shipping clerk-R. L. McCarty; commissary manager-W. M. Wilson; commissary clerk-C. Hickman; physician-Dr. G. H.Spurlock; mill foreman-W. H. Kelley; sawyer-Frank Marion; filer-J. B. Heider; planer foreman-J. F. Pennington; yard foreman-C. H. Dougherty; sawmill engineer-William Hopkins; planer engineer-Frank Martin; woods foreman-T. M. Caderhead; locomotive engineer-William Frisby; and machinist-Joe Olsen. Dr. Spurlock is a graduate of Memphis Hospital Medical College...

Some time afterward, the Doucette sawmill was purchased by the Thompson Brothers Lumber Company, who also operated the sawmills at Willard, New Willard, and Grayburg. A cancelled check in the Forestry Archives indicated that on January 3, 1907, the Thompson Brothers Lumber Company issued a check to 'S. F. Carter, President,' for $500,000 to cover the cost of the Sunset Lumber Company sawmill and several thousand acres of timberlands.[112]

An article of September, 1908 observed that the Thompson mill at Doucette had just ordered a new 45-ton locomotive and thirty new log cars for their tram road. A month later another article noted that:[112]

> ...The Thompson Brothers Lumber Company at Doucette are operating a small transient sawmill at that place with a capacity of 20,000 feet daily. They installed the temporary mill in order to cut the timbers and lumber necessary for the building of the new mill...The Doucette mill burned several months ago, and is being replaced by one of the biggest and finest mills in the state, a double band (saw) affair...

A subsequent article in December, 1908, reported on the completion of the new Doucette mill, adding that:[112]

> ...The plant of the Thompson Brothers Lumber Company is very nearly completed...The big double band mill has been up for some time, and they have been installing the machinery since the first of December. They have completed a log pond that will hold 2,000,000 feet, have received a new Shay locomotive and a number of cars, and are making their logging road standard gauge instead of the old narrow gauge road. They have ordered a great steel water tank which will be installed at the top of the hill to replace the old wooden tank...

The Thompson people eventually disposed of the Doucette mill and timber to Fidelity Lumber Company, whose successor was the Long-Bell Lumber Company of Kansas City. The last purchaser of the Doucette mill and timber was the International Paper Company, but the last sawmill at Doucette has already ceased sawing in 1944, leaving only a certain stillness in the forests there where once a mighty steam whistle reigned and greeted the dawn.[113]

THE SAWMILL AT HILLISTER

One early mill town on the East Texas Railroad in the early 1880's was Hillister, four miles north of Warren, in south central Tyler County. Hillister was originally named 'Hollister' for some early railroad official, but the name soon became corrupted, perhaps when the post office was established on June 8, 1882. The town's first postmaster as well as first sawmiller was the same William R. McCarty, who a year earlier had joined with T. J. Ridley in building one of the first sawmills at Kountze.[114] Just as McCarty did not remain at Kountze very long, neither did he remain at Hillister, and within one or two years, he sold out to W. M. Gilder of the Express Lumber Company. As of June, 1886, the Express sawmill was a very small affair, cutting only 15,000 feet daily. Gilder may also have replaced McCarty at postmaster, since the first store at Hillister was operated by the Gilder Brothers. [115]

The Hillister sawmill thereafter was a struggle in financial woes. About 1888, two Hutchins brothers of Houston purchased the old Gilder sawmill and organized the Tyler County Lumber Company, with Arthur Hutchins as president and L. Hutchins as secretary. Beginning about 1886, Arthur Hutchins had operated the Nona Planing Mills at Nona, Hardin County (of 60,000 feet daily capacity), and until about 1890, the proprietors of the Tyler County Lumber Company prospered.[116]

An 1890 news account of the Hillister sawmill provided the following information:[117]

> ...A. Hutchins, president; L. Hutchins secretary. The sawmill capacity is 50,000 feet per day. This company operates three miles of tram road, with one narrow gauge engine and fifteen tram cars. There are about 150 men employed. At an early date, they contemplate removing the extensive planing mill (at Nona) to this point, Hillister, where they will have one of the best appointed outfits in southeastern Texas. They have in operation a Cople dry kiln with a daily capacity of 25,000 feet. Mr. J. Valentine is in charge of the books...It is with great pleasure that the record of this mill of the Hutchins Bros. is given...

While operated by the Express Lumber Company, the Hillister sawmill shipped 354 box cars of lumber during the year ending on September 1, 1886, but the sawmill lay idle in June, 1888.[118] In October, 1890, while the Hutchins brothers operated the mills, the Hillister sawmill had 1.6 million feet of lumber on the yard, but six months later, their yard inventory had been reduced to 950,000 feet.[119]

During the summer of 1891, however, the Tyler County Lumber Company fell on hard times after the death of its owner, Arthur Hutchins, in May, 1891, as the following quote reveals:[120]

> ...Manufacturers generally express deep regret at the death

> of Mr. Arthur Hutchins, of the Tyler County Lumber Company, which occurred in Houston last Monday. He was a genial, courtly gentleman, and popular with all the mill men...

By July, 1891, the Tyler County Lumber Company was in bankruptcy, and a Galveston editor recounted that:[121]

> ... Mr. John Delaney has been appointed receiver of the Tyler County Lumber Company. His application was heard before Judge Ford at Woodville last Wednesday. This is the mill formerly operated by the Hutchins brothers. ...The company's liabilities are said to be nearly $80,000.

At least for a few months, Delaney attempted to operated the Hillister sawmill in an effort to pay off debts. In January, 1892, it was noted that: "...P. M. Delaney, manager of the Hillister mills...reports logging almost at a standstill owning to the continued rains."[122] The Hillister mill was still sawing 30,000 feet daily in April, 1893, but that was the last news report of that mill ever found. The writer believes that Tyler County Lumber Company was auctioned off on the court house steps, and that it was purchased by the Warren Land and Lumber Company, who wanted the property only for its uncut timber reserve or stumpage, but dismantled and sold the old sawmill. In 1905, when J. I. Campbell Lumber Company, successor to Warren Land and Lumber Company, went into receivership, one of the Campbell Company's assets was the old Tyler County Land and Lumber Company of Hillister, who still owned uncut timber in Tyler and Polk counties. There is evidence as well that the old Tyler County Lumber Company sawmill may have passed to the J. P. Carter Lumber Company between 1904 and April, 1907. At about the same time, J. Press Carter had been involved in the receivership of the J. I. Campbell Lumber Company at Warren. The Carter Company consisted of J. P. Carter, who until 1906 had been vice-president of the old Emporia Lumber Company of Doucette and Emporia, Angelina County. Also his partner, Jonas S. Rice, had been one of the owners of the old Rice sawmill at Hyatt, which also was dismantled in 1906.[123] Although subsequent sawmills probably manufactured lumber at Hillister at later dates, the writer has no further information about them.

THE SAWMILLS AT WARREN

Warren was another of the early Tyler County mill towns with a checkered and uneven history, sometimes enjoying great financial prosperity, and at other times, beset by indebtedness and bankruptcy. Warren was founded on the East Texas Railroad in south-central Tyler County when Alexander Young of Warren, Pennsylvania, arrived there in 1882 to found a sawmill known at the Warren Land and Lumber Company.[124] By 1887, Warren had a second sawmill. By 1886, the Warren Lumber Company's mill was already cutting 35,000 feet daily and had shipped 524 box cars of lumber products during the year.[125]

Before settling at Warren, Young had briefly been associated with J. L. Williams of Beaumont in the operation of Globe Planing Mills. One of the first accounts of his Warren facility was published in 1887, as follows:[126]

> ...At Warren we met Mr. A.(lexander) Young, managing proprietor of the Warren Lumber Company. This company owns and operates a sawmill which turns out daily 50,000 feet of lumber, and a planing mill with a daily capacity of 30,000 feet. Messrs. Brough and Krueger of Grand Rapids, Michigan, are about to erect a 70,000 foot sawmill about 400 yards north of Mr. Young's mill...

The completion of the Brough and Krueger mill gave the town a daily cutting capacity of about 125,000 feet of lumber. A Galveston editor published an excellent account of Warren in 1889, as follows:[127]

> ...One of the largest and finest mills on the Sabine and East Texas Railroad is located at Warren. The plant of the Warren Lumber Company consists of the Warren mill, built in 1882, and Brough and Krueger's mill (the B. and K.), built in 1887. The two mills have a joint daily capacity of 125,000 feet, while the planer is totalling out 50,000 feet of dressed lumber per day.
>
> ...The logs are brought to the mill by tram road, the company having some five miles of tram, two locomotives, and twenty-five (log) cars. The average of lumber carried (in yard) is 4,000,000 feet. This concern employs about 200 hands. On the day of your correspondent's visit, the shipments of cars were sixteen of lumber.
>
> ...There is quite a respectable town built up at this place,

> consisting of about 1,000 inhabitants. This place is continually improving, the mills giving regular employment to many others in addition to the attachees of the company.
>
> ...The establishment has a large store situated between the main line of the Sabine and East Texas Railroad and their siding, which also serves as a depot, telegraph office, and post office. Mr. H. D. Darden, the postmaster, handles 100 letters daily, which is a pretty good showing for a fourth-class post office.
>
> ...Mr. A. Young, to whom Warren and the mills owe their existence, is president and manager of the Warren Lumber Company, and is ably assisted in his duties by Mr. William Brough, Jr., who acts as superintendent of the operating department. Mr. William Peters...is foreman of the mills; Mr. W. N. Curtain of Kansas attends to orders, while Mr. S. F. Tooke...is chief in the mercantile department. Mr. J. Huey...holds the responible position of bookkeeper, while Mr. J. E. Gray handles the checks and attends to the pay rolls.
>
> ...We also found engaged in the store and office two ex-newspaper men, Messrs. I. C. Davis and P. H. Fall, to whom we are indebted for many favors, as well as the accomplished engineer, Mr. J. O. Russell...

It is not entirely clear how the Warren and Brough and Krueger mills interacted with each other. Obviously, Young and Brough were the principally owners, whereas Brough served as superintendent of both mills. Although the sawmills cut timber separately, they maintained only one office, sales force, and commissary, and one tram road logged both mills. In September, 1890, another news account reported that:[128]

> ...Warren, Texas, Sept. 2—Now here is what you may call ...a complete sawmill establishment. Last month they shipped out 260 cars (of lumber), and for this month they are averaging twelve cars a day so far. Their stock is growing less daily, and the orders absorb everything as fast as it comes from the saw. They have a splendid tram road of ten miles of rails, three locomotives, and 35 logging cars. The plant consists of two sawmills with a combined capacity of 130,000 feet per day. The planing

> mill capacity is 40,000 feet per day, and their dry kilns 30,000 feet per day. They have orders in abundance...We return our thanks to Mr. Harry Miller, in charge of the merchandise department...Mr. Charles Martin is the bookkeeper...and Mr. T. H. Plant, timekeeper and cashier, with W. E. Hannon as secretary...

Nonetheless, after a period when all newspaper articles about the mills were so promising and pleasant, the Warren proprietors were already undergoing a period of financial woes, even though a depression or financial panic, perhaps with under-financing for the company; overproduction from too much competition, or a lack of orders did not appear to have played a role in the company's troubles, as the following quote reveals:[129]

> ...Application was filed yesterday with the clerk of the district court of Tyler County for a receiver for the Warren Lumber Company, which will be acted upon by Judge Ford on April 9th. Pending the hearing of the application, S. T. Swinford was appointed temporarily as receiver. The exact condition of the company cannot be obtained, but it is known that the assets are largely in excess of liabilities. The Warren...embarrassment, which is due to slow collections, is greatly regretted...

Although no follow-up article was located, it appeared that the Warren Lumber Company was soon once again able to operate profitably and outside of the receivership.

An article of 1892 also praised the Warren school district immensely, as the following quote relates:[130]

> ...Warren has the largest public school community in Tyler County. It has gone steadily for six months, but will close Tuesday, with an examination and concert. Prof. M. R. Crow, principal; Miss Maude Plant, intermediate (school); and Miss Clara Busby, primary department, have given general satisfaction and deserve great credit for the efficient manner in which they have managed the school...

By 1895, it became apparent that both the Warren and B. and K. sawmills had been greatly enlarged. One article of that year observed that: "...On the 18th the B. and K. (mill) cut 106,000 feet (of lumber), and on the 13th the Warren mill cut 108,000 feet."[131] In fact, during that same period, the B. and K. sawmillers believed they had accomplished a new record cut for a single circular saw, as follows:[132]

> ...The B.(rough) and K.(rueger) mill at this place proved a pace setter last month and made a record believed to be the best ever made on the Sabine and East Texas Railroad. The B. and K. worked 25 days in July and sawed 2,264,000 feet of lumber, an average of nearly 91,000 feet a day. Now 91,000 feet is not a record-breaker for one day, but for 25 consecutive days, it comes near the top. The work force of men are too numerous to be named, but the principals were: R. B. Zimmerman, superintendent; W. L. Cunningham, foreman; John Newton, filer; G. W. Quinn, sawyer; B. M. Cooper, (block) setter; J. D. Brown, engineer; M. J. Reddy, fireman; Charles Richards and John H. Pollitt, yard foremen; and G. L. Justice, tram engineer. Vice president (Wm.) Brough is also charged with being a participant...

At the same moment at the Warren mill, Pat Griffin served as millwright; George Hymers was the sawyer; and H. M. Fleming was superintendent of motive power and machinery (chief machinist).[133]

In addition to Vice President Brough, the Warren Land and Lumber Company officers still included Alexander Young as president and J. N. Johnston as secreatry, but 1896 would be the last year Brough and Young would be present in Warren before the Warren Land and Lumber Company was sold to J. I. Campbell Lumber Company of Houston in 1897. In February, 1896, President Young would make one more championship cut, "claiming the horns" (of the Henry Disston Saw manufacturers) for the month of January, when the B. and K. mill cut 2,422,000 feet of lumber in 27 days on a single circular saw.[134]

It is uncertain whether the J. I. Campbell Company leased the mills briefly or bought the mills outright in May, 1897, but Vice President Johnston went with the new sawmill owners as mill manager at Warren.[135] In 1900, the two Warren sawmills were recorded as cutting 175,000 feet of lumber daily.[136] In 1899, Campbell chartered the Warren and Corsicana Pacific Railroad, a tram road which ran seventeen miles west of Warren to Camp Wood and five miles farther to Stutts in Polk County. The railroad passed within two miles of the Alabama-Coushatta Indian reservation, and "many of the Indians earn money by getting pine knots for the tram road engine."[137]

In about April, 1904, the Campbell sawmill at Warren burned, but was soon rebuilt, one editor soon observing that:[138]

> ...The new mill of J. I. Campbell at Warren on the T. and N. O. (Texas and New Orleans Railroad) in Tyler County will be ready to resume work about the 1st of Nobember. The mill was destroyed by fire last spring and since that time, the planer only has been running. The new mill will have a daily capacity of 80,000 feet, and will give employment to 200 men. The logs will be hauled from Stutts in Polk County, a distance of 22 miles...

(At about the same time, logs for the Keith Lumber Company sawmill at Voth, Jefferson County, were trammed the 22 miles from Stutts to Warren and thence the 32 miles from Warren to Voth, the 54 miles believed to have been a record for tram distance in East Texas during that period.)

On January 15, 1905, the Beaumont *Enterprise* published a history of Warren and the Warren sawmills, particularly of the new Campbell mill just completed, under the caption of "Warren Is Growing." However, financial woes, most likely triggered by the sawmill fire, were soon to plague the Campbell Company as well. In February, 1905, the company asked that a receiver be appointed, and W. H. Norris of the Norris Lumber Company and Samuel Fain Carter of Houston were appointed by the court.[139]

Later the receivers' report to the court confirmed that the value of the Campbell Company's assets were considerably less than its liabilities. The assets included the Campbell properties purchased as part of the Warren Land and Lumber Company; the assets of the Tyler County Land and Lumber Company (mostly timber lands) in Tyler and Polk County; and the assets of the Warren and Corsicana Pacific Railroad. The writer has no further information on the outcome of the receivership. The latter railroad lost its common carrier status in 1908.[140]

One sawmill under the firm name of Tyler County Lumber Company continued to operate at Warren until 1918. And most likely, others or other timber enterprises, to include logging, log debarking and pulpwood, etc., hve been operated at or near Warren in later years. Unlike the ghost towns that many other Tyler County sawmill towns spawned, Warren has survived, even if only a shadow of its former self, to serve a populace consisting of farmers, cattlemen, business men, and even oil drillers who flocked to that area in World War II days. Warren still had a population of 500 and several stores in 1948, but it has long since ceased to exist as an important lumber-manufacturing community.[141]

THE EARLY WOODVILLE SAWMILLS

Woodville, the county seat and the only important town in Tyler County in 1881, does not have an early sawmill history to match many of the other points in the county. The county seat, for reasons unknown, did not attract the sawmillers even though the town had a multitude of marketable timber within easy reach. The principal reason was probably a matter of who owned and controlled the timberlands surrounding the town. The writer found no Woodville sawmills mentioned in the "Letterbook of the East Texas and Louisiana Lumbermen's Association, 1884-1886." However, another source verifies that 1,239 bales of cotton and 68 cars of lumber products were shipped from Woodville during the year ending September 1, 1886. That unknown Woodville sawmill of 1886 was either very small or else cut only a few weeks or months during that year. In other accounts of sawmill information for the years 1888, 1890, and 1891, again no sawmills or lumber shipments were recorded from Woodville.[142] Hence, the writer can readily conclude that, until the erection of the Kirby mill at Woodville in 1902, the contributions of Woodville to the early sawmill history of Tyler County were inconsequential.

There was a hardwood mill at Woodville in 1897, as the following quote records, but no other information about Hodges' sawmill was ever found:[143]

> ...Mr. A. J. Hodges is cutting a bill of 315 sticks (logs) of hardwood timber at his mill in Woodville for export via Sabine Pass...

After losing his Nebraska Lumber Company sawmill at Doucette to fire in October, 1897, P. A. "Pete" Doucette moved to Woodville in 1899 and founded another small mill in the latter town in partnership with L. J. Chapman. After Kirby Lumber Corporation bought out the Doucette and Chapman mill at Woodville in 1901, the Kirby Company assigned Pete Doucette to superintend the Kirby Log Camp No. 2 at Woodville, a duty Doucette was eminently qualified to perform. For the reader desiring to see the size of saw logs at Kirby's Woodville sawmill in 1902 (previously mentioned), a full-page photograph (page 164) in that company's history book of 1902, *Timber Resources of East Texas*, shows logs that appear to be as high as a man's neck, some obviously from 45 to 55 inches in diameter.[144]

The only surviving history of Kirby's new Woodville sawmill of 1902 also appears in that book and is as follows:[145]

> ...The mill operated at this point (Woodville's Kirby Mill M) was a single-circular affair, with an average capacity of 20,000 feet. It was constructed in January, 1900. . . .It was then run steadily until September, 1901, when it was shut down pending negotiations with Kirby Lumber Company, but it was again set in motion in October of the same year.
>
> ...The mill was logged by a tram road, which extends about three miles to the northeast and directly east of the Texas and New Orleans road. It was equipped with one single circular saw; a 2-saw trimmer; edger with four saws and a cut-off saw. The timberlands at this point are exceptionally fine and there is no break in the forest from Woodville to Kirbyville, a distance of thirty miles east. The mill was used mainly to cut ties and the sides were utilized in making yard stocks, which were shipped as fast as a sufficient stock was accumulated.
>
> ...The commissary located here does an average business of $3,500 a month (in 1900 dollars), and this result is accomplished in the face of competition from the other mercantile establishments at Woodville. The checks issued by the company are received by the other stores and are in turn traded out by them at the commissary.
>
> ...The company's pay roll at this point had the names of about sixty men on it and included those in the mill, logging camp and commissary departments. There was only one locomotive used in hauling logs to the saw. The mill was not supplied with a log pond, the water being furnished the engines being filled from a neighboring creek or, when this failed, from a well bored for that purpose.
>
> ...One of the new mills will be located here, as the former plant is not deemed capable of cutting the immense amount of timber lying to the north and east of the plant (all of which belonged to Kirby). It is now nearly ready for

> operation. Woodville is a village of about 1,000 inhabitants and...it is not incorporated. The village is entirely surrounded with pine timber, which stretches to the west a distance of eighteen miles, and the area to the north, east, and south is also grown up in pine trees...

By 1904, the big new mill of Kirby Lumber Corporation at Woodville was cutting 125,000 feet of lumber daily and providing employment to 200 men, all of them in the mills at Woodville. The plant's saw logs were being trammed for 26 miles from Kirby's Silsbee Front or Camp No. 1, twelve miles north of Silsbee, and at least half of the tram mileage to Woodville was over the main line of the East Texas Railroad.[146] In 1907, the Kirby mill at Woodville produced 15,348,000 feet of yellow pine lumber, which was about equal to seven other Kirby mills, but considerably less than that of Kirby's four largest mills at Kirbyville, Call, Browndell, and Bessmay.[147] The last record that the writer has of the Kirby mill at Woodville indicated it was still sawing timber in 1910, but because of the enormous company-owned stumpage then available in north and eastern Tyler County to be sawn only by the Woodville and Mobile sawmills, the writer believes that the Woodville sawmill continued sawing lumber until at least 1920, and perhaps much later.[148]

Lumbering continued to be a major industry at Woodville until about 1968, when the last two sawmills there (believed to have been the Smith-Feagin and Turkey Creek sawmills) closed their doors. Cecil Smith, long-time superintendent of the huge Peavy-Moore sawmill at Deweyville, was the owner of both mills, as well as the Smith-Feagin (formerly Nona-Fletcher) sawmill at Kountze around 1950. Logging and log debarking is still a major industry at a point only four or five miles south of Woodville, and as long as pulpwood and wood chips are in demand at the Lufkin and Evadale paper mills, timber products will continue to supply a sizeable percentage of Woodville's income.

A SUMMARY OF TYLER COUNTY SAWMILLING

For decades to come, Tyler County history will continue to point out the ghost towns in the forest where thriving lumber industries once stood, and where many residents lived, went to school and church, and died and were buried. And it is good to know more of those industries that once existed, then moved away, leaving only the silent pine needles to cover their foot prints. Sometimes only a lost and lonely graveyard survives in the forest underbrush to bear mute witness of that fact and record here or there a name or two of those who once worked or lived there.

Whatever else one can say, those early sawmills built Tyler County into what it is today, as well as those neighboring counties of East Texas in all directions. The mills did leave millions of acres of unsightly "cutover" stump lands, some of which were converted to farms and orchards, but most of which gradually returned to second-growth forests. And on most any still, clear day at sunset, that is, if one is inclined to reminisce of days gone by, one might still hear the faint sounds of yesteryear's sobs and laughter, or catch the faint echoes of the big bandsaws' screech, the same saws that cut the building blocks of Tyler County's foundations and provided the payrolls that sustained that county's earliest citizens.

FOOTNOTES

Camps such as this were used by cypress loggers in Orange County. This camp was probably located on Adams Bayou in the 1920s.

Footnotes

Introduction

[1]Kirby Lumber Co., *Timber Resources of East Texas* (Houston: 1902), p. 164.

[2]Photograph, "Log in Jasper County," (Beaumont) *Enterprise*, Aug. 3, 1927.

[3](Beaumont) *Journal*, Nov. 13, 1904.

[4]P. R. McCoy, *Shelby County Sampler* (Lufkin: 1982), p. 63.

[5]Bob Bowman, *The 35 Best Ghost Towns of East Texas* (Lufkin: 1988), p. 14.

[6]Reputed to have been published in (Nacogdoches) *Texas Republican*, and reprinted in H. P. Easton, "The History of The Texas Lumbering Industry," Ph. D. Dissertation, University of Texas at Austin, 1947, p. 20.

[7]"Oldest Mill in East Texas," (Houston) *Morning Star*, Feb. 1, 1842.

[8]E. W. Winkler (ed.), *Manuscript Letters and Documents of Early Texans* (Austin: 1937), pp. 83, 93.

[9]W. P. Webb et al, *The Handbook of Texas* (Austin: 1952), I, 117.

[10]W. F. Gray, *From Virginia to Texas: The Diary of Col. William Fairfax Gray* (reprint: Houston: 1965), p. 146.

[11](Brazoria, Tx.) *The People*, April 18, 1838.

[12](Houston) *Morning Star*, Jan. 16, 1840.

[13](Houston) *Telegraph and Texas Telegraph*, April 21, 1841.

[14]J. DeCordova, *Texas: Her Resources and Her Public Men* (reprint; Waco: 1979), p. 39.

[15]W. T.Block, "An Early East Texas Captain Of Commerce: David Robert Wingate," *Texas Gulf Historical and Biographical Record*, VIII (Nov. 1977), p. 60.

[16]"Among The Millmen," (Galveston) *Daily News*, July 11, 1896.

[17]*Ibid.*

[18]"The Page Circular Sawmill," (Houston) *Telegraph and Texas Register*, Oct. 12, 1848, p. 1. In 1867-1868, a number of advanced model circular sawmills were advertised in Galveston and Houston papers.

[19]Block, "David Robert Wingate," pp. 59-80; Letter, J. H. to E. H. Cushing, Oct. 24, 1862, and "Night Attack at Sabine Pass," (Houston) *Tri-Weekly Telegraph*, Nov. 5, 1862.

[20]Webb, *Handbook of Texas*, I, 275-276.

[21]"Lumber and Lumbermen," (Galveston) *Daily News*, Aug. 14, 1889; "History of The Thompson-Tucker Lumber Company," (Beaumont) *Enterprise*, April 2, 1905.

[22]"Beaumont's Big Business," (Galveston) *Daily News*, Feb. 15, 1888.

[23]W.T. Block, "Henry J. Lutcher: Sawmiller and Architect of Orange," in W.T. Block, *Frontier Tales of The Texas-Louisiana Borderlands* (Nederland: 1988), pp. 324-327.

[24]"Southern Pine From Forest to Retailer," *American Lumberman* (April 16, 1904), pp. 35-42.

[25]From the writer's personal count from each county's 1860 Sched. IV, Products of Agriculture, for his two books on the Trinity and Sabine rivers, entitled *Cotton Bales, Keelboats and Sternwheelers.*

[26]"Timbered Texas," (Galveston) *Weekly News*, April 13, 1893.

[27]Ibid.

[28]G. C. and Santa Fe Railroad *Circular No. 2090*, Galveston, March 28, 1904.

[29]"Lumber Mills of Texas," *Southern Industrial and Lumber Review*, (Sept. 6, 1906), pp. 27-29.

[30]Webb, *Handbook of Texas*, I, 745-746, 966; "Kirby Sawmills," (Beaumont) *Enterprise*, March 26, 1905.

[31]R. S. Maxwell, *Sawdust Empire: The Texas Lumber Industry* (College Station: 1983), p. 102; "Kirby Lumber Company, Statement of Production and Payroll, December 31, 1918," Houston, R. F. Ford, Auditor, in Kirby Papers.

[32]"Lumber Still Big Industry," (Beaumont) *Enterprise*, Aug. 2, 1927.

[33]"Sawmills Still Do Big Business," (Beaumont) *Enterprise*, Aug. 2, 1927, page 31-A.

[34](Beaumont) *Journal*, March 5, 19, 1905.

[35]*Ibid.*, Aug. 24, 26, 1911; Letter, C. P. Myer to John S. Bonner, Oct. 23, 1914, in the Kirby Papers, SFASU Forestry Archives.

[36]W.T. Block, "Documents of The Early Sawmilling Epoch" *Texas Gulf Historical and Biographical Record*, IX (Nov., 1973), 47-63, copied from the 1880 Jefferson and Orange County

censuses, Scheds. IV, Products of Industry, Microfilm Reel #48, Texas State Archives.

[37]George Creel, "Feudal Towns of East Texas,"*Harper's Weekly*, LX (Sept. 15, 1915), pp. 76-78.

[38]John C. Gee, Jr., "The Call of Texas," in Pauline Hines et al, *Crosscuts: An Anthology of Memoirs of Newton County Folk* (Austin: 1984), pp. 3-6.

Angelina County

[1]W.P. Webb et al (eds.), *The Handbook of Texas* (Austin: 1952), I, 49.

[2]"Texas Timbered Wealth," (Galveston) *Daily News*, Jan. 1, 1900.

[3]"Angelina County," (Beaumont) *Journal*, January 18, 1904; 1904 *Texas Almanac* (Dallas: Belo and Company, 1904), p. 204.

[4]Effie Maddox Boon, "The History of Angelina County," MA thesis, University of Texas at Austin, 1937, p. 83.

[5]*Ibid.*

[6]R. S. Maxwell, *Whistle In The Piney Woods: Paul Bremond and The Houston East and West Texas Railroad* (Texas Gulf Coast Historical Assn., 1963), pp. 1-77.

[7]Webb, *Handbook of Texas*, I, 50, 341; Boon, "History of Angelina County," pp. 93-95; Lita Mayberry, "Keltys - An East Texas Sawmill Town," MA Thesis, Stephen F. Austin University, 1948, pp. 12-13.

[8]Webb, *Handbook of Texas*, I, p. 937, and II, p. 530.

[9]*Ibid.*, I, 536.

[10]Boon, "History of Angelina County", p. 112.

[11]"Sawmill Vicissitudes: The Passing of Michelli," (Beaumont) *Journal*, Nov. 27, 1904.

[12]Webb, *Handbook of Texas*, II, 760.

[13](Houston) *Telegraph and Texas Register*, April 21, 1841.

[14]J. DeCordova, *Texas; Her Resources and Her Public Men* (Waco: 1969), p. 39.

[15]Boon, "History of Angelina County," p. 86.

[16]"Oldest Mill in Texas," (Houston) *Morning Star*, Feb. 1, 1842.

[17](Beaumont) *Journal*, June 12, 1914.

[18]A. Williams and E. Barker (eds), *The Writings of Sam Houston* (Austin: Pemberton Press, 1970). IV, pp. 32-34.

[19](San Augustine) *Redlander*, Jan. 20, 1844; L. F. Blount, "The Story of Old Pattonia," *East Texas Historical Journal,* V (Mar., 1967), pp. 14-16.

[20]"Contract With Steamboat*Angelina*, Moses L. Patton, Master," 1849, Vol. A, p. 69, Personal Property Records, Jefferson County Archives; (Galveston) *Weekly News*, March 11, 1850; (Houston) *Telegraph and Texas Register*, May 2, 1850; (Nacogdoches) *Times*, Jan. 13, 1849; A. P. McDonald (ed.), *Hurrah For Texas! The Diary of Adolphus Sterne* (Waco: 1969), pp. 194-199.

[21]Boon, "History of Angelina County," p. 88.

[22]Manuscript Census Returns of 1860, Angelina County, Texas, Schedule V, Products of Industry, on microfilm, Lamar University Library.

[23]Boon, "History of Angelina County, p.88; B. Bowman et al (eds.), *Land of The Little Angel: A History of Angelina County, Texas* (Lufkin: 1976), pp. 113-114, 137.

[24]Bowman, *Land of The Little Angel*, pp. 29, 162; "Timbered Texas," (Galveston) *Weekly News*, April 13, 1893.

[25]"Texas' Timbered Wealth," (Galveston) *Daily News*, Jan. 1, 1900; Maxwell, *Whistle in The Piney Woods*, p. 9.

[26] Bowman et al, *Land of The Little Angel*, p. 162.

[27](Galveston) *Weekly News*, April 13, 1893; "Texas' Timbered Wealth," (Galveston) Daily News, Jan. 1, 1900; Maxwell, *Whistle in The Piney Woods*, pp. 69-71; Manuscript Census Returns of Angelina County, Texas, 1900, Sch. I, Precinct 1, res. 477-479.

[28]Boon, p. 118; Bowman, *Land of The Little Angel*, p. 180.

[29]"Angelina County," (Beaumont) *Journal*, Jan. 8 and Nov. 13, 1904.

[30]Webb, *Handbook of Texas*, II, 73; Manuscript Census Returns of 1910, Logtown, Newton County, Texas, Sch. I, Pct. 1, res. 101ff.

[31]No. 1992-009-0004—Technical Drawing of Lufkin Foundry and Machine, dated Aug. 2, 1918, for Boynton Bros. sawmill, Sturgis, Texas; Webb, *Handbook of Texas*, II, 682; "Lufkin Thrives With Pine Mills," (Beaumont) *Enterprise*, Aug. 3, 1927.

[32]Manuscript Census Returns of 1900, Angelina County, Texas, Sch. I, Pct. 2, res. 218; Bowman et al, *Land of The Little Angel*, p. 178.

[33](Beaumont) *Journal*, April 9, 1905; (Beaumont) *Enterprise*, April 3, 1905.

[34]"Sawmills on The Sante Fe," *Southern Industrial and Lumber Review*, Sept., 1906, p. 27.

[35]"Jasper Peavy Passes," (Lufkin) *Daily News*, Nov. 16, 1942; "A. J. Peavy Dies," (Shreveport) *Times*, Nov. 16, 1942.

[36]"The Passing of Michelli," (Beaumont) *Journal*, Nov. 27, 1904.

[37]Boon, "History of Angelina County," pp. 102-103.

[38]"Timbered Texas," (Galveston) *Weekly News*, April 13, 1893.

[39]"Texas' Timbered Wealth," (Galveston) *Daily News*, Jan. 1, 1900.

[40]Letter, Johnathan Gerland to W. T. Block, Lufkin, July 1, 1993.

[41]Plat of Michelli Land Grant, Vol. 29, pp. 366-67, Angelina County Deed Records; "Wm. Cameron and Co.: The Angelina Plant," *American Lumberman* (Nov. 17, 1900), pp. 50-53; "The Passing of Michelli," (Beaumont) *Journal*, Nov. 27, 1904.

[42]Earnest Balch, "Biographical Sketches of Industrial Leaders in Lufkin and Angelina County," M. A. thesis, Stephen F. Austin University, 1949, pp. 6-7; "German Immigrant Is Factor in East Texas Development," (Beaumont) *Enterprise*, Aug. 3, 1927; F. Johnson and E. Barker (eds.), *Texas and Texans* (New York: 1914), IV, p. 2,413; J. Dione, "Fifty Years in Lumber," *Gulf Coast Lumberman* (July, 1937), p. 5.

[43]Balch, "Biographical Sketches," pp. 6-7.

[44]"Lumber Business at Keltys," (Galveston) *Daily News*, April 10, 1890.

[45]"Angelina County Lbr. Co. Incorporation," (Galveston) *Daily News,* October 28, 1890.

[46]"Lumber and Lumbermen," (Galveston) *Daily News*, May 23 and Nov. 14, 1896.

[47]Manuscript Census Returns of 1900, Sched. I, Keltys, Angelina County, Texas, Precinct 1, res, 292-407.

[48](Beaumont) *Journal*, Jan. 18, 1904.

[49]"German Immigrant Is Factor In Development," (Beaumont) *Enterprise*, Aug. 3, 1927; Earnest Balch, "Biographical Sketches of Industrial Leaders of Lufkin," pp. 6-7.

[50]*Southern Industrial and Lumber Review*, Dec., 1908, p. 54, and June, 1909, p. 72.

[50a]Direct Quote, Letter, Jonathan Gerland, East Texas Sawmill Database, Texas Forestry Museum, Lufkin, to W. T. Block, Aug. 19, 1993.

[51]"Keltys," (Galveston) *Daily News*, Aug. 31, 1895.

[52]Manuscript Census Returns of 1900, Keltys, Angelina County, Texas, Sched. I, Pct. 2, res. 331-428.

[53]Boon, "History of Angelina County," pp. 92-97; Webb, *Handbook of Texas*, I, 50; Bowman et al, *Land of The Little Angel*, pp. 167-168.

[54]Boon, "History of Angelina County," pp. 97-98; Lita Mayberry, "Keltys - An East Texas Sawmill Town" M. A. Thesis, Stephen F. Austin State University, 1948, pp. 16-20; J. Dionne, "Fifty years in Lumber," *Gulf Coast Lumberman* (July, 1937), p. 5.

[55]Boon, "History of Angelina County," pp. 99-100; L. Mayberry, "Keltys - An East Texas Sawmill Town," pp. 23-29.

[56]Boon, p. 91; Bowman et al, *Land of The Little Angel*, p. 169.

[57]Letter, Jonathan Gerland, Texas Forestry Museum, to W. T. Block, August 19, 1993; J. Dionne, "Fifty years in Lumber," *Gulf Coast Lumberman* (July, 1937), pp. 5ff; (Beaumont) *Journal*, Jan. 18, 1904.

[58]"Biography of J. Press Carter," *Southern Industrial and Lumber Review* (March, 1909), pp. 51, 80.

[59]Manuscript Census Returns of 1910, Baber, Angelina County, Texas, Sched. I, Pct. 5, residences 234-267.

[60]Boon, "History of Angelina County," p. 117; Bowman et al, *Land of The Little Angel*, p. 143.

[61]Webb, *Handbook of Texas*, I, 867; Bowman et al, *Land of The Little Angel*, pp. 163-165.

[62](Beaumont) *Journal*, Jan. 15, 1904.

[63]"From Beaumont to Nacogdoches," *Southern Industrial and Lumber Review* (June, 1904), pp. 24-25.

[64]"Lumber Mills of Texas," *Southern Industial and Lumber Review* (Sept., 1906), pp. 26-28; Bowman et al, *Land of The Little Angel*, p. 177; B. Bowman, *35 Best Ghost Towns in East Texas* (Lufkin: 1988), p. 198.

[65]Boon, "History of Angelina County," p. 118; Bowman, *Land of The Little Angel*, p. 178.

[66]"From Beaumont to Nacogdoches," *Southern Industrial and Lumber Review* (Sept., 1904), p. 24.

[67]"Timbered Texas," (Galveston) *Weekly News*, April 13, 1893.

[68]Letters, Lufkin Manufacturing Company to Angelina County Lumber Company, dated June 24 and Nov. 25, 1896, Forestry Archives, SFA University.

[69]Order for lumber, Lufkin Land and Lumber Company to Angelina County Lumber Company, Dec. 1, 1899, Forestry Archives; also "Texas Timbered Wealth," (Galveston) *Daily News*, Jan. 1, 1900.

[70]R. S. Maxwell, "History of Three Lumber Families," *Texas Forests and Texans* (May-June, 1964), pp. 21, 27; "Life of G. A. Kelley," *American Lumberman*, January 16, 1915.

[71]Manuscript Census Returns of 1900, Lufkin, Angelina County, Texas, Sched. I, Pct. 1, res. 48-222.

[72]"History of Lufkin Land and Lumber Company," (Beaumont) *Enterprise*, April 9, 1905.

[73](Beaumont) *Journal*, Nov. 13, 20, 27, 1904; Boon, "History of Angelina County, p. 110; Bowman, *Land of The Little Angel*, pp. 138-139.

[74](Beaumont) *Journal*, Nov. 20, 1904; Jan. 22, 1905.

[75]*Ibid.*, Jan. 22, 1905.

[76]*Ibid.*, June 25 and July 9, 1905.

[77]Manuscript Census Returns of 1910, Lufkin, Angelina County, Texas, Sched. I, Pct. 1, res. 91-146.

[78]Boon, "History of Angelina County," pp. 110-111; Bowman et al, *Land of The Little Angel*, p. 179.

[79]Bowman et al, *Land of The Little Angel*, pp. 120, 315.

[80]Manuscript Census Returns of 1910, Lufkin, Angelina County, Texas, Sched. I, Pct. 1, res. 10-137.

[81]"Lufkin Thrives On Pine Mills," (Beaumont) *Enterprise*, Aug. 3, 1927.

[82]Bowman, *Land of The Little Angel*, p. 144; "Mill at Clark's Ferry," *Southern Industrial and Lumber Review*, Oct., 1908, p. 41.

[83]B. Bowman, *The 35 Best Ghost Towns*, p. 200; Webb, *Handbook of Texas*, I, p. 4 and II, p. 258.

[84]Bowman, *35 Best Ghost Towns*, p. 202; Bowman et al, *Land of The Little Angel*, p. 178; Angelina County Census of 1900, Sched. I, Pct. 2, res. 265.

[85]Webb, *Handbook of Texas*, II, 138; Bowman, *35 Best Ghost Towns*, p. 198; (Beaumont) *Journal*, Feb. 5, 1905.

[86]Manuscript Census Returns of 1910, Durant, Angelina County, Texas, Sched. I, Pct. 3, res. 19-88; Webb, *Handbook of Texas*, I, 527; Bowman, *35 Best Ghost Towns*, p. 186; Bowman et al, *Land of The Little Angel*, p. 176.

[87]Bowman, *Land of The Little Angel*, p. 176; (Galveston) *Weekly News*, April 13, 1893; *Daily News*, Jan. 1, 1900.

[88]"Lumber Mills of Texas," *Southern Industrial and Lumber Review*, Sept., 1906, pp. 26-28; Bowman et al, *Land of The Little Angel*, p. 177.

[89]Boon, "History of Angelina County," pp. 110-111.

[90]Bowman, *35 Best Ghost Towns*, p. 193; Webb, *Handbook of Texas*, I, 841.

[91]Webb, *Handbook of Texas*, I, 472; Bowman, *Land of The Little Angel*, p. 174; "Plat of Portion of Michelli Grant Showing Davisville and Davisville Lumber Company," Angelina County Deed Records, Vol. 29, pp. 366-367, dated Oct., 1909.

[92]Webb, *Handbook of Texas*, I, 248; R. W. Haltom, *History and Description of Angelina County* (reprint: Austin, 1961), pp. 59-60; Bowman, *Land of The Little Angel*, pp. 299, 309; (Galveston) *Weekly News*, April 13, 1893; *Daily News*, Jan. 1. 1900.

[93]Bowman et al, *Land of The Little Angel*, p. 176.

[94]Webb, *Handbook of Texas*. II, 242

[95]Bowman, *35 Best Ghost Towns*, p. 203; Webb, *Handbook of Texas*, II, 385.

[96]Bowman et al, *Land of The Little Angel*, p. 176.

[97]*Ibid*, p. 180.

[98]See Footnote 40.

[99]Haltom, *History and Description of Angelina County*, p. 55.

[100]Jonathan Gerland, Texas Forestry Museum, to W. T. Block, July 1, 1993.

[101]"Timbered Texas," (Galveston) *Weekly News*, April 13, 1893; "Texas Timbered Wealth," (Galveston) *Daily News*, Jan. 1, 1900.

[102]Copies of five letterheads, Clawson Lumber Co., Clawson, Texas, dated April 20 and Sept. 27, 1897; and Mar. 3, July 2, and Dec. 1, 1899, in Angelina County Lumber Company Papers.

[103]Manuscript Census Returns of 1900, Clawson, Angelina County, Texas, Sched. I, Pct. 1, res. 477-535.

[104]Bowman, *Land of The Little Angel*, p. 171.

[105]Price List B-22, Foster Lumber Company, Jan. 21, 1903, Forestry Archives.

[106]"Jasper Peavy Dies," (Lufkin) *Daily News*, Nov. 16, 1942.

[107]Bowman et al, *Land of The Little Angel*, p. 171.

[108]Manuscript Census Returns of 1910, Clawson, Angelina County, Texas, Sched. I, Pct. 3, res. 361-371.

[109]Webb, *Handbook of Texas*, II, 391.

[110]Boon, "History of Angelina County, Texas," pp. 103-104.

[111]Bowman et al, *Land of The Little Angel*, p. 162.

[112]B. Bowman, *35 Best Ghost Towns*, p. 203.

[113]R. W. Haltom, *History and Description of Angelina County*, p. 61.

[114]"Timbered Texas," (Galveston) *Weekly News*, April 13, 1893; "Texas' Timbered Wealth," (Galveston) *Daily News*, Jan. 1, 1900.

[115]"Notes From Keltys," (Galveston) *Daily News*, Jan. 30, 1897.

[116]Hattie Roach, A History of Cherokee County (Dallas: 1934), p. 158; *Land of The Little Angel*, p. 162.

[117]Dorothy Young, "The Pollok Sawmill," in Angelina County Genealogical and Historical Society, *The History of Angelina County, Texas, 1840-1991* (Dallas: 1992), p. 114; Letter, J. Gerland, Texas Forestry Museum, to W. T. Block, Lufkin, Aug. 24, 1993; Angelina County Deed Records, Vols. Q: 394-397; 1: 513-514; 5: 310-316; 41:72-78; 43: p. 1.

[118]Manuscript Census Returns of 1910, Pollok, Angelina County, Texas, Sched. I, precinct 3, res. 445 to 526; Bowman et al, *Land of The Little Angel*, p. 163.

[119]Webb et al, "S. F. Carter," *Handbook of Texas*, I, 303; "J. Press Carter," *Southern Industrial and Lumber Review* (March, 1909), pp. 51, 80; "Beaumont's Big Business," (Galveston) *Daily News*, Feb. 15, 1888.

[120]East Texas Sawmill Database Report on Emporia Lumber Company, prepared by J. Gerland, Aug. 11, 1993, Texas Forestry Muesum; "Timbered Texas," (Galveston) *Weekly News*, April 13, 1893; Angelina County Deed Records, Vols. U, pp. 1-7; V, 350-352; and Vol. 22, p. 490.

[121](Galveston) *Daily News*, Aug. 1 and Sept. 26, 1896; Bowman et al, *Land of The Little Angel*, p. 176; Manuscript Census of 1900, Emporia, Angelina County, Texas, Sched. I, Pct. 4, res. 280-323.

[122]Boon, "History of Angelina County," p. 101; "Texas' Timbered Wealth," (Galveston) *Daily News*, Jan. 1, 1900; Letterhead, Emporia Lumber Company, dated Aug. 28, 1897, Angelina County Lumber Company Papers, Forestry Collections.

[123]"J. Press Carter," *Southern Industrial and Lumber Review* (March 1909), pp. 51, 80; "History of The Doucette Sawmill," (Beaumont) *Enterprise*, March 25, 1905.

[124]E. Boon, "History of Angelina County," pp. 101-1-2.

[125](Beaumont) *Journal*, November 13, 1904; Webb, *Handbook of Texas*, I, 303, "Biography of Press Carter," *Southern Industrial and Lumber Review* (March 1909) p. 51.

[126]Boon, "History of Angelina County," pp. 114-116; Bowman et al, *Land of The Little Angel*, p. 176; *The 1915 Directory of American Sawmills* (Washington D. C.: Govt. Prtg. Office, 1915), p. 210; Angelina County Lumber Co. records in SFASU Forestry Archives; and technical drawing archives of Lufkin Industries.

[127]*Ibid.*; Webb, "Ewing," *Handbook of Texas*, I, 578.

[128]Boon, p. 116.

[129]*Ibid.*

[130]Cynthia Jones, "The Builders of Manning," 1934, in R. L. Poland, *Were You At Manning?* (Aug., 1985), p.5.

[131]Jones, "The Builders of Manning," p. 5; "Biography of W. T. Carter," in Boon, "History of Angelina County," pp. 131-134; F. T. Beck, "Development of the W. T. Carter and Brother Lumber Industry," M. A. Thesis, S.- F. Austin State University, 1950, pp 1-8; J. and J. Wheat, "W. T. Carter," in *Sketches of Tyler County History* (Bevil Oaks: 1986), pp. 147-149.

[132]"G. A. Kelley," in Boon, "History of Angelina County," pp. 134-135; R. Maxwell, "History of Three Lumber Families," *Texas Forests and Texans* (May-June, 1964), pp. 21, 27; "Biography of G. A. Kelley," *American Lumbermen* (Jan. 16, 1915).

[133]Jones, "The Builders of Manning," in *Were You At Manning?* p. 5-6.

[134]V. M. Pride, "The Origin of Manning," p. 4-5, in Poland, *Were You At Manning?*

[135]Boon, "History of Angelina County," p. 112.

[136]"Carter-Kelley Runs Nights," *Southern Industrial and Lumber Review* (Oct. 1908), p. 40.

[137]Boon, p. 112.

[138]"Sawmill Race Troubles," *Southern Industrial and Lumber Review* (Oct. 1908), p. 40.

[139]Manuscript Census Returns of 1910, Manning, Angelina County, Texas, Sched. I, Precinct 5, residences 64-170.

[140]D. Laird, "Important Events in Manning," pp. 10-11, in Poland, *Were You At Manning?* (1985).

[141]R. L. Poland, "Manning in General," p. 22, in *Were You At Manning?*

[142]M. M. Flournoy, "The History of Manning," in the pamphlet *Were You At Manning?*

[143]Boon, "History of Angelina County," p. 113; R. L. Poland, "Manning in General," pp. 21-23, in *Were You at Manning*?

[144]H. Richmond, "Present Facts About Manning," p. 13, in *Were You At Manning?* (1985)

[145]Chapters I through XVI, pp. 1-29, of *Were You At Manning?* (1985); for the Turner-Cummings mill, see Nellis and Pierson, Directory of American Sawmills-Series No. 27 (Washington: Dept. of Agriculture, Govt. Printing Office, 1915), p. 211.

[146]Webb et al, "Diboll," *Handbook of Texas*, I, pp. 499-500, and II, p. 760.

[147]Bowman et al, *Land of The Little Angel*, p. 156.

[148]M. Biesele, *The Cornbread Whistle: Oral History of A Texas Timber Company Town* (Diboll: 1986), p. 2; "Southern Pine From Forest To Retailer," *American Lumberman* (April 16, 1904), pp. 35-42.

[149]Letter, Southern Pine Lumber Co. to Angelina County Lbr. Co., Keltys, Jan. 21, 1896, Angelina County Lbr. Co. Archives, SFAS University.

[150]*Ibid.*, letterhead dated Nov. 12, 1909.

[151]*Ibid.*, letterhead dated April 18, 1923.

[152]M. Biesele, *Cornbread Whistle*, pp. 9-11; Boon, "History of Angelina County," p. 106; *American Lumberman* (April 16, 1904), p. 36; "History of Lufkin Land and Lbr. Co.," (Beaumont) *Enterprise*, April 9, 1905.

[153]"Southern Pine," *American Lumberman* (April 16, 1904), p. 37.

[154]Manuscript Census Returns of 1900, Diboll, Angelina County, Texas, Sched. I, Pct. 4, res. 178-823.

[155]*American Lumberman* (April 16, 1904), pp. 38-39.

[156]*Ibid.*, pp. 39-40.

[157]Manuscript Census Returns of 1910, Diboll, Angelina County, Texas, Sched. I, pct. 4, res. 244-356.

[158]"The Three Southern Pine Lumber Company Mills," *Southern Industrial and Lumber Review* (July, 1912), p. 45. The Central Coke and Coal Company (the 4-C) sawmill, which existed at Kennard, Houston County, from 1902 until 1917, cut 350,000 feet daily and was the second largest sawmill in the United States.

[159]"Diboll Mill Burns," *Gulf Coast Lumberman* (April 15, 1915), p. 18.

[160]Webb, "Texas South-Eastern," *Handbook of Texas*, II, 760; *American Lumberman* (April 16, 1904), pp. 37-38.

[161]Boon, "History of Angelina County," p. 106.

[162]Biesele, *Cornbread Whistle*, p. 27.

[163]Hattie J. Roach, *A History of Cherokee County, Texas* (Dallas: 1934), p. 88.

[164]Boon, "History of Angelina County," pp. 106-108.

[165]"Diboll Distinctly on Lumber Map," (Beaumont) *Enterprise*, Aug. 3, 1927.

[166]*Ibid.*

[167]H. P. Easton, "The History Of The Texas Lumbering Industry," Ph. D. dissertation, University of Texas at Austin, 1947, pp. 383-384.

[168]Boon, "History of Angelina County," p. 108.

[169]H. B. Ellison, "The Lumber Industry at Pineland, Texas," p. 15, Grad. Paper, 1947, Forestry Collections, S. F. Austin State University.

[170]M. Biesele, *The Cornbread Whistle*, p. 3.

[171]Easton, "The History Of The Texas Lumbering Industry," pp. 384-385.

[172]Boon, "Diboll," in "History of Angelina County," p. 106.

[173]Easton, p. 385.

[174]M. Biesele, *The Cornbread Whistle*, p. 10.

[175]*Ibid.*

[176]"Southern Pine From Forest to Retailer," *American Lumberman* (April 16, 1904), pp. 36-40.

[177]Bowman et al, *Land of The Little Angel*, p. 157.

[178]Biesele, *The Cornbread Whistle*, p. 58.

[179]"Diboll Distinctly On Lumber Map," (Beaumont) *Enterprise*, Aug. 3, 1927.

[180]M. Biesele, *The Cornbread Whistle*, pp. 67-71.

[181]Boon, "History of Angelina County," pp. 105-109.

[182]Biesele, *The Cornbread Whistle*, p. 116.

[183]"Sawmills Still Do Big Business," (Beaumont) *Enterprise*, Aug. 3, 1927.

Chambers County

[1]J. H. Harry, *A History of Chambers County, Texas* (Dallas: 1981), pp. 140-141.

[2](Houston) *Morning Star*, Feb. 19, 1842; W. P. Webb et al (eds.), "Biography of T. F. McKinney," *The Handbook of Texas* (Austin: 1952), II, 117.

[3]Volume G, pp. 44-47, Deed Records, Chambers County, Texas Archives.

[4]M. S. Henson and K. Ladd, *Chambers County: A Pictorial History* (Norfolk: 1988), p. 136, quoting (Galveston) *Weekly News*, June 3, 1851.

[5](Galveston) *Weekly News*, Dec. 20, 1853.

[6]*Ibid.*, May 12, 1855.

[7]Manuscript Census Returns of 1860, Liberty County, Texas, Sched. V, Products of Industry; M. Partlow, *Liberty, Liberty County and The Atascosita District* (Austin: 1974), pp. 61, 329.

[8]Webb, *Handbook of Texas*, II, 241.

[9]Henson and Ladd, *Chambers County: A Pictorial History*, p. 136.

[10]Vol. C, pp. 705-706, Chambers County Deed Records.

[11]*Ibid.*, Vol. D, p. 57.

[12]Vol. D, p. 101, Chambers County Deed Records.

[13]*Ibid.*, Vol. D, pp. 53, 112-113; Henson and Ladd, Chambers County, p. 136.

[14]*Ibid.*, Vol. D, pp. 113-117.

[15](Galveston) *Daily News*, Dec. 15, 1884.

[16]*Ibid.*, December 10, 1896.

[17](Wallisville, Tx.) *Age*, March 15, 1897; October 5, 1898; Manuscript Census Returns of 1900, Chambers County, Texas, p. 10.

[18](Liberty, Tx.) *Vindicator*, July 23, 1897.

[19](Galveston) *Daily News*, December 10, 1896.

[20](Wallisville) *Age*, June 14, 15, 1899; Oct. 5, 1898.

[21]*Ibid.*, Dec. 22, 1897.

[22]Henson and Ladd, *Chambers County*, p. 136; (Liberty) *Vindicator*, April 23, 1896.

[23](Liberty) *Vindicator*, Jan. 10, 1896.

[24]*Ibid.*

[25]Henson and Ladd, *Chambers County*, quoting J. H.Harry, *A History of Chambers County*, p. 131; (Wallisville) *Age*, Dec. 22, 1897.

[26](Liberty) *Vindicator*, Nov. 26, 1897; (Wallisville) *Age*, March 29, 1899.

[27]Harry, *A History of Chambers County*, p. 141; Vols. G, p. 429; K, p. 470; and L, p. 795, Chambers County Deed Records.

[28]Henson and Ladd, *Chambers County*, p. 138, quoting (Wallisville) *Age*, April 19, 1899.

[29]Coastal Environments, Inc., prepared for Corps of Engineers, *Cultural Resources Evaluation of the J. J. Mayes Farm and The Cummings Lumber Mill.....* (Baton Rouge: July, 1986), p. 28; (Liberty) *Vindicator*, October 8, 1897.

[30](Wallisville) *Age*, Dec. 22, 1897; (Liberty) *Vindicator*, Sept. 1, 1893.

[31]"Liberty County," (Galveston) *Daily News*, April 10, 1897.

[32](Liberty) *Vindicator*, July 31, 1903.

[33]"Chambers County," (Galveston) *Daily News*, July 4, 1896.

[34](Liberty) *Vindicator*, Jan. 10, 1896.

[35]*Ibid.*

[36]Henson and Ladd, *Chambers County*, p. 136.

[37](Liberty) *Vindicator*, Sept. 17, 1897; (Galveston) *Daily News*, Sept. 15, 1897.

[38](Wallisville) *Age*, Oct. 5, 1898.

[39](Galveston) *Daily News*,Dec. 10, 1896.

[40]Hensonand Ladd, *Chambers County*, p. 136.

[41]*Cultural Resources Evaluation etc.*, pp. 29-30.

[42]*Ibid.*, p. 30; Vol. W, pp. 370-371, 457, Liberty County Deed Records.

[43]*Cultural Resources Evaluation Etc.*, p. 30.

[44](Beaumont) *Journal*, Jan. 15, 1904.

[45]*Ibid.*, Nov. 13, 1904.

[46]*Cultural Resources Evaluation Etc.*, p. 36.

[47]*Ibid.*, pp. 27, 36.

[48]*The Photographic Heritage of Wallisville*, p. 1.

[49]*Cultural Resources Evaluation Etc.*, p. 36.

[50]Vols. 14, pp. 417-419; 15, p. 508; 16, pp. 205-207, 314-316, Liberty County Deed Records.

[51](Liberty) *Vindicator*, Feb. 12, 1904

[52](Beaumont) *Journal*, February 26, 1905.

[53]*Ibid.*

[54](Liberty) *Vindicator*, Feb. 5 and March 12, 1909.

[55]"New Cummings Mill Starts up," (Beaumont) *Enterprise*, July 21, 1907.

[56]"Lumber Mills of Texas...on Texas and New Orleans Railroad," *Southern Industrial and Lumber Review* (Sept., 1906), p. 28.

[57]Harry, *History of Chambers County*, p. 141.
[58]*Cultural Resources Evaluation Etc.*, p. 29.
[59]*Ibid.*, pp. 29-37.
[60]Henson and Ladd, *Chambers County*, p. 138.
[61]*Ibid.*, p.138; *Cultural Resources Evaluation Etc.*, p. 37.
[62]Henson and Ladd, *Chambers County*, p. 138; (Anahuac) *Progress*, Jan. 14, 1916.
[63]*Ibid.*, p. 138; W. O. Work's "Map of The Lost Lake Area" of 1930.
[64]Harry, *History of Chambers County*, p. 142.
[65]*Ibid.*
[66](Anahuac) *Progress*, July 23, 1942.
[67]*Ibid.*, Jan. 27, 1949.

Jefferson County

[1]"Timbered Texas Wealth," (Galveston) *Daily News*, January 1, 1900.
[2](Houston) *Telegraph and Texas Register*, March 17, 1838.
[3]"Quarterly Return," R. C. Doom, collector, to the Secretary of Treasury, June 30 and Sept. 30, 1839, Port of Sabine Customs Records, File 4-21/10, Texas State Archives.
[4]*Texas Almanac*, 1859 (Galveston: Richardson and Co., 1860), p. 150.
[5]H. P. N. Gammel, *Laws of Texas*, 1822-1897 (Austin: Gammel Book Co, 1898), II, 13.
[6]Vol. D, pp. 40-47, Deed Records, Jefferson County Archives.
[7]Vol. F, pp. 166-167, Jefferson County Deed Records.
[8]*Ibid.*, Vol. F, pp. 168-169; (Nacogdoches) *Times*, January 20, 1849.
[9]Vol. G, pp. 157, 164, Jefferson County Deed Records.
[10]Manuscript Census Returns of 1850, Jefferson County, Texas, Schedule IV, Products of Industry.
[11]Vol. L, pp. 571-572, Jefferson County Deed Records.
[12]*Ibid.*, pp. 573-574.
[13]Manuscript Census Returns of 1860, Jefferson County, Texas, Schedule IV, Products of Industry.
[14](Galveston) *Weekly News*, August 14, 1860.
[15]"Biography of Judge David Wingate," (Galveston) *Daily News*, February 18, 1899.
[15a]Article, "Night Attack at Sabine Pass," and letter, J. H. to E. H. Cushing, (Houston) *Tri-Weekly Telegraph*, November 5, 1862.
[16](Galveston) *Weekly News*, May 26, 1857.
[17]Manuacript Census Returns of 1860, Jefferson County, Texas, Schedule IV, Products of Industry.
[18]Vol. C, pp. 120, 126, 141-142, Commissioners Court Minutes, Jefferson County, Texas Archives.
[19](Galveston) *Weekly News*, July 26, 1876; File 45-B, Estate of C. H. Alexander, Probate Files, and Vols. P, pp. 246, 354, and Q, pp. 249, 352, 404, 545, Deed Records, Jefferson County, Texas Archives.
[20]Vol. L, pp. 8, 77, Jefferson County Deed Records; K. D. Keith, "The Memoirs of Captain K. D. Keith," *Texas Gulf Historical and Biographical Record*, X (Nov. 1974), p. 51.
[21]Vol. L, p. 292, and M, pp. 134, 483, Jefferson County Deed Records.
[22]Manuscript Census Returns of 1860, Jefferson County, Texas, Schedule IV, Products of Industry.
[23]Vols. N, pp. 276-277, and T, p. 210, Jefferson County Deed Records.
[24](Galveston) *Tri-Weekly News*, September 19, 1873.
[25](Beaumont) *Journal*, August 1, 1908.
[26]"Letter from Hal," (Galveston) *Weekly News*, December 29, 1857; W. T. Block (ed.), "Beaumont in the 1850s: The Letters of Henry R. Green," *Texas Gulf Historical and Biographical Record*, XI (Nov. 1975)., p. 58.
[27]Vol. B, pp. 226-229, Personal Property Records, Jefferson County, Texas Archives.
[28]"Letter from Hal," (Galveston) *Weekly News*, March 1, 1859; Block (ed.), "Beaumont in the 1850s," p. 60.
[29]Vol. M, p. 113, Jefferson County Deed Records.
[30]Manuscript Census Returns of 1870, Jefferson County, Texas, Schedule IV, Products of Industry, Microfilm Reel No. 46, Texas State Archives.
[31](Galveston) *Tri-Weekly News*, June 23, 1873.
[32]"History of the Long Families," (Beaumont) *Journal*, November 4, 1906.
[33]S. G. Reed, *A History of The Texas Railroads* (Houston: 1941), pp. 85-86.
[34]*Ibid.*, 87-88.

[35](Galveston) *Weekly News*, October 29, 1877; *Daily News*, October 24, 1877.

[36](Galveston) *Weekly News*, November 11, 1877; *Daily News*, November 6, 1877.

[37](Galveston) *Weekly News*, January 14, 1878; *Daily News*, January 8, 1878.

[38]Manuscript Census Returns of 1880, Jefferson County, Texas, Schedule IV, Products of Industry, Microfilm Reel No. 48, Texas State Archives.

[39]"Beaumont-What The Tribune Man Saw In The Mill City," (Orange, Tx.) *Tribune*, March 5, 1881, reprinted (Beaumont) *Enterprise*, March 12, 1881; (Galveston) *Daily News*, June 23, 1895.

[40]"Beaumont's Big Business," (Galveston) *Daily News*, February 15, 1888, reprinted in *Texas Gulf Historical and Biographical Record,* XIII (Nov. 1977), pp. 84-90.

[41](Beaumont) *Journal*, April 23, 1905.

[42]Manuscript Census Returns of 1860, Jefferson County, Texas, Schedule IV, Products of Industry.

[43]Vols. M, p. 375, and P, p. 313, Jefferson County Deed Records.

[44]Vols. P, pp. 317-320; Q, p. 265; and S, pp. 191-195, 417-423, Jefferson County, Texas Deed Records.

[45]Manuscript Census Returns of 1880, Jefferson County, Texas, Schedule IV, Products of Industry, Microfilm Reel No. 48, Texas State Archives.

[46]Vol. U, p. 213, Jefferson County Deed Records; (Beaumont) *Journal*, April 23, 1905.

[47](Beaumont) *Enterprise*, March 12, 1881; July 2, 1910; (Beaumont) *Journal*, July 2, 1910.

[48](Beaumont) *Enterprise*, October 1, 1881.

[49]*Ibid.*, November 6, 1880.

[50](Beaumont) *Enterprise*, March 12, 19, 1881; (Beaumont) *Journal*, April 23, 1905; W. T. Block (ed.), *Emerald Of The Neches: The Chronicles of Beaumont, Texas From Reconstruction To Spindletop* (MSS: Nederland, Tx., 1880), p. 191.

[51](Beaumont) *Journal*, April 9, 1905.

[52](Galveston) *Weekly News*, June 30, 1887.

[53]*Ibid.*, September 24, 1891.

[54]"Beaumont's Big Business," (Galveston) *Daily News*, February 15, 1888; also "Beaumont, Texas, 1889," (New Orleans) *Times-Democrat*, March 29, 1889.

[55]"Reliance Sawmill Again in Operation," (Beaumont) *Journa*l, April 23, 1905, reprinted in *Texas Gulf Historical and Biographical Record* (Nov. 1973), pp. 52-53.

[56]"Biography of Sidney C. Olive," *Memorial and Biographical History of McLennan, Falls, Bell and Coryell Counties* (Chicago: 1893), p. 619.

[57]Vol. R, p. 138, Jefferson County Deed Records.

[58](Galveston) *Weekly News*, December 13, 1877; *Daily News*, December 13, 1877.

[59](Galveston) *Weekly News*, January 14, 1878; *Daily News*, January 8, 1878.

[60](Galveston) *Weekly News*, September 23, 1878; *Daily News*, September 20, 1878.

[61](Galveston) *Daily News*, September 15, 1878.

[62]Manuscript Census Returns of 1880, Jefferson County, Texas, Schedule IV, Products of Industry, Microfilm Reel No. 48, Texas State Archives.

[63]Book of Log Brands, p. 5, 1879, Jefferson County, Texas Archives.

[64]"What The Tribune Man Saw in The Mill City," (Orange, Tx.) *Tribune*, March 5, 1881, reprinted (Beaumont) *Enterprise*, March 12, 1881.

[65]"In The Pineries," (Beaumont) *Enterprise*, October 1, 22, 1881.

[66]Vol. N, pp. 470, 472, 480, 507, Hardin County, Texas Deed Records.

[67]Manuscript Census Returns of Jefferson County, Texas, 1880, Schedule I, Res. 77.

[68]Vols. R, pp. 433, 548; S, pp. 11, 21, 343; V, pp. 106, 144-145, 260, Jefferson County Deed Records.

[69]Manuscript Census Returns of 1880, Jefferson County, Texas, Schedule IV, Products of Industry, Microfilm Reel No. 48, Texas State Archives.

[70]"What The Tribune Man Saw In The Mill City," (Orange) *Tribune*, March 5, 1881, reprinted in (Beaumont) *Enterprise*, March 12, 1881.

[71]"Lumber Trade," (Beaumont) *Enterprise*, June 4, 1881, (Galveston) *Daily News*, June 9, 1881.

[72]"Jefferson County," (Galveston) *Daily News*, June 7, 1884.

[73]Manuscript Census Returns of 1880, Jefferson County, Texas, Schedule IV, Products of Industry, Microfilm Reel No. 48, Texas State Archives.

[74](Orange, Tx.) *Tribune*, March 5, 1881, reprinted in (Beaumont) *Enterprise*, March 12, 1881.

[75]"Blaze at Beaumont," (Galveston) *Daily News*, March 23, 1882; W. T. Block (ed.), *Emerald of The Neches: The Chronicles of Beaumont, Texas From Reconstruction to Spindletop* (MSS: Nederland, 1880), p. 265.

[76]Vol. V, p. 222, Jefferson County Deed Records.

[78](Galveston) *Weekly News*, January 18, 1883.

[79](Galveston) *Daily News*, June 21, 1884.
[80]Vols. R, pp. 244-246, 313; S, p. 368; and T, p. 130, Jefferson County Deed Records.
[81](Galveston) *Weekly News*, January 14, 1878; *Daily News*, January 8, 1878; Block (ed.), *Emerald of The Neches*, p. 61.
[82]"Timbered Texas," (Galveston) *Weekly News*, April 13, 1893.
[83]Manuscript Census Returns of 1880, Jefferson County, Texas, Schedule IV, Products of Industry, Microfilm Reel No. 48, Texas State Archives.
[84]"Jefferson County," (Galveston) *Weekly News*, September 17, 1877.
[85]"Beaumont Lumberman," (Galveston) *Weekly News*, April 8, 1878; *Daily News*, April 2, 1878.
[86]"Buna, Texas," in W. P. Webb et al, *Handbook of Texas* (Austin: 1952), I, 245.
[87]"Beaumont's Big Business," (Galveston) *Daily News*, February 15, 1888; see also "Beaumont, Texas, 1889," (New Orleans) *Times-Democrat*, March 29, 1889.
[88]"Among the Lumbermen," (Galveston) *Daily News*, January 26, 1898: also Charlsie E. Berly, "Records of the Nona Mills Company, Inc.," *Texas Gulf Historical and Biographical Record*, XX (Nov. 1984), p. 71.
[89]"Timbered Texas," (Galveston) *Weekly News*, April 13, 1893.
[90]"Among the Lumbermen," (Galveston) *Daily News*, March 14, 1896.
[91]*Ibid.*, May 2, 1896.
[92](Beaumont) *Enterprise*, July 4, 1902.
[93]Vol. Q, p. 459, Jefferson County Deed Records.
[94]*Ibid.*, p. 619.
[95]W. T. Block, "A Tale of King Lumber, Godparent of Beaumont," in W. T. Block, *Frontier Tales of The Texas-Louisiana Borderlands* (MSS: Nederland, 1988), pp. 377-382.
[96]Vols. S, p. 100, and W, pp. 146-147, Jefferson County Deed Records.
[97]Vol. S, p. 396, Jefferson County Deed Records.
[98]*Ibid.*, p. 263.
[99]Manuscript Census Returns of 1880, Jefferson County, Texas Schedule IV, Products of Industry, Microfilm Reel No. 48, Texas State Archives.
[100]Vols. S, p. 559, and T, pp. 122, 170, 423-424, 599, Jefferson County Deed Records.
[101]Vol. W, p. 143, Jefferson County Deed Records.
[102]*Ibid.*, Vol. T, p. 335.
[103]Vol. W, p. 120, Jefferson County Deed Records.
[104](Galveston) *Daily News*, August 7, 1883.
[105](Galveston) *Daily News*, March 21, 1888.
[106]"Beaumont's Big Business," (Galveston) *Daily News*, February 15, 1888; also reprinted in *Texas Gulf Historical and Biographical Record*, XIII (Nov. 1977), pp. 88-89.
[107]"Beaumont, Texas, 1889," (New Orleans) *Times-Democrat*, March 29, 1889.
[108]"State Press,"(Galveston) *Daily News*, March 27, 1888.
[109]"Beaumont Sawmill Company," (Beaumont) *Enterprise*, October 8, 15, 1905.
[110]*Ibid.*, Ad of Miller-Vidor Lumber Company, September, 29, 1910.
[111]"Vidor, Texas," W. P. Webb et al, *Handbook of Texas*, II, 841.
[112]Ad of Miller-Vidor, (Beaumont) *Enterprise*, September 29, 1910.
[113]"Miller Vidor Sawmill Burns," *Ibid.*, January 6, 1918.
[114]"The Industrial Company," (Beaumont) *Enterprise*, October 15, 1905.
[115]"Helbig, Texas," Webb et al, *Handbook of Texas*, I, 793.
[116]"Mill about Completed,"(Beaumont) *Journal*, January 15, 23, 1905.
[117]*Ibid.*, January 29; February 12, 1905.
[118]C. E. Berly, "Records of The Nona Mills Company, Inc." *Texas Gulf Historical and Biographical Record*, XX (Nov. 1984), p. 71; also American Lumberman, May 12, 1906, p. 68.
[119]Telephone Conversation with Charlsie E. Berly, May 1, 1992.
[120]"Voth, Texas," in Webb, *Handbook of Texas*, II, 847.
[121]"At Voth," (Beaumont) *Journal*, November 13, 1904.
[122]*Ibid.*, June 18, 1905.
[123]"Voth, Texas," in Webb, *Handbook of Texas*, II, 847; for the Ryder sawmill, see Nellis and Pierson, *Directory of American Sawmills, Series No. 27* (Washington: Dept. of Agriculture, Govt. Prtg. Office, 1915), p. 212.
[124]"Cummings Mill Starts," (Beaumont) *Enterprise*, July 21, 1907.
[125]Ad of Neches Lumber Company, (Beaumont) *Journal*, Deep Water Edition, September 3, 1910.
[126]C. E. Pearson et al, *Cultural Resources Evaluation of the...Cummings Lumber Mill...*, Coastal Environmental Inc., for the U. S. Corps of Engineers, MSS, April, 1985, pp. 1-65.

[127]"Cummings Mill Starts," (Beaumont) *Enterprise*, July 21, 1907.

Nacogdoches County

[1](Galveston) *Daily News*, January 1, 1900; Elva A. Cox, "Lumbering in Nacogdoches County," grad. paper, 1947, Forest Collections, SFA State University.

[2]Cox, "Lumbering in Nacogdoches County," pp. 1-2.

[3]R. Maxwell, *Whistle in The Piney Woods: Paul Bremond and The Houston East and West Texas Railroad* (Texas Gulf Coast Historical Association: 1963), pp. 9-37.

[4]"From Beaumont To Nacogdoches," *Southern Industrial and Lumber Review*, June, 1904, p. 24.

[5]W. P. Webb et al, *The Handbook of Texas* (Austin: 1952), II, 257; "Historical Nacogdoches," (Galveston) *Daily News*, Aug. 16, 1903; "Inauguration of A Great Texas Enterprise," *American Lumberman* (Nov. 5, 1904), p. 34.

[6]Webb, *Handbook of Texas*, I, 298.

[7]E. W. Winkler, "The First Newspaper in Texas," *Southwestern Historical Quarterly*, VI, 2 (Oct., 1902), 163, quoted in H. P. Easton, "A History of The Texas Lumbering Industry," Ph. D. dissertation, University of Texas at Austin, 1947, p. 20; "Oldest Sawmill in Texas," (Houston) *Morning Star*, Feb. 1, 1842.

[8]Manuscript Census Returns of 1850, Nacogdoches County, Texas, Sched. V, Products of Industry, on microfilm, Lamar University Library.

[9]*Ibid.*, Manuscript Census of 1860, Sched. V, Lamar Univ. Library.

[10]"Lumber," (Beaumont) *Journal*, Nov. 13, 1904.

[11]Manuscript Census Returns of 1860, Nacogdoches County, Texas, Sched. V, Products of Industry, on microfilm, Lamar University Library.

[12]*Ibid.*

[13]Manuscript Census Returns of 1870, Nacogdoches County, Texas, Sched. IV, Manufacturers-Products of Industry, Microfilm Reel No. 46, Texas State Archives.

[14]Manuscript Census Returns of 1880, Nacogdoches County, Texas, Special Schedules V and VI, Lumber Mills and Sawmills, for year ending May 31, 1880, on Microfilm Reel No. 47, Texas State Archives.

[15]East Texas Sawmill Data Base, Texas Forestry Museum, Lufkin; *Daily Phone*, Nacogdoches (1899); (Galveston) *Weekly News*, April 13, 1893.

[16]"Timbered Texas," (Galveston) *Weekly News*, April 13, 1893; "Texas' Timbered Wealth," (Galveston) *Daily News*, Jan. 1, 1900; R. Maxwell, "HE & WT List of Rail Stops," in *Whistle in The Piney Woods*, p. 8; (Beaumont) *Journal*, April 9, 1905; "Lumber Mills of Texas," *Southern Industrial and Lumber Review* (Sept. 6, 1907), p. 27.

[17](Nacogdoches) *News-Chronicle* (1884); ET Sawmill Database, Texas Forestry Museum, Lufkin.

[18](Beaumont) *Journal*, Jan 15, 1904; (Nacogdoches) *Sentinel* (1905); "Lumber Mills of Texas," *Southern Industrial and Lumber Review* (6 Sept. 1906), p. 27.

[19](Beaumont) *Journal.*, Nov. 20, 1904.

[20](Beaumont) *Enterprise*, Feb. 26, 1905.

[21]Webb, *Handbook of Texas*, I, 362, and II, 806; "From Beaumont to Nacogdoches," *Southern Industrial and Lumber Review* (June, 1904), p. 24.

[22](Beaumont) *Journal*, Nov. 13, 1904 and Jan. 15, 1905;"Lumber Mills of Texas," *Southern Industrial and Lumber Review* (6 Sept. 1906), p. 27; (Nacogdoches) *Daily Phone*, (1899-1900); also *Daily Sentinel*, (1900-1905); also ET Sawmill Database, Texas Forestry Museum, Lufkin.

[23]Beaumont Journal., Nov. 13, 1904; *Southern Industrial and Lumber Review* (June, 1904), p. 24; also *Southern Industrial Etc.....*, Sept., 1906, p. 28; Manuscript Census Returns of 1880, Nacogdoches County, Texas, Scheds. V and VI, Lumber Mills and Saw Mills, for year ending May 31, 1880, Microfilm Reel No. 47, Texas State Archives; also ET Sawmill Database, Texas Forestry Museum, Lufkin.

[24](Beaumont) *Enterprise*, March 15, 1905.

[25](Beaumont) *Journal*, Feb. 19, 1905; (Nacogdoches) *Daily Sentinel* (1901-1905); ET Sawmill Database Texas, Forestry Museum, Lufkin.

[26]"Historical Nacogdoches," (Galveston) *Daily News*, Aug. 16, 1903.

[27]Robert Maxwell, "History of Three Lumbering Families Is Told," *Texas Forests and Texans* (May-June, 1964), p. 21.

[28]Book 55, pp. 73-75, Nacogdoches County Deed Records.

[29]"Inauguration of a Great Texas Enterprise," *American Lumberman* (Nov. 5, 1904), p. 34; see also "The Hayward Mill," *Southern Industrial and Lumber Review* (4 Nov. 1904), p. 14.

[30](Nacogdoches) *Daily Sentinel*, Nov. 1, 1904.
[31](Beaumont) *Journal*, Oct. 23, 1904.
[32]Webb et al (eds.), "Nacogdoches and Southeastern,"*Handbook of Texas*, II, 257.
[33](Beaumont) *Journal*, Nov. 27, 1904 and Jan. 8, 1905.
[34]*Ibid.*, March 12 and June 11, 1905.
[35]*Ibid.*, March 5, 19, 1905.
[36]"Accident Closes Mill," *Southern Industrial and Lumber Review* (Feb. 1909), p. 45.
[37]*Ibid.*, "Hayward Orders New Sawmill," (June, 1909), p. 72.
[38]*Ibid.*, "Wind Storm Damages Mill," p. 73.
[39]Maxwell, "History of Three Lumber Families," pp. 20, 27.
[40]Manuscript Census Returns of 1910, Nacogdoches County, Texas, Sched. I, Prec. 1, res. 69-126.
[41]Maxwell, "History of Three Lumber Families," p. 27.
[42]Webb et al, *Handbook of Texas*, I, 298; Whiteman-Decker sawmill in ET Sawmill Database.
[43]Charter Records of the Secretary of State, Austin, Texas.
[44]East Texas Sawmill Database, Texas Forestry Museum, Lufkin.
[45](Beaumont) *Journal*, Nov. 13, 17, 1904.
[46]*Ibid.*, Jan. 15. 23, 1905.
[47]*Ibid.*, May 14 and June 11, 1905; also *American Lumberman* (May 5, 1906), p. 53; also *Southern Industrial and Lumber Review* (May 26, 1906), p. 14.
[48]"Lumber Mills of Texas," *Southern Industrial and Lumber Review* (Sept. 1906), p. 28.
[49]*Ibid.*, p. 29; ET Sawmill Database, Lufkin; Technical Drawings 1990.015.0049 through 0052, all dated 1922, Lufkin Foundry and Machine Co. for Whiteman-Decker Lumber Company, Farmersville, Louisiana.
[50]"Saner-Whiteman Suffers Fire," *Southern Industrial and Lumber Review* (June, 1909), p. 73.
[51]Manuscript Census Returns of 1910, Caro, Nacogdoches County, Tx., Sched. I, Prec. 5, res. 220-305.
[52](Beaumont) *Journal*, Nov. 13, 1904, and Apr. 9, 1905; (Nacogdoches) *Daily Phone* (1899); also Sentinel (1901-1902); ET Sawmill Database, Lufkin.
[53](Beaumont) *Journal*, Jan. 15, 22, 1905; (Nacogdoches) *Daily Sentinel* (1901-1905); "Lumber Mills of Texas," *Southern Industrial and Lumber Review* (Sept. 6, 1906), p. 27.
[54]"Harrington Mill," (Beaumont) *Journal*, May 14, 1904, quoting (Nacogdoches) *Daily Sentinel.*
[55]"Harrington Puts In Planer," (Beaumont) *Journal*, January 18, 1905; also (Beaumont) *Enterprise*, June 11, 1905.
[56]*Ibid.*, "Moved To Dallas,"Feb. 12, 1906.
[57]"Lumber Mills of Texas," *Southern Industrial and Lumber Review* (Sept. 6, 1906), p. 28; also *American Lumberman* (Feb. 18, 1905); also *Reference Book of The Lumbermen's Credit Association* (Jan., 1905), p. 37.
[58]East Texas Sawmill Database; (Nacogdoches) *Daily Sentinel* (1901-1905).
[59]ET Sawmill Database; *Daily Sentinel*, May 29, 1901; also "Lumber Mills of Texas," *Southern Industrial and Lumber Review* (Sept. 6, 1906), p. 28.
[60]*Ibid.*, "Lumber Mills of Texas," (Sept. 1906), p. 28.
[61]*Ibid.*; also (Beaumont) *Journal*, Jan. 18, 1905; also *Reference Book of The Lumbermen's Credit Association* (Jan., 1905), p. 37.
[62]ET Sawmill Database; "Twelve Years in Yellow Pine," *American Lumberman* (April 21, 1906), pp. 47-94.
[63]*Southern Industrial and Lumber Review* (6 Sept. 1906), p. 28; (Nacogdoches) *Daily Sentinel* (1905).
[64](Beaumont) *Enterprise*, Dec. 13, 1931; (Nacogdoches) *Daily Sentinel*, May 31, 1934; (Lufkin) *Daily News*, Oct. 3, 1943; Elva A. Cox, "Lumbering in Nacogdoches County," 1947, grad. paper, S. F. A. State University Forestry Archives.
[65]"To Utilize Hardwood," (Beaumont) *Enterprise*, June 11, 1905.
[66]ET Database; *Gulf Coast Lumberman* (Apr. 14, 1946), p. 37.
[67]ET Sawmill Database; (Beaumont) *Enterprise*, Dec. 13, 1931, p. 8-C; (Nacogdoches) *Daily Sentinel*, May 11, 1934.
[68]Manuscript Census Returns of 1910, Cushing, Nacogdoches County, Sched. I, Pct. 7, res. 28, 30, 123, 170, 390.
[69]Webb et al (eds.), "S. F. Carter," in *Handbook of Texas*, I, 303; "Biography of J. P. Carter," in *Southern Industrial and Lumber Review* (March, 1909), p. 51; ET Sawmill Database, Lufkin; photocopy, Thompson Bros. check to S. F. Carter for $500,000, in Thompson-Tucker Mill Records, Box 27, Folder 12, SFA University Forestry Archives.
[70]Charter Records of the Secretary of State, Austin; Webb et al, "Biography of J. S. Rice,"

Handbook of Texas, II, 467.

[71]The J. P. Carter mills, East Texas Sawmill Database, Lufkin.

[72]"Biography of J. Press Carter," *Southern Industrial and Lumber Review* (March, 1909), pp. 51, 80.

[73]ET Sawmill Database; *Southern Industrial and Lumber Review* (Feb., 1906), pp. 18, 41; also (Jan. 20, 1907), p. 44; and also (March 20, 1907), p. 3; also *American Lumberman* (March 17, 1906), p. 62.

[74]"Lumber Interests at Sacul," (Beaumont) *Journal*, June 18, 1905; (Nacogdoches) *Daily Sentinel* (1905).

[75]W. T. Block, "Olive, Hardin County, Texas: An Extinct Sawmill Town and The Olive-Sternenberg Partnership That Built It," *Texas Gulf Historical and Biographical Record*, XXVI (November, 1990), p. 32; 4 letterheads, Clawson Lumber Co., Clawson, Texas, to Angelina County Lumber Co., dated Sept. 27, 1897 and March 3, July 3, and Dec. 1, 1899, in SFA University Forestry Records.

[76](Beaumont) *Journal*, Nov. 13, 20, 1904; also June 18, 1905; ET Sawmill Database; also (Nacogdoches) *Daily Sentinel* (1905).

[77]ET Sawmill Database, Lufkin; "Lumber Interests at Sacul," (Beaumont) *Journal*, June 18, 1905.

[78]"Lumber Mills of Texas," *Southern Industrial and Lumber Review* (6 Sept. 1906), p. 27; ET Sawmill Database.

[79]"Around Cushing," (Beaumont) *Enterprise*, March 12, 1905; (Beaumont) *Journal*, Oct. 23, 1904.

[80]ET Sawmill Database, Lufkin; (Nacogdoches) *Daily Sentinel* (1905); "Lumber Mills of Texas," *Southern Industrial and Lumber Review* (6 Sept. 1906), p. 27.

[81]Manuscript Census of 1910, Cushing, Nacogdoches County, Sched. I, Pct. 7, res. 26-30, 281-390.

[82]"Cariker's New Tram," (Beaumont) *Journal*, Feb. 26, 1905.

[83]"Lumber Mills of Texas," *Southern Industrial and Lumber Review* (6 Sep. 1906), p. 27; (Nacogdoches) *Daily Sentinel*, (1901-1905); (Beaumont) *Journal*, Nov. 13, 1904.

[84]"Bad at Cushing," (Beaumont) *Enterprise*, Mar. 12, 1905; (Beaumont) *Journal*, Feb. 12, 1905.

[85]Manuscript Returns of 1910, Cushing, Nacogdoches Co., Tx., Sched. I, Pct. 7, res. 119-172.

[86]"Cushing Sawmill Burned," (Beaumont) *Enterprise*, Nov. 8, 1909; 1910 Cushing census, Pct. 7, res. 38; (Nacogdoches) *Daily Sentinel*, (1901-1905); "Lumber Mills of Texas," *Southern Industrial and Lumber Review* (6 Sep. 1906), p. 27.

[87](Beaumont) *Journal*, January 8, 1905; *American Lumberman* (Feb. 17, 1906), p. 63; also (Feb. 10, 1906), p. 61.

[88]"J. H. Summers Moves Sawmill," *Southern Industrial and Lumber Review* (Jan. 1909), p. 69.

[89]"Lumber Mills of Texas," *Southern Industrial and Lumber Review* (6 Sep 1906), p. 27.

[90]*American Lumberman* (March 17, 1906), p. 62.

[91]"Lumber Mills of Texas," *Southern Industrial and Lumber Review* (6 Sept. 1906), p. 28.

[92](Nacogdoches) *Daily Phone*, Nov. 4, 1899.

[93]"Lumber Mills of Texas," p. 28; "Blount Sawmill," *American Lumberman* (June 16, 1906), p. 43; also *American Lumberman* (May 26, 1906), p. 47.

[94](Beaumont) *Journal*, Jan. 15, 1905.

[95]*Ibid.*, April 9, 1905.

[96]*Ibid.*, Nov. 27, 1904; see also 1910 Nacog. County census, Garrison, Sched. I, Pct. 5, res. 45.

[97](Nacogdoches) *Daily Sentinel* (1904); "Lumber Mills of Texas," *Southern Industrial and Lumber Review* (6 Sept. 1906), p. 28; ET Sawmill Database.

[98]"East Texas Lumber District," *American Lumberman* (March 17, 1906), p. 62.

[99]"Town of Rush," (Beaumont) *Enterprise*, April 3, 19; *American Lumberman*, March 17, 24, 1906.

[100]"Lumber Mills of Texas," p. 28; ET Sawmill Database, Lufkin.

[101]"New Mill Being Built," (Beaumont) *Enterprise*, July 30, 1905.

[102]J.C. Nellis and A. H. Pierson (compilers), *Directory of American Sawmills* (Washington: Dept. of Commerce, Gov. Printing Office, 1915), p. 209.

[103]ET Sawmill Database, Lufkin; *Gulf Coast Lumberman* (Apr. 15, 1946), p. 37; " Trinity Valley Section," (Lufkin) *Daily News*, Oct. 3. 1943.

[104]R. Maxwell, *Whistle In The Piney Woods*, p. 2; "Lumber Mills of Texas," Southern Industrial and Lumber Review (Sept. 1906), p. 28.

[105]"Cavin Makes Sawmill Deal," *Southern Industrial and Lumber Review (Ma*r. 1909), p. 36.

[106]"Lumber Mills of Texas," p. 28; Manuscript Census Returns of 1910, city of Nacogdoches, Sched. I, Pct. 1, res. 5.

[107](Beaumont) *Journal*, April 9, 1905; (Beaumont) *Enterprise*, April 3, 1905.

[108]ET Sawmill Database; (Nacogdoches) *Daily Sentinel*, June 4, 1901.

[109](Nacogdoches) *Daily Sentinel*, Oct. 16, 1902; ET Sawmill Database, Lufkin.

[110]"Sullivan Mill Burned," *Southern Industrial and Lumber Review* (Aug. 1914), p. 45; ET Sawmill Database.

[111]ET Sawmill Database; (Nacogdoches) *Daily Sentinel*, Feb. 19, 1901.

[112]*Southern Industrial and Lumber Journal* (Oct. 1905), p. 12; also (Sept. 20, 1908), p. 29.

[113]ET Sawmill Database, Lufkin; (Nacogdoches) *Daily Sentinel*, May 11,1902.

[114]"Nacogdoches County," (Beaumont) *Enterprise*, Dec. 13, 1931, all of page 8-C.

[115]*Southern Industrial and Lumber Review* (Jan. 20, 1907), p. 75; also (Mar. 20, 1907), p. 73.

[116]"Sawmills Still Do Big Business," (Beaumont) *Enterprise*, Aug. 3, 1927.

[117]Elva A. Cox, "Lumbering in Nacogdoches County," 1947 grad. paper, S. F. A. State University Forest Collections, p. 11.

[118]*Ibid.*, p. 12.

Newton County

[1]W. P. Webb et al, "Newton County," *The Handbook of Texas* (Austin: 1952), II, 277; (Galveston) *Daily News*, "Texas' Timbered Wealth," January 1, 1900; Josephine C. Peavy, "The History of Newton County, Texas," M. A. Thesis, University of Texas at Austin, 1942, p.p. 1-8.

[2]Peavy, "History of Newton County," thesis, pp. 79-80; Webb, *Handbook of Texas,* II, 317; C. S. Potts, *Railroad Transportation in Texas* (Austin: 1909), p. 70; "History of the Orange and Northwestern Railroad," *Southern Industrial and Lumber Review*, November, 1902, p. 25.

[3]Webb et al, *Handbook of Texas*, I, 907; Peavy, "History of Newton County," p. 79; F. W. Johnson et al, *A History of Texas and Texans* (Chicago: 1914), II, 1059.

[4]Webb, *Handbook of Texas*, I, 937, and II, 514, 524; Potts, p. 70; Peavy thesis, pp. 65, 78-79.

[5]"Peavy Interests Buy Big Holdings," *Gulf Coast Lumberman*, VII (September 15, 1919), p. 64.

[6]Kirby Lumber Company, *Timber Resources of East Texas* (N. P.: 1902), pp. 82, 89, 145.

[7]Census Returns of 1860, Newton County, Texas, Schedule II, Slaves, and Schedule IV, Products of Agriculture, p. 7, No. 19; W. T. Block, "An Early East Texas Captain of Commerce: David Robert Wingate," *Texas Gulf Historical and Biographical Record*, XIII (November, 1977), pp. 63-64; "Life of Judge D. R. Wingate," (Sabine Pass) *News*, February 23, 1899.

[8]Pauline Hines et al, *Glimpses of Newton County History* (Nortex Press: 1982), pp. 46, 95-97.

[9]Block, "An Early East Texas Captain of Commerce: David R. Wingate," pp. 64-65.

[10]Dean Tevis, "Old East Texas Water Mills," (Beaumont) *Enterprise*, November 17, 1935.

[11]J. Peavy, "History of Newton County," p. 85.

[12]Census Returns of Newton County, Texas, 1860, Schedule V, Products of Industry.

[13]Peavy, "History of Newton County," p. 85; Tevis, "Old East Texas Water Mills," (Beaumont) *Enterprise*, November 17, 1935.

[14]Census Returns of Jasper County, Texas, 1860, Schedule V, Products of Industry.

[15]Hines, *Glimpses of Newton County*, p. 24; Tevis, "Water Mills," *Enterprise*, November 17, 1935.

[16]Webb, *Handbook of Texas*, II, 45; "Newton County," (Beaumont) *Enterprise*, June 30, 1935; Tevis, "Old Water Mills," *Enterprise*, November 17, 1935.

[17]"Old East Texas Water Mills," (Beaumont) *Enterprise*, November 17, 1935.

[18]Manuscript Census Returns of Orange County, Texas, 1860, Schedule V, Products of Industry.

[19](Galveston) *Daily News*, March 9 and September 15, 1878; "Orange, Texas," (New Orleans) *Times-Democrat*, March 22, 1889; (Galveston) *Weekly News*, November 19, 1877, and "The Long Leaf Belt," *Ibid.*, November 26, 1877.

[20](Galveston) *Weekly News*, August 31; September 17, and December 10, 1877.

[21]*Ibid.*, November 5, 1877.

[22]*Ibid.*, November 19, 1877.

[23]*Ibid.*, December 3, 1877.

[24]*Ibid.*, December 24, 1877.

[25]*Ibid.*, April 1, 1878.

[26]*Ibid.*, August 12, 1878.

[27]*Ibid.*, September 2, 1878.

[28]*Ibid.*, October 28, 1878.

[29]*Ibid.*, February 4, 1878

[30]*Ibid.*, January 6, 1879.

[31]*Ibid.*, May 26, 1879.
[32]*Ibid.*, April 15, 1880.
[33](Galveston) *Weekly News*, February 24, 1879.
[34]Webb, *Handbook of Texas*, I, 271.
[35](Galveston) *Daily News*, December 18, 1890.
[36](Galveston) *Weekly News*, September 17, 1891; W. T. Block, *Emerald of the Neches: The Chronicles of Beaumont, Texas From Reconstruction to Spindletop* (Nederland: 1980), p. 451, to be hereinafter cited as 'Emerald,'
[37]"The Orange Budget," (Galveston) *Daily News*, July 13, 1891; *Emerald*, pp. 475-476.
[38]"Lumber News From Salem," (Galveston) *Daily News*, September 16, 1891; Emerald, p. 481.
[39]"Through The Pineries," (Galveston) *Daily News*, March 3, 1892; Emerald, p. 513.
[40]"Lumber News From Salem," (Galveston) *Daily News*, September 28, 1895.
[41]*Ibid.*, January 4 and April 11, 1896.
[42]*Ibid.*, November 14, 1896.
[43]*Ibid.*, June 19, 1897.
[44]"Call Sawmill Burns," (Beaumont) *Weekly Enterprise*, October 7, 1899.
[45]Kirby Lumber Company, *Timber Resources of East Texas* (N. P.: 1902), pp. 145-149.
[46]"History of The Call, Texas Sawmill," (Beaumont) *Enterprise*, March 12, 1905.
[47]R. Maxwell et al, *Sawdust Empire: The Texas Lumbering Industry* (College Station: 1983), p. 102.
[48]*Ibid.*, p. 40.
[49]Webb, *Handbook of Texas*, I, 271.
[50]John C. Gee, Jr., "The Call of Texas," in Pauline Hines et al, *Crosscuts: An Anthology of Memoirs By Newton County Folk* (Austin: Eakin Press, 1984), pp. 3-6.
[51]Webb, *Handbook of Texas*, I, 271
[52]Webb, *The Handbook of Texas*, II, 804.
[53]*Ibid.*
[54]Hamilton P. Easton, "A History of The Texas Lumbering Industry," Ph. D. dissertation, University of Texas at Austin, 1947, p. 264.
[55]Kirby, *Timber Resources of East Texas*, p. 82.
[56]Webb, *Handbook of Texas*, I, 855.
[57]Manuscript Census Returns of 1910, Howard's Switch, Newton County, Texas, Precinct 1, res. 49-132.
[58](Beaumont) *Journal*, January 8 and November 13, 1904.
[59]Webb, *Handbook of Texas*, II, 73.
[60]Manuscript Census returns of 1910, Logtown, Newton County, Texas, Precinct 1, residences 101ff.
[61]Webb, *Handbook of Texas*, I, 10; also "Bought Small Mill," *Southern Industrial and Lumber Review* (Sept., 1908), p. 28.
[62]*Ibid.*; Manuscript Returns of 1910, Adsul, Newton County, Texas, Precinct 5, residences 16ff.
[63]Webb, *Handbook of Texas* I, 782; Manuscript Census returns of 1910, Hartburg, Newton County Texas, Precinct 5, residences 76ff.
[64]Webb, *Handbook of Texas*, I, 782; P. Hines et al, *Glimpses of Newton County History*, p. 47; Nellis and Pierson, *Directory of American Sawmills-Series No. 27* (Washington: Dept. of Agriculture, Govt. Prtg. Office, 1915), p. 210.
[65]Webb, *Handbook of Texas*, I, 587; P. Hines et al, *Glimpses of Newton County History*, p. 23.
[66]Hazel Wilborn, "Living in a Box Car at Old Woodmeyer," in Hines et al, Crosscuts, pp. 4-6; Webb, *Handbook of Texas*, II, 932; and Kirby, *Timber Resources of East Texas*, p. 139.
[67]Webb, *Handbook of Texas*, II, 309, 533.
[68]*Ibid.*, I, 773.
[69]*Ibid.*, I, 174; and Hines, *Glimpses of Newton County*, p. 24; Hines, *Crosscuts*, pp. 27-28.
[70]*Ibid.*, I, 185; Hines et al, *Glimpses of Newton County History*, p. 24; and J. Peavy, "History of Newton County, Texas," thesis, p. 85; also "Will Build New Sawmill," *Southern Industrial and Lumber Review* (Dec., 1908), p. 54.
[71]Webb, *Handbook of Texas*, I, 248; Hines, *Glimpses*, pp. 25-30; Peavy thesis, "History of Newton County," pp. 62-64.
[72](Galveston) *Daily News*, February 20, 1897.
[73]*Ibid.*, May 15 and June 19, 1897.
[74]Webb, *Handbook of Texas*, II, 514.
[75]*Ibid.*, I, 287.

[76]Hines et al, *Glimpses of Newton County*, pp. 85-86.
[77]Webb, *Handbook of Texas*, II, 327.
[78]*Ibid.*, II, 133.
[80]*Ibid.*, II, 424.
[81]P. Hines et al, *Glimpses of Newton County History*, pp. 22-23; Manuscript Census Returns of 1910, Newton County, Texas, Schedule I, Precinct 2, residence 244.
[82]Webb, "Bon Wier," *Handbook of Texas*, I, 185; Kirby Lumber Company, *Timber Resources of East Texas,* p. 139.
[83]Webb, "Texla," *Handbook of Texas*, II, 768; "Miller-Link Building Texla Mill," *Gulf Coast Lumberman*, VI (December 15, 1918), p. 36; R. S. Maxwell et al, *Sawdust Empire: The Texas Lumbering Industry, 1830-1940* (College Station: 1983), pp. 195-196.
[84]"Wier Long Leaf Lumber Company," *The Gulf Coast Lumberman*, VI (Dec. 15, 1918), p. 36.
[85]Earl Hines, "Prime Time," and Sylvia Larkin, "Wiergate," in P. Hines,*Crosscuts*, pp. 23, 125.
[86]Webb, *Handbook of Texas*, I, 787; Hines, *Glimpses of Newton County*, p. 154; *Gulf Coast Lumberman,* December 15, 1918.
[87]Sylvia Larkin, "Wiergate," in P. Hines et al, *Crosscuts*," pp. 23-24.
[88]H. P. Easton, "A History of The Texas Lumbering Industry," Ph. D. dissertation, U. T. Austin, 1947, pp. 359-360.
[89]"Largest Sawmill in Texas Is At Wiergate," (Beaumont) *Enterprise*, August 3, 1927.
[90]H. P. Easton, "The Log Pond and Unloading Rig," in Ph. D. dissertation, "A History of The Texas Lumbering Industry," pp. 337-338.
[91]"Largest Sawmill in Texas," (Beaumont) *Enterprise*, August 3, 1927.
[92]Easton, "History of the Texas Lumbering Industry," pp. 342-350.
[93]"Wier Long Leaf Lumber Company," (Beaumont) *Enterprise*, April 3, 1932; and "Wiergate," (Beaumont) *Enterprise*, March 21, 1937.
[94]"With Wier at Wiergate," *Gulf Coast Lumberman*, XXV (Nov. 15, 1937), pp. 16-24.
[95]*Ibid.*, "With Wier at Wiergate," pp. 16-18.
[96]Easton dissertation, p. 353.
[97]"With Wier at Wiergate," *Gulf Coast Lumberman*, Nov. 15, 1937, pp. 17-24.
[98]Easton, pp. 355-356.
[99]*Ibid.*, p. 357.
[100]"Wiergate Still Has Long Leaf," (Beaumont) *Enterprise*, June 30, 1940, p. 11-G.
[101]"Wiergate To Join Other Ghost Towns of East Texas," (Beaumont) *Enterprise*, January 1, 1943, p. 8-E.
[102]Earl S. Hines, "Prime Time," in P. Hines et al, *Crosscuts*, pop. 125-127.
[103]*Ibid.,* pp. 128-129.
[104]B. Bowman, "Wiergate," *The 35 Best Ghost Towns in East Texas* (Lufkin: 1988), pp. 45-46.
[105]H. P. Easton dissertation, "History of the Texas Lumbering Industry," pp. 360-363. This work contains two extremely long and detailed chapters about all facets of Wiergate history.
[106]"Deserted Wiergate," (Beaumont) *Enterprise*, November 6, 1955.
[107](Beaumont) *Enterprise*, November 6, 1955.
[108]"Laurel's Laurels," (Galveston) *Daily News*, November 22,1890; *Emerald*, pp. 443-444.
[109](Galveston) *Daily News*, December 18, 1890; *Emerald*, pp. 447-448.
[110]*Emerald*, p. 451; "A Great Day's Work," (Galveston) *Weekly News*, September 17, 1891.
[111]"A Fine Logging Tram," (Galveston) *Daily News*, May 12, 1891; *Emerald*, pp. 469-470.
[112]H. P. Easton interview with W. A. Smith, Nov. 28, 1940, in Easton, "A History of The Texas Lumbering Industry," pp. 262-268.
[113]*Ibid.*
[114](Galveston) *Daily News*, February 8, 1896.
[115]*Ibid.*, January 7, 1899.
[116]*Gulf Coast Lumberman*, September 15, 1919; (Beaumont) *Journal*, November 20, 27, 1904.
[117](Beaumont) *Journal*, November 13, 20, 27, 1904.
[118](Beaumont) *Journal*, November 20, 1904.
[119]"A Sawmill Town," *Ibid.*, November 20, 1904.
[120]"Peavy Interests Buy Big Holdings," Gulf Coast Lumberman, VII (Sept. 15, 1919), p. 64.
[121]"Sabine Tram To Manufacture Hardwoods," *Gulf Coast Lumberman*, April 15, 1917.
[122]"Well-Known Trio in Lumber Circles," *The Lufkin Line* (publication of Lufkin Foundry and Machine Company), April, 1921, p. 1
[123]Obituaries of A. Jasper Peavy, (Shreveport) *Times* and (Lufkin) *Daily News*, November 16, 1942; "Well-Known Trio," *Lufkin Line*, April, 1921, p. 1; also "Sabine Tram To Erect Two Big Mills," (Beaumont) *Journal*, July 27, 1917.

[124]"Peavy Interests Buy Big Holdings," *Gulf Coast Lumberman*, VII (September 15, 1919), p. 64.

[125]Webb, *Handbook of Texas*, II, 524.

[126]*Ibid.*, II, 768.

[127]"The New Mill at Deweyville, Texas," *The Lufkin Line*, II, No. 8 (April-May, 1923), pp. 1-3; "Peavy-Moore at Deweyville Rebuilt," *The Lufkin Line*, May-June, 1926, pp. 8-9.

[128]*Ibid.*, p. 3.

[129]J. C. Peavy, "A History of Newton County, Texas," thesis, personal biography of Josephine Cochrum Peavy; also letters, Mrs. Virginia Peavy, Lufkin, Texas, to W. T. Block, March 13, 20, 1993.

[129a]Peavy thesis, pp. 86-87.

[130]*Ibid.*, p. 86.

[131]*Ibid.*, p. 65.

[132]*Ibid.*, p. 98.

[133]"Jasper Peavy Honored," *Lufkin Line*, V (May-June, 1926), p. 1; R. Maxwell, *Sawdust Empire: The Texas Lumbering Industry* (College Station: 1983), p. 204; (Shreveport) *Times*, November 16, 1942.

[134]Webb, *Handbook of Texas*, I, 496.

[135]Peavy thesis, p. 85.

[136]Manuscipt Census Returns of 1910, Newton County, Schedule I, Precinct 1, residence 309FF.

[137]"History of the Miller-Link Lumber Company," *Where The Port of Orange Is Located* (1916), p. 50; also "History of the Miller-Link Lumber Company,"*The Orange Leader*, The Holiday Issue, (December, 1910), p. 3; also "Miller-Link Buys Newton Sawmill," *Southern Industrial and Lumber Review* (Sept., 1908), p. 28; also *Ibid.* (Jan., 1909), p. 68.

[138]Webb, *Handbook of Texas*, I, 496, 782.

[139]"Newton County, Texas," (Beaumont) *Enterprise*, July 30, 1930.

[140]Webb, *Handbook of Texas*, II, 277. 1932; and "Wiergate," (Beaumont) *Enterprise*, March 21, 1937.

Orange County

[1]"Texas' Timbered Wealth," (Galveston) *Daily News*, January 1, 1900.

[2](Galveston) *Weekly News*, November 21, 1876.

[3]"More About The Mills," *Ibid.*, August 6, 1891.

[4](Houston) *Telegraph and Texas Register*, October 7, 1840.

[5]"Mills At Orange," (Orange, Tx.) *Leader*, 1895 Special Edition.

[6]"Description of Orange County," (Nacogdoches, Tx.) *Chronicle*, August 16, 1853: also (New Orleans) *Time-Democrat*, March 22, 1889.

[7]"Entrances and Clearances," J. D. Swain, Collector, March 31, 1840, File 4-21/10, Sabine Bay Customs Records, Texas Statge Archives; (Richmond, Tx.) *Telescope*, April 4, 1840.

[8]"Quarterly Return," R. C. Doom, Collector, June 30 and Sept. 30, 1839; and "Quarterly Return," N. F. Smith, Collector, December 31, 1842, Sabine Bay Customs Records, File 4-21/10, copied by W. T. Block at Texas State Archives.

[9]Manuscript Census Returns of 1850, Jefferson County, Texas, Schedule V, Products of Industry.

[10]Manuscript Census Returns of 1860, Orange County, Texas Schedule I, residences 50, 147; Schedule II, Slaves.

[11]"Description of Orange County," (Nacogdoches) *Chronicle*, August 16, 1853.

[12]"Mills at Orange," (Orange) *Leader*, 1895 Special Edition.

[13]"Empire Mills Burn," (Galveston) *Weekly News*, June 7; July 15, 1856; also *Tri-Weekly News*, July 15, 1856.

[14]Dr. Howard Williams, *Gateway To Orange: A History of Orange and Orange County* (Orange: 1988), p. 108.

[15](Galveston) *Weekly News*, May 26, 1857.

[16] "Letter From Hal," (Galveston) *Weekly News*, June 30, 1857; also reprinted in W. T. Block (ed.), "Extracts From The Writings of Henry R. Green," *Texas Gulf Historical and Biographical Record*, XI (Nov. 1975), p. 72.

[17]"Letter From Hal," *Ibid.*, May 10, 1859, reprinted in *Texas Gulf Historical and Biographical Record*, XI (Nov. 1975), p. 76.

[18]Manuscript Census Returns of 1860, Orange County, Texas, Schedule V, Products of Industry.

[19]*Texas Almanac*, 1859, (Galveston: 1860), p. 150.

[20]R. E. Russell, "A History of Orange," published in (Beaumont, Tx.) *Enterprise*, April 23, 1922.

[21]"Mills at Orange," (Orange) *Leader*, 1895 Special Edition; "Obituary of E. T. Chenault, (Beaumont) *News-Beacon*; "Among the Lumbermen: (Galveston) *Daily News*, March 21, 1896.

[22]Manuscript Census Returns of 1860, Orange County, Texas, Schedule I, res. 8.

[23]"Obituary of A. Gilmer," (Beaumont) *Enterprise*, July 31, 1906.

[24]*Ibid.*

[25]"Memoirs of Alexander Gilmer," (Galveston) *Daily News*, February 11, 1905.

[26](Sabine Pass, Tx.) *Beacon*; quoted in (Houston) *Tri-Weekly Telegraph*, November 12, 1871.

[27]"A. Gilmer Memoirs," (Galveston) *Daily News*, February 11, 1905.

[28]*Ibid.*

[29]"The Sabine Trade," (Galveston) *Daily News*, February 1, 1877.

[30](Galveston) *Weekly News*, August 13, 1877, quoting the *Tribune*.

[31]*Ibid.*, September 10, 1877, quoting the *Tribune*.

[32]Manuscript Census Returns of 1880, Orange County, Texas, Schedule V, Products of Industry, copied by W. T. Block from Microfilm Reel No. 48, Texas State Archives.

[33]"The Orange Riot," (New Orleans) *Democrat*, August 21, 1881.

[34]"Orange Fire," (Galveston) *Weekly News*, December 7, 1882.

[35](Galveston) *Weekly News*, December 22, 1887.

[36]"Orange County Prospects," (Galveston) *Weekly News*, February 9, 1888.

[37]"Orange, Texas—The Sawmill City of The Lone Star State," (New Orleans) *Times-Democrat*, March 22, 1889.

[38]"Laurel's Laurels," (Galveston) *Daily News*, November 22, 1890.

[39]"Orange Marine Matters," *Ibid.*, September 14, 1891.

[40]"Texas Fires," (Galveston) *Daily News*, February 19, 1891.

[41]"A. Gilmer Mill Burns," (Galveston) *Daily News*, March 16, 1893; "Lumber Mills Burned," (Galveston) *Weekly News*, March 23, 1893.

[42]"Gilmer Mill Destroyed," (Beaumont) *Journal*, October 6, 1899.

[43](Beaumont) *Enterprise*, July 31, 1906.

[44]"Memoirs of A. Gilmer," (Galveston) *Daily News*, February 11, 1905.

[45]S. G. Thigpen, *Pearl River: Highway to Gloryland* (privately printed: 1965), pp. 83-86; N. Hickman, *Mississippi Harvest: Lumbering In The Long Leaf Belt* (University, Ms., 1962), pp. 18-19.

[46]Letter, J. H. to E. H. Cushing, and article, "Night Attack at Sabine Pass," (Houston) *Tri-Weekly Telegraph*, November 5, 1862.

[47]Vol. B, pp. 302-303, and C, p. 43, Orange County, Texas Deed Records.

[48](Galveston) *Tri-Weekly News*, February 5, 1872; *Weekly News*, September 19, 1873; July 16, 1891.

[49](Galveston) *Daily News*, February 18, 1899; *Weekly News*, October 29: November 5, 12, 1877; April 1; August 26, 1878, each quoting the *Tribune*.

[50](Galveston) *Weekly News*, August 20 and November 5, 1877; August 26 and October 7, 1878, each quoting *Tribune*.

[51](Beaumont) *Enterprise*, December 4, 1880; *Weekly News*, October 7, 1878.

[52]Manuscript Census Returns of 1880, Orange County, Texas, Schedule V, Products of Industry, copied by W. T. Block from Microfilm Reel No. 48, Texas State Archives.

[53](Beaumont) *Enterprise*, December 4, 1880; (Galveston) *Weekly News*, December 2, 1880.

[54](Galveston) *Daily News*, February 18, 1899.

[55]"Prospects of Orange County," (Galveston) *Weekly News*, February 9, 1888.

[56](Galveston) *Daily News*, June 2, 1890; Vol. L, pp. 351-355, Orange County Deed Records.

[57](Galveston) *Daily News*, March 10 and April 14, 1895.

[58]"Life of Judge Wingate," (Sabine Pass) *News*, February 23, 1899; also "Obituary of Judge Wingate," (Galveston) *Daily News*, February 18, 1899.

[59]"Among The Lumbermen," (Galveston) *Daily News*, January 10, 1903, p. 6, c. 6-7.

[60](Galveston) *Weekly News*, April 9, 1877.

[61]*Ibid.*, April 20, 1877.

[62]"T and N O Railroad," (Galveston) *Daily News*, July 15, 1879.

[63]"Orange County," *Ibid.*, October 22, 1879.

[64](New Orleans) *Democrat*, November 20, 1881.

[65](Galveston) *Weekly News*, April 9, 1877, quoting *Tribune*.

[66]*Ibid.*, August 13, 1877.

[67]*Ibid.*, October 22, 1877, quoting *Tribune*.

[68]Manuscript Census Returns of 1880, Orange County, Texas, Schedule V, Products of Industry, Microfilm Reel No. 48, Texas State Archives, copied by W. T. Block.

[69]"History of the Lutcher and Moore Lumber Company," (Beaumont) *Enterprise*, October 15, 1905.

[70]"Orange," (Galveston) *Daily News*, March 28, 1885.

[71]*Ibid.*, April 5, 1885.

[72]"Among The Lumbermen," (Galveston) *Daily News*, June 16, 1895; April 25 and August 15, 1897.

[73]"History of the Lutcher and Moore Lumber Company," (Beaumont) *Enterprise*, October 15, 1905.

[74]"Among the Millmen," (Galveston) *Daily News*, December 20, 1887.

[75]"Prospects of Orange," (Galveston) *Weekly News*, February 9, 1888.

[76]*Ibid.*

[77]"Orange and Beaumont," (Galveston) *Daily News*, September 1, 1890.

[78]"Orange, Texas," (New Orleans) *Times-Democrat*, March 22, 1889.

[78a]"History of The Lutcher and Moore Lumber Company," (Beaumont) *Enterprise,* October 15, 1905; "The Lutcher and Moore Lumber Company of Port Orange," *Where The Port Of Orange Is Located On The World's Map* (Orange *Leader*, 1916), pp. 24-26; "Lumber History," *Southern Industrial and Lumber Review* (Nov. 1902), p. 36.

[79]W. T. Block, *Cotton Bales, Keelboats, and Sternwheelers: A History of The Sabine River Trade* (Nederland: 1976), p. 54.

[80](Beaumont) *Enterprise*, "History of The Lutcher and Moore Lumber Company," October 15, 1905.

[81]*Ibid.*

[82]Dr. Howard Williams, *Gateway to Texas: A History of Orange and Orange County*, p. 113; "Orange: The Gateway to Texas," *Southern Industrial and Lumber Review*, November, 1902, p. 37.

[83]R. Maxwell et al, *Sawdust Empire: The Texas Lumber Industry* (College Station: 1983), pp. 32-33; Dr. Williams, *Gateway to Texas*, pp. 111-114; "Orange: Gateway City To Texas," *Southern Industrial and Lumber Review*, November, 1902, pp. 36-39.

[84]Williams, *Gateway to Texas*, p. 114.

[85]"Lumber History," *Southern Industrial and Lumber Review*, November, 1902, p. 32.

[86]"History of Orange," by Robert E. Russell, January 11, 1911, published (Beaumont) *Enterprise*, April 23, 1922; also (Orange) *Leader*, January 21, 1918.

[87](Galveston) *Weekly News*, October 29, 1877.

[88]*Ibid.*, October 29, 1877.

[89]*Ibid.*, August 12, 1878.

[90]*Ibid.*, August 26, 1878.

[91]*Ibid.*, December 3, 1877; January 21, 1878.

[92]Manuscript Census Returns of 1880, Orange County Texas, Schedule V, Products of Industry, Microfilm Reel No. 48, copied by W. T. Block at the Texas State Archives and published by Block in *Texas Gulf Historical and Biographical Record*, IX (November, 1973), p. 58.

[93]"Death of a San Jacinto Veteran," (Beaumont) *Enterprise*, December 4, 1880; see also (Galveston) *Daily News*, December 9, 1880.

[94]"Orange, Texas: The Sawmill City of The Lone Star State," (New Orleans) *Times-Democrat*, March 22, 1889.

[95]"Some Sawmill History," (Galveston) *Weekly News*, July 16, 1891.

[96]*Ibid.*, "More About The Mills," August 6, 1891.

[97]"Orange: Gateway To Texas," *Southern Industrial and Lumber Review*, November, 1902, p. 34; (Galveston) *Weekly News*, August 20, 1877.

[98](Galveston) *Weekly News*, April 1, 1878.

[99]Census Returns of 1880, Orange County, Texas, Sched. V, Products of Industry, Microfilm Reel No. 48, Texas State Archives, copied by W. T. Block and published in *Texas Gulf Historical and Biographical Record*, IX (Nov., 1973), p. 59.

[100]"Orange, Texas," (New Orleans) *Times-Democrat*, March 22, 1889; "Lumber History," *Southern Industrial and Lumber Review* (Nov. 1901), p. 34.

[101]*Ibid.*, (Nov. 1902), p. 35.

[102]George Bancroft, "Retrospective and Memorandum of Business Events" (undated 21-page unpublished manuscript of family business history: Orange, Ca. 1920), pp. 2-3, to be hereinafter cited as "Bancroft."

[103]None of the writer's research ever noted a "Sims" mill at Orange, and apparently Bancroft's memory may have failed him. However, in 1873, Eberle Swinford operated the "Phoenix" or

"Upper" shingle mill on the Louisiana side at Phoenix Lake. In 1874 the Phoenix mill became Wingate and Swinford, and in 1877 it became Moore and Swinford, when D. R. Wingate sold out to Charles H. Moore of Galveston. Moore and Swinford also operated the Conway's Bayou or "Lower" shingle mill, both mills being in Louisiana, and hence not a part of this story. The "Lower" mill burned out on March 25, 1884, but was soon rebuilt, and the "Upper" mill was eventually dismantled around 1892 when the cypress timber was cut out. See "Shingle Mill Burns," (Galveston) *Daily News*, March 27, 1884; also "Orange, Texas," (New Orleans) *Times-Democrat*, March 22, 1889.

[104]Bancroft, p. 3; also Vol. F, pp. 397-398, Orange County Deed Records.

[105](Galveston) *Weekly News*, Dec. 24, 1877.

[106]*Ibid.*, March 17, 1879.

[107]The 1880 Orange County census, Sched. IV, Products of Industry, copied by W. T. Block from Microfilm Reel No. 48, Texas State Archives, and published in *Texas Gulf Historical and Biographical Record* (Nov. 1973), p. 58.

[108]Bancroft, p. 4.

[109]"Orange, Texas," (New Orleans) *Times-Democrat*, March 22, 1889.

[110]Bancroft, pp. 5, 12.

[111]*Ibid.*, p. 8.

[112]*Ibid.*, p. 9.

[113]Bancroft, pp. 13-18.

[114]*Ibid.*, p. 10

[115]*Ibid.*, pp. 12-21.

[116]Kirby Lumber Co., *Timber Resources of East Texas* (Houston: Nov. 1902), p. 131.

[117]Kirby, *Timber Resources of East Texas*, p. 132.

[118]Typescript, "List of Kirby Lumber Company Plants—Mill D," Houston, Feb. 9, 1903, p. 3, Kirby Papers, SFA State Unviversity Forestry Archives.

[119]*Ibid.*; Bancroft, p. 21; *Timber Resources of East Texas*, p. 136.

[120]"List of Kirby Lumber Company Plants-Mill D," p. 3; Santa Fe Railroad *Circular No. 2090*, Galveton, March 28, 1904.

[121]"Mill D Sold to Miller-Link," (Beaumont) *Enterprise*, July 4, 1905.

[122]"List of Kirby Lbr. Co. Plants-Mill D," p. 3; Bancroft, p. 21; "Lumber History," *Southern Industrial and Lumber Review* (Nov. 1902), p. 37.

[123]"Lumber History," *Southern Industrial and Lumber Review* (Nov. 1902), pp. 34, 37.

[124]"Orange, Texas," (New Orleans) *Times-Democrat*, March 22, 1889.

[125]"Orange and Northwestern Railroad," *Southern Industrial and Lumber Review* (Nov. 1902), p. 25.

[126]"A Great Lumber Manufacturing Enterprise-Miller-Link," *The Orange Leader*, Holiday Number (Nov. 1910), pp. 3-4; "Miller-Link Buys Machine," *Southern Industrial and Lumber Review* (June, 1909), p. 73.

[127]Webb, "Quigley, Texas," *Handbook of Texas*, II, 424.

[128]*The Orange Leader*, Holiday Edition, Nov. 1910, p. 3; also Webb, "Bunker Hill, Texas," *Handbook of Texas*, I, 245.

[129]*Orange Leader*, Holiday Number, Nov. 1910, p. 3; also "Miller-Link Buys Newton Sawmill," *Southern Industrial and Lumber Review* (Sept. 1908), p. 28; also *Ibid.* (Jan. 1909), p. 28;l also *Ibid.*, (Feb. 1909), p. 45.

[130]"A Great Lumber Manufacturing Enterprise," *Orange Leader*, Holiday Number (Nov. 1910), p. 3.

[131]Nellis and Pierson, *Directory of American Sawmills* (Washington: 1915), p. 211.

[132]"History of the Miller-Link Lumber Company," in *Where The Port Of Orange Is Located On The World Map (Orange Leader*: 1916), p. 50.

[133]Webb, *Handbook of Texas*, II, 841; also *American Lumberman* (December 15, 1918); also J. E. Johnson, "The Economic Development of Orange County," *Texas Gulf Historical and Biographical Record*, XIII (Nov. 1977), p. 25; also *Gulf Coast Lumberman* (Nov. 15, 1917), p. 30.

[134]Bancroft, p. 21; Dr. H. Williams, *Gateway to Orange*, p. 117.

[135]"Lumber History," *Southern Industrial and Lumber Review* (Nov. 1902), pp. 34, 37; also (Galveston) *Weekly News*, Apr. 29; Aug. 26; Oct. 21; and Nov. 11, 1878.

[136]Manuscript Census of 1880, Orange County, Tx., Sched. IV, Products of Industry, copied by W. T. Block from Microfilm Reel No. 48, Texas State Archives.

[137]*Southern Industrial and Lumber Review* (Nov. 1902), p. 37; "Orange County Prospects," (Galveston) *Weekly News*, Feb. 9, 1888.

[138]"Orange, Texas," (New Orleans) *Times-Democrat*, March 22, 1889.

[139](Galveston) *Daily News*, March 10 and April 14, 1895.

[140]"Lumber," (Galveston) *Daily News*, Jan. 11, 1902; Jan. 10, 1903; (Beaumont) *Journal*, April 9, 16, 1905; also (Beaumont) *Journal*, Oct. 23, 1904.

[141]"Lumber Mills of Texas," *Southern Industrial and Lumber Review* (Sept. 1906), p. 28.

[142]Dr. H. Williams, *Gateway to Orange*, p. 117.

[143]"Miller-Vidor Lumber Company," *American Lumberman* (Oct. 8, 1910), p. 52.

[144]Webb, *Handbook of Texas*, II, 841; "Beaumont Sawmill Co." (Beaumont) *Enterprise*, Oct. 15, 1905; "Miller-Vidor Mill Burns," (Beaumont) *Enterprise*, Jan. 6, 1918.

[145]Webb, *Handbook of Texas*, II, 768; "R. W. Weir Lumber Co." *Orange Leader*, Holiday Number, Dec. 1910; *American Lumberman*, Dec. 15, 1918; "Lumber Mills of Texas," *Southern Industrial and Lumber Review* (Sept. 1906), p. 27.

[146]Webb, *Handbook of Texas*, II, 160.

[146a]*Southern Industrial and Lumber Review* (Feb., 1909), p. 45.

[147]*Handbook of Texas*, II, 727.

[148]Webb, *Handbook of Texas*, I, 515.

[149]Nellis and Pierson, *Directory of American Sawmills* (Washington: 19150, p. 211.; also Webb, *Handbook of Texas*, II, 308.

[149a] *Southern Industrial and Lumber Review* (Jan., 1909).

[150]*Handbook of Texas*, II, 48.

[151]G. C. and Santa Fe *Circular No. 2090*, Galveston, Mar. 28, 1904; (Beaumont) *Enterprise*, Oct. 30, 1904.

[152](Beaumont) *Enterprise*, Oct. 15, 1905.

[153]"Gilmer Obituary," (Beaumont) *Enterprise*, July 31, 1906.

[154]Webb, *Handbook of Texas*, II, 48.

[155]"Lumber History," *Southern Industrial and Lumber Review* (Nov., 1902), p. 36; "Orange, Texas," (New Orleans) *Times-Democrat*, March 22, 1889.

[156]"Lumber Mills of Texas," *Southern Industrial and Lumber Review* (Sept., 1906), p. 28.

[157]"Miller-Vidor Lumber Company," *American Lumberman* (Oct. 8, 1910), p. 52.

[158]Company Advertisement, (Beaumont) *Enterprise*, September 29, 1910.

[159]Webb, "Vidor," *Handbook of Texas*, II, 841.

[160]"Miller-Vidor Lumber Co.," *American Lumberman* (Oct.8, 1910), pp. 84-85; "Orange Sawmill Burns" (Beaumont) *Enterprise*, Aug. 4, 1910, p. 2.

[161]*Ibid.*

[162]Dr. H. Williams, *Gateway to Orange*, p. 117.

[163]L Miller, "Lumber-The Great Asset," in *Where The Port Of Orange Is Located On The World Map* (1916), pp. 9-10; "Paper From Yellow Pine Waste," *Southern Industrial and Lumber Review* (——, 1909, month blurred), p. 47.

[164]J. E. Johnson, "The Economic Development of Orange County, Texas, 1900-1920," *Texas Gulf Historical and Biographical Record*, XIII (Nov. 1977), 10-27.

[165]*Ibid.*, W. T. Block, "An Early East Texas Captain of Commerce: David Robert Wingate, XIII (Nov. 1977), p. 72.

[166]Dr. H. Williams, *Gateway to Orange*, pp. 109-119; (Galveston) *Weekly News*, June 24 and July 22, 1878.

Polk County

[1]W. P. Webb et al, *The Handbook of Texas* (Austin: 1952), II, 390; also "Texas Timbered Wealth," (Galveston) *Daily News*, January 1, 1900.

[2]Webb, *Handbook of Texas*, I, 852, and II, 848.

[3]*Ibid.*, II, 241; R. Peebles et al, *The Pictorial History of Polk County, Texas* (N. P.: 1976), p. 122.

[4]*Ibid.*, II, 117, 170: (Houston) *Morning News*, February 19, 1842; Peebles et al, *Pictorial History of Polk County*, p. 103.

[5]Mrs. Aline Thompson Rothe, "A Brief History of Polk County," research paper, University of Texas at Austin, 1934, cited by H. P. Easton, "A History of the Texas Lumbering Industry," Ph. D. dissertation, University of Texas at Austin, 1947, p. 36.

[6]Emma Haynes, *A History of Polk County, Texas* (Livingston: 1968), p. 16; Peebles et al, *Pictorial History of Polk County*, p. 114.

[7]Ralph A. Wooster, "The Battle of Galveston," *Texas Gulf Historical and Biographical Record*, XXVIII (November, 1992), p. 34.

[8]E. Haynes, *History of Polk County* (Livingston: 1968), p. 46: Webb, *Handbook of Texas*, II, 143.

[9]Interview with Tom Hickman, cited by Easton, "History of Texas Lumbering" p. 136; Peebles,

"History of Moscow," *Pictorial History of Polk County*, p. 278.

[10]Webb, *Handbook of Texas*, I, 197; Peebles et, *Pictorial History of Polk County*, p. 66.

[11](Galveston) *Daily News*, August 18, 1889.

[12]"Timbered Texas," (Galveston) *Weekly News*, April 13, 1893.

[13](Galveston) *Daily News*, April 13, 1893.

[14]"Texas' Timbered Wealth," (Galveston) *Daily News*, January 1, 1900.

[15]*Ibid.*, also (Galveston) *Daily News*, September 11, 1897.

[16]Webb, *Handbook of Texas*, I, 112; Peebles, *Pictorial History of Polk County*, p. 63.

[17]Peebles et al, p. 63.

[18]"In the Pineries," (Galveston) *Daily News*, August 18, 1889.

[19](Galveston) *Weekly News*, March 14; and December 25, 1891.

[20](Galveston) *Daily News*, April 13, 1893.

[21]*Ibid.*, November 27, 1896.

[22](Galveston) *Daily News*, March 14, 1896.

[23]*Ibid.*, September 11, 1897.

[24]"Barnum News Notes," (Galveston) *Daily News*, February 11, 1898.

[25]Webb, *Handbook of Texas*, I, 147.

[26]"Benford Mill Purchased," (Beaumont) *Journal*, January 23, 1904.

[27]"History of the Benford Sawmill," (Beaumont) *Enterprise*, April 2, 1905.

[28]Webb, *Handbook of Texas*, I, 147; R. Peebles, *Pictorial History of Polk County*, pp. 63-64; E. Haynes, *History of Polk County, Texas* (typescript: 1937), p. 145.

[29]Webb, *Handbook of Texas*, II, 852.

[30](Galveston) *Weekly News*, March 14, 1891.

[31]"Timbered Texas," (Galveston) *Weekly News*, April 13, 1893; "Texas' Timbered Wealth," (Galveston) *Daily News*, January 1, 1900.

[32]Webb, *Handbook of Texas*, I, 850; and II, 400, 852; Peebles et al, *Pictorial History of Polk County*, pp. 108, 119.

[33]Peebles, p. 119; P. B. Hensley et al, "History of P. E. Josserand Family," *History of Trinity County, Texas* (N. P.: 1986), pp. 507-508.

[34](Galveston) *Daily News*, August 18, 1889.

[35]"Timbered Texas," (Galveston) *Weekly News*, April 13,1893.

[36]*Ibid.*; also "Texas Timbered Wealth," (Galveston) *Daily News*, January 1, 1900.

[37]"Lumber and Lumbermen," (Galveston) *Daily News*, August 31, 1889.

[38](Galveston) *Daily News*, December 9, 1890; March 14, 1891.

[39]*Ibid.*, April 13, 1893.

[40]*Ibid.*, September 11, 1897.

[41]Peebles et al, *Pictorial History of Polk County*, p. 62.

[42]Webb, *Handbook of Texas*, II, 386; Peebles et al, *Pictorial History of Polk County*, pp. 109, 115; E. Haynes, *History of Polk County* (typescript: 1937), pp. 143-144.

[43](Galveston) *Daily News*, August 20, 1889.

[44]"Lumber and Lumbermen," (Galveston) *Daily News*, July 23, 1891.

[45]"Timbered Texas," (Galveston) *Weekly News*, April 13, 1893.

[46]Webb, *Handbook of Texas*, I, 284-285; Robert Maxwell et al, *Sawdust Empire: The Texas Lumbering Industry* (College Station: 1983), pp. 65, 152.

[47]E. Haynes, "The History of Polk County" (typescript: 1937), p. 146.

[48]Webb, *Handbook of Texas*, I, 285.

[49]Personal Interview with Jim Whiteside, Kountze, Texas, February 10, 1993.

[50]Webb, *Handbook of Texas*, II, 864; "At Voth," (Beaumont) *Journal*, November 13, 1904; July 19, 1905.

[51]Webb, *Handbook of Texas*, I, 276-76, 298; Peebles et al, *Pictorial History of Polk County*, p. 68.

[52]"Lumber and Lumbermen," (Galveston) *Daily News*, December 9, 1890.

[53]"Timbered Texas," (Galveston) *Weekly News*, April 13, 1893.

[54](Galveston) *Daily News*, April 13, 1893.

[55]"Lumber and Lumbermen," (Galveston) *Daily News*, March 12, 1897.

[56]"Texas' Timbered Wealth," (Galveston) *Daily News*, January 1, 1900; "The Cameron Plant at Carmona, Texas," *American Lumberman* (Nov. 17, 1900), p. 50.

[57]Webb, *Handbook of Texas*, I, 298.

[58]W. T. Chambers, "Life in A Southern Sawmill Community," *Journal of Geography*, XXX (Mar. 1931), pp. 181-182.

[59]Chambers, "Life in a Sawmill Community," pp. 182-183.

[60]Webb, *Handbook of Texas*, I, 298; Peebles et al, *Pictorial History of Polk County*, pp. 68-69.

[61]George Creel, "The Feudal Towns of Texas," *Harper's Weekly* (January 23, 1915), pp. 76-78.

[62]H. P. Easton, "A History of the Texas Lumbering Industry," Ph.D. dissertation, University of Texas at Austin,1947, p. 265.

[63]Webb, *Handbook of Texas*, I, 132, and II, 313; Peebles et al, *Pictorial History of Polk County*, p. 107.

[64]E. Haynes, "History of Polk County" (typescript: 1937), pp. 147-149

[65]Maxwell et al, *Sawdust Empire*, p. 139; E. Haynes, "History of Polk County," typescript, 1937, p. 149; *Lumber Trade Journal*, January 1, 1910, cited in D. H. Hardy et al, "Biography of James M. West," *The Historical Review of Southeast Texas* (Chicago: 1910), II, pp. 1009-1011; for the Ryan hardwood mill, see Nellis and Pierson, *Directory of American Sawmills* (Washington: Dept. of Agriculture, Govt. Prtg. Office, 1915). p. 211.

[66]Peebles, *Pictorial History of Polk County*, p. 90.

[67]"Timbered Texas," (Galveston) *Weekly News*, April 13, 1893; "Texas' Timbered Wealth," (Galveston) *Daily News*, January 1, 1900; Nina Smith et al, "Biography of the C. A. Bender Family," *The History of the Humble, Texas Area* (Humble: 1976), pp. 20-21, 72-74.

[68]Peebles, pp. 90, 121; Webb, *Handbook of Texas*, II, 229; (Beaumont) *Journal*, February 5, 1905.

[69]*Polk County Enterprise*, March 10, 1910, quoted in Peebles, *Pictorial History of Polk County*, pp. 131, 136.

[70]Peebles, pp. 92, 121; "Texas Timbered Wealth," (Galveston) *Daily News*, January 1, 1900.

[71]Peebles et al, p. 90; E. Balch, "Biographical Sketches of Industrial Leaders in Lufkin and Angelina County," MA Thesis, 1949, pp. 14-19; L. M. Mayberry, "Keltys-An East Texas Sawmill Town," MA Thesis, 1948, pp. 1-5.

[73]E. Haynes, "History of Polk County,"typescript, 1937, p. 143; R. Peebles, *Pictorial History of Polk County*, p. 64; Webb, Handbook of Texas, I, 149; "Timbered Texas,' (Galveston) *Weekly News*, April 13, 1893.

[74](Galveston) *Daily News*, September 26, 1896.

[75]Peebles et al, pp. 119-121; "Timbered Texas," (Galveston) *Weekly News*, April 13, 1893; "Texas Timbered Wealth," (Galveston) *Daily News*, January 1, 1900; Webb, *Handbook of Texas*, II, 827; (Beaumont) *Journal*, October 23, 1904.

[76](Galveston) *Daily News*, August 20, 1889.

[77]Webb, *Handbook of Texas*, I, 416; (Galveston) *Daily News*, December 9, 1890.

[78]"Timbered Texas," (Galveston) *Weekly News*, April 13, 1893.

[79]Webb, *Handbook of Texas*, I, 416; also E. Haynes, "History of Polk County," typescript, 1937, p. 150.

[80]Webb, *Handbook of Texas*, I, 416.

[81]J. and J. Wheat, *Sketches of Tyler County History* (Bevil Oaks: 1986), pp. 147-148; (Galveston) *Daily News*, February 11, 1898.

[82]Haynes, "History of Polk County," typescript, p. 146.

[83]Maxwell, *Sawdust Empire*, pp. 137-138.

[84]*Ibid.*, pp. 138-139.

[85]*Ibid.*, pp. 45-46.

[86]J. Wheat, *Sketches of Tyler County History*, pp. 147-149.

[87]Peebles et al, *Pictorial History of Polk County*, p. 698.

[88]R. Maxwell, *Sawdust Empire*, p. 111.

[89]Peebles, *Pictorial History of Polk County*, p. 114; E. Haynes, "History of Polk County", typescript, p. 150.

[90]Webb, *Handbook of Texas*, II, 69.

[91](Beaumont) *Enterprise*, June 11, 1905.

[92]"East Mayfield: A Part of Our Past," *In Touch* (Temple-Inland Magazine), Fall, 1992, pp. 16-17.; Maxwell, *Sawdust Empire*, pp. 111-115.

[93]Peebles et al, *Pictorial History of Polk County*, pp. 91, 121; Webb et all, *Handbook of Texas*, II, 46.

[94]*Ibid.*; (Galveston) *Weekly News,* March 14, 1891.

[95]*Ibid.*; (Galveston) *Weekly News*, April 13, 1893.

[96]Peebles, pp. 91, 121; (Galveston) *Daily News*, September 26, 1896.

[97]Peebles, pp. 92, 121.

[98]*Ibid.*; "Texas' Timber Wealth," (Galveston) *Daily News*, January 1, 1900.

[99]Webb et al, *Handbook of Texas*, II, 46.

[100]*Ibid.*, II, 274; Peebles et al, pp. 105, 121.

[101]"History of the Thompson and Tucker Lumber Company," (Beaumont) *Enterprise*, April 2,

1905; Maxwell et al, *Sawdust Empire*, p. 88.

[102]Maxwell, *Sawdust Empire*, p. 144.

[103]Charles E. Wilson, "The Big Mill," in P. B. and J. W. Hensley, *The History of Trinity County, Texas* (N. P.: 1986), No. T-36; Webb, *Handbook of Texas*, II, 274; Peebles, pp. 105, 121.

[104] Webb et al, *Handbook of Texas*, I, 237; E. Haynes, "History of Polk County," typescript, p. 143; R. Peebles et al, *Pictorial History of Polk County*, pp. 66-67, 121.

[105](Beaumont) *Enterprise*, June 11, 1905.

[106](Galveston) *Weekly News*, April 13, 1893; and *Daily News*, January 1, 1900; Peebles, Pictorial *History of Polk County*, pp. 112, 121; Webb, *Handbook of Texas*, II, 593.

[107]"Will Build New Mill,": (Beaumont) *Enterprise*, June 11, 1905.

[108]E. Haynes, "History of Polk County," typescript, p. 145; Peebles et al, *Pictorial History of Polk County*, pp. 111, 121.

[109]Peebles et al, *Pictorial History*, p. 121.

[110]*Ibid.*, pp. 62, 92; (Galveston) *Daily News*, January 1, 1900.

[111]Peebles, pp. 121, 278.

[112]*Ibid.*, p. 65.

[113]*Ibid.*, p. 102.

[114]*Ibid.*, pp. 76, 119-121.

[115]Peebles, pp. 79, 121.

[116]*Ibid.*, p. 81.

[117]*Ibid.*, pp. 81, 101-102, 119-121.

[118]*Ibid.*, pp. 104-121.

[119]*Ibid.*, p. 106.

[120]*Ibid.*, p. 121.

[121](Galveston) *Weekly News*, April 13, 1893; *Daily News*, January 1, 1900.

[122]Peebles, p. 121; R. Fincher, "The History of Liberty County," MA thesis, U. T. Austin, 1937, p. 74.

[123]Peebles, p. 121.

[124]E. Haynes, "History of Polk County," typescript, p. 150; Webb, "Livingston," *Handbook of Texas*, II, 69.

TYLER COUNTY

[1]W. P. Webb et al, *The Handbook of Texas* (Ann Arbor: 1986), II, 814.

[2]*Ibid.*, I, 871, and II, 467.

[3]"Letter from Hyatt," (Galveston) *Daily News*, October 29 and November 6, 1884.

[4]"Letterbook of the East Texas and Louisiana Lumbermen's Association, 1884-1886," published in *Texas Gulf Historical and Biographical Record*, IX (Nov. 1973), p. 60.

[5](Galveston) *Daily News*, September 1, 1886.

[6]*Ibid.*, June 6, 1888; W. T. Block (ed.), *Emerald of The Neches: The Chronicles of Beaumont, Texas, From Reconstruction to Spindletop* (Nederland: 1980), p. 287, hereinafter cited as *Emerald.*

[7](Galveston) *Daily News*, August 14, 1889; *Emerald*, p. 290.

[8]*Emerald*, pp. 432-433; (Galveston) *Daily News*, August 27, 1890.

[9](Galveston) *Weekly News*, March 29, 1891; *Emerald*, p. 452.

[10]Emerald, p. 296; (Galveston) *Daily News*, April 24, 1892.

[11](Galveston) *Daily News*, March 14, 1896.

[12]*Ibid.*, May 23, 1896.

[13]*Ibid.*, October 17, 1896.

[14](Galveston) *Daily News*, March 12 and May 19, 1897.

[15]*Ibid.*, September 11, 1897.

[16](Beaumont) *Journal*, January 15, 1904.

[17]*Ibid.*, July 9, 1905.

[18]*Ibid.*, November 13, 1904; "Among The Hyatt Pines," *Southern Industrial and Lumber Review*, June, 1904, p. 26.

[19]Webb, *Handbook of Texas*, I, 871.

[20]Webb, *Handbook of Texas*, II, 467; J. and J. Wheat, *Sketches of Tyler County History* (Bevil Oaks: 1986), pp. 149-150.

[21]Webb, *Handbook of Texas*, II, 220.

[22](Galveston) *Daily News*, August 31, 1889.

[23]*Ibid.*, December 7, 1890.

[24]*Ibid.*, April 13, 1893.

[25]Wheat, *Sketches of Tyler County History*, p. 151.

[26]Kirby Lumber Company, *Timber Resources of East Texas* (n. p., 1902), p. 166; Letterhead, Southwestern Lumber Company, Mobile, Texas, to F. M. Aldridge, June 16, 1902, Forestry Collections, SFA State University.

[27](Beaumont) *Journal*, November 20, 1904.

[28](Beaumont) *Enterprise*, March 26, 1905.

[29]*American Lumberman*, April 4, 1908.

[30]*Enterprise*, September 20, 1910.

[31]Webb, *Handbook of Texas*, II, 220.

[32]*Ibid.*, II, 494; Wheat, *Sketches of Tyler County History*, p. 151.

[33]"Letterbook of the East Texas and Louisiana Lumbermen's Association," 1884-1886, in *Texas Gulf Historical and Biographical Record*, IX (Nov. 1973), p. 60.

[34](Galveston) *Daily News*, September 1, 1886; *Emerald*, p. 296.

[35]*Emerald*, p. 282; (Galveston) *Daily News*, June 6, 1888.

[36](Galveston) *Daily News*, August 14, 1889; June 6, 1888; *Emerald*, p. 291.

[37]*Emerald*, pp. 452, 454; (Galveston) *Daily News*, October 9, 1890; March 29, 1891.

[38](Galveston) *Daily News*, September 13, 1891; *Emerald*, p. 465.

[39]*Emerald*, p. 526; (Galveston) *Daily News*, April 26, 1892.

[40](Galveston) *Daily News*, August 31, 1895.

[41]*Ibid.*, January 30, 1897.

[42](Beaumont) *Journal*, November 20, 1904; "The Rockland, Texas Mill," *American Lumberman* (Nov. 17, 1900), p. 52.

[43]*Ibid.*, October 23, 1904.

[44](Beaumont) *Journal*, November 13, 1904.

[45](Beaumont) *Enterprise*, June 11, 1905.

[46]*Ibid.*, September 10, 1910; R. Maxwell et al, *Sawdust Empire: The Texas Lumber Industry* (College Station: 1983), p. 102.

[47]"History of the Cameron Mill at Rockland," (Beaumont) *Enterprise*, January 23, 1905.

[48]Webb, *Handbook of Texas*, II, 494.

[49]Wheat, *Sketches of Tyler County History*, p. 151.

[50]Webb, *Handbook of Texas*, I, 377.

[51](Galveston) *Daily News*, September 1, 1886; June 6, 1888; *Emerald*, pp. 282, 296.

[52]*Emerald*, pp. 287-289; (Galveston) *Daily News*, July 27, 1889.

[53]*Emerald*, pp. 431-432; (Galveston) *Daily News*, August 27, 1890.

[54](Beaumont) *Enterprise*, June 11, 1905.

[55]Brenda J. Goode, "History of Colmesneil, Texas," *Texas Gulf Historical and Biographical Record*, XXVII (Nov. 1991), p. 45, quoting (Galveston) *Weekly News*, April 20, 27, 1893.

[56](Galveston) *Daily News*, April 14, 1891.

[57]*Ibid.*, September 13, 1901.

[58]*Ibid.*, May 17, 1891.

[59]*Ibid.*, March 22, 1892.

[60]Wheat, *Sketches of Tyler County History*, p. 151.

[61](Galveston) *Daily News*, March 17, 1896.

[62]*Emerald*, pp. 290-291; (Galveston) *Daily News*, August 14, 1889.

[63](Galveston) *Daily News*, May 23, 1896.

[64]*Ibid.*, August 15, 1896.

[65]*Ibid.*, January 30 and April 17, 1897; May 1, 1897.

[66](Galveston) *Daily News*, January 9, 1897.

[67](Beaumont) *Journal*, January 22, 1905.

[68]*Ibid.*, June 11, 18, 1905.

[69]B. Goode, "The History of Colmesneil, Texas," p. 55.

[70]Webb, *Handbook of Texas*, II, 202; Wheat, *Sketches of Tyler County History*, p. 152.

[71]"Kountze," (Galveston) *Daily News*, October 16, 1886.

[72]"Letterbook of the East Texas and Louisiana Lumbermen's Association, 1884-1886," in *Texas Gulf Historical and Biographical Record*, IX (Nov. 1973), p. 60; "Lumber Shipments," (Galveston) *Daily News*, September 1, 1886; *Emerald*, p. 296.

[73](Galveston) *Daily News*, March 28 and June 6, 1888; *Emerald*, pp. 281-282.

[74]"Seneca Mills," (Galveston) *Daily News*, June 6, 1888; *Emerald*, pp. 281-282.

[75](Galveston) *Weekly News*, October 9, 1890; *Daily News*, March 29, 1891; *Emerald*, pp. 452, 454.

[76](Beaumont) *Journal*, June 18; July 9, 1905; (Beaumont) *Enterprise*, June 11, 1905.

[77](Galveston) *Daily News*, August 14, 1888; *Emerald*, p. 289.

[78]*Ibid.*, June 6, 1888; *Emerald*, p. 282.

[79]*Emerald*, p. 282; (Galveston) *Daily News*, June 6, 1888.
[80](Galveston) *Daily News*, August 14, 1889; *Emerald*, pp. 290-291.
[81]Wheat, *Sketches of Tyler County History*, p. 152; Webb, *Handbook of Texas*, II, 864.
[82]Wheat, *Sketches of Tyler County History*, p. 152.
[83]Webb, *Handbook of Texas*, II, 628.
[84]*Ibid.*, I, 4.
[85]*Ibid.*, I, 805.
[86]*Ibid.*, I, 786.
[87](Galveston) *Daily News*, March 29, 1891.
[88]*Emerald*, p. 291; (Galveston) *Daily News*, August 18, 1889.
[89]"Timbered Texas," (Galveston) *Weekly News*, April 13, 1893; also March 14, 1891.
[90](Galveston) *Daily News*, August 15, 1896.
[91]*Ibid.*, January 30, 1897.
[92]Webb, *Handbook of Texas*, I, 763.
[93]"History of the Hampton, Texas Sawmill," (Beaumont) *Enterprise*, January, 28, 1905.
[94]Webb, *Handbook of Texas*, I, 763; Wheat, *Sketches of Tyler County History*, pp. 148-149.
[95]"Timbered Texas," (Galveston) *Weekly News*, April 13, 1893.
[96]Webb, *Handbook of Texas*, I, 515; Wheat, *Sketches of Tyler County History*, p. 152; Kirby, *Timber Resoures of East Texas*, p. 165.
[97]Vol. K, p. 30, Hardin County Texas Deed Records.
[98](Galveston) *Daily News*, September 11, 1893.
[99]*Ibid.*, March 12, 1897.
[100]*Ibid.*, September 11, 1897.
[101]*Ibid.*, August 15, 1896; January 30, 1897.
[102](Liberty) *Vindicator*, October 29, 1897.
[103](Liberty) *Vindicator*, November 5, 1897.
[104]Kirby, *Timber Resources of East Texas*, p. 165.
[105]Ruth Peebles et al, *A Pictorial History of Polk County, Texas* (n. p.: 1976), p. 79; "Texas Timbered Wealth," (Galveston) *Daily News*, January 1, 1900.
[106]Webb, "Samuel F. Carter," *The Handbook of Texas*, I, 303 Wheat, *Sketches of Tyler County History*, p. 152.
[107]"Doucette," (Beaumont) *Enterprise*, March 25, 1905.
[108]"Will Enlarge Mill," (Beaumont) *Journal*, January 8, 1904.
[109]"May Not Be Sold," (Beaumont) *Journal*, February 19, 1905.
[110]"S. F. Carter, *Handbook of Texas*, I, 303; Wheat, *Sketches of Tyler County History*, p. 152.
[111]"Town of Doucette," (Beaumont) *Enterprise*, January 15, 1905.
[112]Maxwell, *Sawdust Empire*, p. 72; Copy of cancelled check, Thompson Brothers to S. F. Carter, January 3, 1907, in the Thompson Brothers papers, Box 27, Folder 12, Forestry Archives, SFA University; *Southern Industrial and Lumber Review*, Sept. 1908, p. 29; Oct., 1908, p. 41; Dec., 1908, p. 53, and Jan. 1908, p. 69.
[113]Wheat, *Sketches of Tyler County History*, p. 152.
[114]Webb, *Handbook of Texas*, I, 815; Wheat, *Sketches of Tyler County History*, p. 151; (Beaumont) *Enterprise*, September 17 and October 1, 1881.
[115]Webb, *Handbook of Texas*, I, 815; "Letterbook of the East Texas and Louisiana Lumbermen's Association, 1884-1886," in *Texas Gulf Historical and Biographical Record*, IX (Nov. 1973), p. 60.
[116](Galveston) *Daily News*, October 5, 1887 and September 2, 1890.
[117]*Emerald*, p. 439; (Galveston) *Daily News*, September 2, 1890.
[118](Galveston) *Daily News*, September 1, 1886; June 6, 1888; *Emerald*, pp. 282, 296.
[119]*Emerald*, pp. 452, 454; (Galveston) *Daily News*, March 29, 1891; *Weekly News*, October 9, 1890.
[120](Galveston) *Daily News*, May 17, 1891.
[121](Galveston) *Weekly News*, July 23, 1891.
[122]"Another Receiver," (Galveston) *Daily News*, January 21, 1892.
[123](Beaumont) *Journal*, February 19, 26, 1905; also Southern Industrial and Lumber Review, June, 1904, p. 25; also issues for January and March, 1907.
[124]Webb, *Handbook of Texas*, II, 864; Wheat, *Sketches of Tyler County History*, p. 151.
[125](Galveston) *Daily News*, September 1, 1886; *Emerald*, p. 296; *Texas Gulf Historical and Biographical Record*, IX (Nov. 1973), p. 60.
[126]*Emerald*, p. 280; (Galveston) *Daily News*, October 5, 1887.
[127](Galveston) *Daily News*, July 21, 1889; *Emerald*, p. 283.
[128]*Emerald*, p. 439; (Galveston) *Daily News*, September 4, 1890.

[129](Galveston) *Daily News*, March 23, 1891; *Emerald*, p. 449.
[130]*Emerald*, p. 509; (Galveston) *Daily News*, February 23, 1892.
[131](Galveston) *Daily News*, August 18, 1895.
[132]*Ibid.*, August 4, 1895.
[133]*Ibid.*
[134](Galveston) *Daily News*, February 8, 1896.
[135]*Ibid.*, July 4, 1897.
[136]"Texas Timbered Wealth," (Galveston) *Daily News*, January 1, 1900.
[137]Webb, *Handbook of Texas*, II, 864; (Beaumont) *Journal*, January 15, 1904.
[138](Beaumont) *Journal*, October 23, 1904.
[139]*Ibid.*, February 19, 1905.
[140]Webb, *Handbook of Texas*, II, 864.
[141](Galveston) *Daily News*, September 1, 1886.
[142]*Ibid.*, June 6, 1888; March 29, 1891; *Emerald*, pp. 282, 452, 454.
[143](Galveston) *Daily News*, April 17, 1897.
[144]Kirby Lumber Company, *Timber Resources of East Texas*, p. 165.
[145]*Ibid.*, pp. 165-166.
[146](Beaumont) *Journal*, November 13, 20, 1904; July 9, 1905.
[147]R. Maxwell et al, *Sawdust Empire*, p. 102.
[148](Beaumont) *Enterprise*, September 10, 1910.